W9-CHF-365

TRANSFORMING INSTITUTIONS

TRANSFORMING INSTITUTIONS
Undergraduate STEM Education for the 21st Century

Edited by
Gabriela C. Weaver, Wilella D. Burgess,
Amy L. Childress, and Linda Slakey

Purdue University Press
West Lafayette, Indiana

TABLE OF CONTENTS

FOREWORD

Carl Wieman

There is a growing awareness of both the need to improve STEM education at the undergraduate level and the opportunities for doing so. The importance of achieving improved educational results in STEM is recognized across the political spectrum as an important element in preserving a vibrant competitive economy. It is also increasingly seen as important for a democracy faced with numerous major decisions involving technical issues, such as addressing climate change and energy sources, novel medical care, and genetically modified foods. This "gathering storm" of factors that were discussed in the 2007 National Research Council report as threatening America's long-term competitiveness and security is now leading to increasingly strong winds of change blowing through higher education. What was missing from that 2007 call to action, however, was the recognition of a large body of research on the teaching and learning of undergraduate STEM, a body of research showing that there exist far more effective ways to teach than the widely used traditional lecture. That research indicates that the 2000-year-old format of a professor standing in front of a large group of students and dispensing knowledge in the form of a lecture is not very effective. It dispenses knowledge, but neither knowledge that sticks nor wisdom—the wisdom to know when, where and how to apply that knowledge to make decisions and new discoveries, or solve real-world problems.

These research results have put STEM education in somewhat the same situation that medicine was in 150 years ago. The conventional treatments, such as bloodletting, had their origins in superstition and tradition and had been in use for many centuries. Their effectiveness was "proven" by after-the-fact confirmation, based on the fact that some patients who received such treatments survived. While such tradition was still the basis of treatment at that time, scientific advances revealed a new understanding of disease, with corresponding indications of more effective treatments, and a growing sense of a more scientific approach for evaluating the effectiveness of treatment. While many questions remained, it was clear to the researchers that there were more effective methods of treatment and an entirely new type of expertise that doctors should have, if they were to be effective. We now exist in an era where institutions are practicing pedagogy based primarily on tradition, with well-meaning faculty that are largely unaware of the dramatic advances that have been made in the past few decades in understanding the learning of STEM and best practices for teaching.

They lack the knowledge and expertise to teach in ways that the research shows are highly effective.

As described in this volume, many individuals and organizations are joining this effort to transform STEM education. The collective hope is to see the teaching of STEM undergo a metamorphosis, transforming into an effective research-based expert practice, as medicine has done. However, in spite of all these efforts and all the potential for improvement, change is slow and far from certain. Changing large well-established institutions and their associated cultures is a very formidable task. It took medicine many decades to change, and their failures died; ours usually just end up switching majors. One can argue that culture is what humans develop to establish stability, and hence inhibit change, and it is very effective at serving this goal of preserving the status quo. Perhaps this is truer for universities than for any other social institutions, as they are among the oldest and most stable.

There are many inter-connected pieces in the modern college or university. All of them are part of this culture and must be involved at some level if large-scale transformation is to take place. The articles in this volume reflect the different players, approaches and ideas for working at multiple levels to bring about change.

To successfully change a complex institution, one must develop a model of change that takes into account all the pieces that are relevant to the change being made and how those pieces connect. A number of the articles here consider different change models. For smaller scale change, such a model need only consider a subset of these issues, but for large-scale institutional change everything is relevant, every piece must be addressed, and all the elements of the model need to fit together—a very formidable task! With such a large, complex system and the individual differences across institutions, a realistic model of change must have considerable flexibility and adaptability built in. It is impossible to know how to get everything right ahead of time. That said, it is also important to recognize that there is a high degree of similarity across institutions of the same type, so much of the basic foundation of the model can and must be the same. One is not starting from scratch with each new institution; the similarities are much greater than the differences. This is true within all types of institutions, but the similarities are particularly large when one is considering large universities with an international presence and representation in both the faculty and student body. Large research universities have a high degree of similarity across their structures and incentive systems and cultures, including their belief that they are each rather unique and special. They are particularly similar with regard to how they approach STEM education.

Although post-secondary institutions believe they operate quite differently from industry, health care, or government, and to some extent that is true, fundamentally they are organizations of people. Many of the studies and principles of organizational change done in other contexts, such as Kotter (Kotter 1996) and others have done in industry or health care, are rooted in basic human organizational behaviors, and so have considerable relevance. One of those findings is a result that is counter-intuitive to many people: namely that it is hardest to make change when times are flush. It is actually much easier to transform an institution when resources are shrinking and times are difficult. One of the reasons for this is that when times are good it is extremely difficult to convince people that there is a problem, and that they should change what they are doing in order to fix that problem. Another general finding about organizational change is that, whenever change is proposed, people in every organization respond by arguing that, "The change might be good but we cannot afford it." Usually this is quickly followed by, "And if we could, we are all so busy we do not have the time." I hear these same arguments from institutions of all types and all levels of resources. I also see some faculty from across all these different institutions who have found ways to make large and impactful changes in their teaching that benefit many students. A number of such examples are given in this book.

While this volume presents many enterprising ways to bring about change, a big issue that is discussed in some of the chapters of Section A but is not explicitly addressed in its own chapter, or in the case studies, is the formal incentive system under which instructors work. This remains the 500-pound gorilla standing in front of the doorway that leads to widespread improvement in teaching methods. And this is not just any 500-pound gorilla; it is a particularly muscular and unforgiving one! Although it is often claimed that one cannot tell faculty what to do, in fact most of them are doing exactly what they are being paid to do, or more precisely, what they are being held accountable for and rewarded for doing.

This shows up in many different ways, but most frequently in the choices they make about allocation of time. Invariably, in discussions about improving teaching methods, the concern is raised that the faculty are already far too busy and overworked to put any more time into their teaching or into learning to use better methods. It is important to remember that no one ever feels they have enough time to do everything they might like to. When someone says they do not have time to do something, they are not making a statement about how much time they have; they are making a statement about their priorities. Learning to teach differently is simply not a high priority in comparison with the other aspects of their job for most faculty members. And there is good reason for this. At every research university the incentive system measures their research output and rewards them accordingly. It penalizes a faculty member (or a department chair) for anything that reduces that productivity, which spending even a small amount of time to become a better teacher will necessarily do. There is great value to this system that carefully measures and rewards research productivity. It is responsible for the establishment of the remarkably beneficial social institution that is the modern research university. The problem is that research productivity is the *only* thing that is measured and rewarded, but universities are expected to serve the dual function of research and teaching.

To have any hope of achieving widespread change in undergraduate STEM teaching, an incentive system must be established that recognizes and rewards contributions in teaching to a meaningful degree. I doubt that a very large change will be needed. Teaching well is inherently rewarding and enjoyable for everyone, so we do not have to massively change the incentive system, only provide a nudge to faculty to take the time to learn better teaching methods and a little professional support to minimize the required time. That is one lesson that has been learned by the Science Education Initiatives at the University of Colorado and the University of British Columbia that have changed the way large numbers of faculty teach through novel department-based incentives and support. One does not have to kill off the 500-pound gorilla, just provide a little space to slip around him.

However, that modest change in the incentive system will never be accomplished without having a better way to measure teaching quality. Currently, the methods of measuring teaching contributions and quality are not remotely close to the thoroughness and effectiveness with which research productivity is measured. The almost universally used method for evaluating teaching is student evaluations. While student evaluations have their value, they are not a good measure of the amount of learning being produced by the teaching. They also provide little guidance for improvement and are sensitive to many confounding variables outside of the instructor's control. In terms of supporting the adoption of better teaching methods, student evaluations are a clear barrier. Many instructors believe that changing to more active learning techniques will cause their student evaluations to go down. In my Science Education Initiatives, we have seen that more effective teaching methods do not cause evaluations to go down, if they are introduced in the right manner. However, the evaluations also do not go up after new teaching methods are introduced that produce demonstrably greater learning and student success.

To be successful in achieving large-scale improvements in teaching, future efforts will need to focus considerable attention on the institutional adoption of better measures for teaching quality, and have those measures become part of the formal incentive system. Without that, the many efforts described in this volume will always be limited in the scope they can achieve. One does not have to be a psychologist to recognize the futility of hoping that many people will choose to take time away from activities for which they are rewarded to invest time and effort into pursuing goals that are never measured and never rewarded.

The Teaching Practices Inventory (Wieman, 2014) and Classroom Observation Protocol for Undergraduate Science (COPUS) (Smith et al., 2013) are new tools that I have worked on developing to address this problem in measuring teaching quality. While these are not the direct measures of learning and enthusiasm for learning that would be ideal as a measure of teaching quality, they are proxies for those measures. The cognitive psychology research and discipline-based education research shows that they are much better proxy measures for those goals than are student evaluations. The widespread use of practical, fair, and valid measures of teaching effectiveness, such as these tools, must be part of any model of large-scale change if it is to be successful.

If you are an individual or part of an institution that is considering launching or joining a transformation effort intended to improve STEM instruction, this volume provides both inspiration and guidance for you at many different levels. It contains many examples that illustrate the opportunities, successes, challenges, wisdom, and lessons that have been learned by the authors. These come from a variety of institutions and organizations and so have a variety of different perspectives and speak to a variety of audiences. They discuss everything from models and theories of large-scale institutional change in section A, through a variety of examples of changes that have been carried out in Sections B and C. The examples in Sections B and C illustrate the issues and successes encountered in contexts ranging from changing multiple departments and how an institution operates, down to modifying individual courses and curriculum, and the faculty development needed to support such efforts. These examples illustrate different types of both top-down and bottom-up implementations. All these scales of change are important, and the models of change presented in these sections give a necessary overall perspective on understanding how to best carry out such efforts. Section E focuses on the special area of metrics and assessment. As we move forward with change, it is essential that it be guided by and supported by data. The cases in the section provide examples of types of data that can be collected and how to collect it. Finally, the last section steps back and offers some broad lessons for moving forward on institutional change.

This volume shows what a varied and energetic enterprise is underway in transforming institutions toward more effective STEM education and foreshadows great progress in the years to come. It will inspire and guide the reader in joining this enterprise.

REFERENCES

Kotter, John P. (1996). *Leading change*. Cambridge MA: Harvard Business School Press.

Smith, Michelle K., Jones, Francis H. M., Gilbert, Sarah L., and Wieman, Carl E. (2006). The classroom observation protocol for undergraduate STEM (COPUS): A new instrument to characterize university STEM classroom practices. *CBE—Life Sciences Education*, Vol. 12, 618–627.

Wieman, Carl & Gilbert, Sarah. (2014). The teaching practices inventory: A new tool for characterizing college and university teaching in mathematics and science. *CBE—Life Sciences Education*, Vol. 13, 552–569.

ABOUT THE AUTHOR

Carl Wieman is a recipient of the Nobel Prize in Physics in 2001, with Eric Cornell (University of Colorado, Boulder) and Wolfgang Ketterle (MIT), for the production of the first Bose–Einstein condensate. His intellectual focus is now on undergraduate physics and science education. He has pioneered the use of experimental techniques to evaluate the effectiveness of various teaching strategies for physics and other sciences. He launched the Science Education Initiative, aimed at improving undergraduate science education, at both the University of Colorado, Boulder, and later at the University of British Columbia, in Vancouver, Canada. Wieman served as founding chair of the Board of Science Education of the National Academy of Sciences in 2004. He was nominated to The White House's Office of Science and Technology Policy, as Associate Director of Science, and served from 2010 until 2012. He currently holds a joint appointment as Professor of Physics and of the Graduate School of Education at Stanford University in Stanford, California.

Introduction

1

Why Now is the Time for Institution-level Thinking in STEM Higher Education

Gabriela C. Weaver

What is the practice that makes a difference?
It is the notion that we can be much better than we are.
F. Hrabowski III, October, 23, 2014

On an October evening in 2014, over 100 people gathered for dinner and to prepare for a day and a half of thought-provoking and challenging ideas and questions. The event was the second conference on Transforming Institutions, aimed at reform in the STEM disciplines in higher education. The attendees were a mix of higher education leaders, STEM faculty, educational researchers and funding agency or educational association representatives. The shared objective of their work, whether recently or for many years, was to achieve an approach to educating STEM undergraduate students that would lead to deeper understanding, larger diversity, increased graduation rates and greater long-term success for students. Like the blind men exploring an elephant (JGRC, 2011), each participant brought an extensive understanding of some component of the challenge at hand, and hoped that the joint dialogue at this conference would help provide a clearer description of the entire system.

There is an abundance of information about how students learn STEM concepts and best practices, techniques and pedagogies that are based on this research (e.g., Ambrose, 2010; Kuh, et al., 2005; Kober, 2015; Kuh, 2008; Labov, et al., 2009; NRC 1999, 2011, 2012; Svinicki, 2004; Weimer, 2013). Much of this knowledge has been supported by decades of investment from sources such as the Department of Education's Fund for the Improvement of Post-Secondary Education (FIPSE) and the programs in the National Science Foundation's Education and Human Resources (EHR) directorate. Programs like Course, Curriculum and Laboratory Improvement (CCLI) and Transforming Undergraduate Education in Science (TUES) at NSF have awarded about $25 million per year in grants to educators to develop, implement and test innovations in teaching (NSF, 2015). Over the last two to three fiscal years, about half of that has been shifted to projects that intentionally target wider, institutional-level transformation, rather than individual course improvement, as many course

3

improvements have failed to result in sustained adoption and widespread dissemination of evidence-based educational practices.

Although many successful innovations have been funded, and much has
been learned, the national-level metrics for success are still not where we would
hope to see them. The recent report on undergraduate STEM education from
the President's Council of Advisors on Science and Technology (PCAST, 2012)
declares that there is evidence for a decline in production of STEM graduates
relative to a rising need in the workplace. A recent publication of the National
Research Council, (Kober, 2015), synthesizes current research on best practices
in STEM education into a practitioner guidebook for undergraduate teaching
in science and engineering. But application of best practices such as these has,
for too long, been localized to individual efforts and to a brief timeframe, after
which the approach often disappears upon the departure or capitulation of its
champion.

Henderson and Dancy (2010) argue that, "the biggest barrier to improving
undergraduate STEM education is that we lack knowledge about how to effectively spread the use of currently available and research tested instructional
ideas and strategies." Part of what makes the knowledge about institutionalization so elusive is that there are many interconnected components that are acting
simultaneously on the people involved and the instructional choices that are
made. Ann Austin (2011) details in a white paper commissioned by the Board
on Science Education of the National Research Council the many influences
that impact faculty decision making about teaching (Fig. 1).

Austin (2011) explains that each faculty member's actions are embedded in
multiple layered contexts, beginning with the department and discipline, then
including the institution and external factors. Some of the influences can support faculty engagement in research-based teaching approaches, while some
factors, sometimes even the same ones, can act as barriers, depending on how
they are leveraged. For example, if the reward system is based primarily on research, then faculty will be disincentivized to take time or creative energy away
from that work. But if substantive recognition and compensation is attached to
specific types of instructional attainment, and if these can be assessed in a reliable way, then faculty will see this as an implicit statement of the institutional
leadership's priorities. In fact, reward and promotion practices are concrete
ways in which institutions inform their faculty member's choices for how to
allocate their efforts, regardless of what is said in mission statements or similar
platforms (Fairweather, 2005; Fairweather & Beach, 2002). This is one example
of the interdependence of the levers and barriers shown in Figure 1: Resources
are dependent on institutional funding, which can be increased by placing
greater emphasis on obtaining external research support, an effort that could

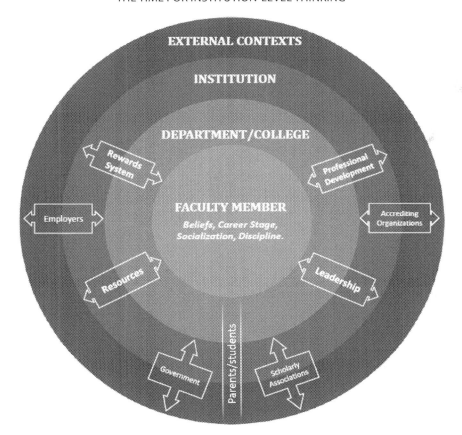

FIGURE 1. Multiple influences on an individual faculty member's approaches to teaching. (Figure adapted, with permission, from Austin [2011] and Sorcinelli [2014]).

have a detrimental impact on the quality of the student experience, potentially affecting enrollments or student success rates, which in turn have negative impacts on state or tuition-based resources. As a result, realizing sustainable institutional transformation must proceed from a systems approach, an assertion echoed by numerous scholars (e.g. Austin, 2011; Henderson, Beach & Finkelstein, 2010; Lemke & Sabelli, 2008; Scileppi, 1988).

The complex interrelationships of the influences on faculty teaching shown in Figure 1 are fundamentally all driven by human relationships and behavioral norms that form the culture in each department, discipline, and institution. The systems approach to change is thus overlaid on the need to achieve cultural change, and the recognition that the goals of the higher education system, whether stated explicitly or as perceived by the participants and beneficiaries of the educational process, are shaped by how knowledge and learning within a discipline are conceptualized, how structures and leadership support the educational process, and how decisions regarding priorities and actions are made. Maton, et al., (2008), emphasize that there are multiple dimensions of cultural change to attend to, including everything from the student experience to organizational behavior. The possibility of encountering resistance is high, and the likelihood of everyone beginning with the same viewpoints is vanishingly small. As a result, institutional transformation requires organizational learning (see Chapters A2 and A4) because "change is seen as a learning process affected by organizational and environmental conditions and by theories of action held by the organization's members" (Kezar, 2011).

Because the stakes are high (PCAST, 2012; National Academies, 2007), and the process is challenging, many of the organizations that have been prime movers in investigating and supporting instructional innovations, are now coming together to do the same for institution level scale-up and sustainability. The Division of Undergraduate Education at the National Science Foundation has become increasingly explicit about their interest in funding projects that will lead to wider dissemination, implementation and, especially, sustainability once their support ends. The American Association of Universities, which ventures into teaching and learning projects only infrequently, convened an expert panel to develop a framework for institutional STEM transformation (AAU, 2012; Chapter A3) that takes into account pedagogy, scaffolding/support, and cultural change. That framework then became the basis for their AAU STEM Initiative competitive proposal process that resulted in eight institutions being named as project sites to study the feasibility and modalities of working with the framework. A recently formed alliance of universities from the United State and Canada, the Bay View Alliance (BVA, 2015), is a consortium of research universities carrying out applied research on the leadership of cultural change for increasing the adoption of improved teaching methods at universities. And the Coalition for Reform of Undergraduate STEM Education (CRUSE) is a group of national organizations that have initiatives aimed at bringing about widespread implementation of evidence-based practice. Coalition members include the American Association for the Advancement of Science (AAAS), the Association of American Universities (AAU), the Association of Public and Land Grant Universities (APLU), Project Kaleidoscope and the Association of American Colleges and Universities (AAC&U), and the Board on Science Education of the National Research Council (NRC). Leaders within these organizations work together on mutual interests, share data and approaches, monitor progress nationally on metrics and models for institutional change, analyze for gaps, encourage action on gaps, and work to attract funding to the agenda to advance the adoption of evidence-based STEM practices at a wide array of college and university campuses (Fry, 2014).

It is clear that achieving sustainable institutional transformation around the widespread adoption of evidence-based teaching practices is not a simple process, and is additionally confounded by cultural norms that have a strong, though often unperceived, grip on the thinking of academic communities. Additionally, there are numerous external forces acting on educational institutions that further complicate the prioritizing of this type of work (as detailed in the following chapter). The efforts will need to involve collaboration, buy-in, patience and mutual support on the part of all the stakeholders in the community. Setting off on the path to transformation is one that requires the willingness to explore which routes lead to the greatest success, sometimes having to double back and try a different approach, because no perfectly reliable map exists yet. Perhaps this is the most challenging notion of all for institutions of higher education.

Those gathered in October 2014 began their engagement that evening listening to a keynote presentation at the second *Transforming Institutions* conference by Dr. Freeman Hrabowski III, president of the University of Maryland, Baltimore County (see Appendix 2). He is widely recognized as one of the most effective academic leaders in the country, having raised UMBC to be among those lauded for offering the best undergraduate education as well as world-class research. As an academic leader who has been down the road of institutional transformation, he admonished the audience that evening, reminding them that the principles that apply to high quality educational experiences for our STEM students mirror those that institutions themselves need to embrace as they strive for instructional excellence:

> If you're going to talk about transformation, you have to be willing to take risks. You have to be willing to be wrong. . . . Any campus that makes progress understands that often we learn more from the failure than we do from the success. It's when you fail and then take the time to understand what went wrong that you learn.

The chapters of this book derive from the presentations given at the two *Transforming Institutions* conferences, held in October 2011 and 2014 (DLRC, 2011 and 2014). They are extended versions of the data presented by the authors at the conferences, in some cases with updates showing developments that took place after they first presented their ideas. These have been arranged in sections that address different considerations for institutional transformation, resulting in varied examples and viewpoints. Section A provides an overview of foundational work on the theories and recent models for institutional transformation.

Sections B and C comprise a variety of examples of institutional transformation efforts in the form of case studies. These are divided into efforts that start out as institution-wide efforts (Section B) and those that are at the course or departmental levels (Section C). These case studies are meant to demonstrate the realities of how efforts at transformation are conceived, launched and put into practice. There are successes, and there are challenges. In some cases the results are different than expected. The case studies represent efforts at various levels of development, some quite nascent and others mature. After the case studies, two sections provide focused attention in areas that are critical for sustaining institutional changes: faculty development (Section D) and assessment (Section E). The concluding Section F provides a big picture overview of the nature of transformation and of threads that run through the work of the authors represented in this book, with the goal of leaving the reader with organizing principles for undertaking what is truly a complex and multifaceted, but worthwhile, undertaking.

REFERENCES

Ambrose, S. A., Bridges, M. W., DiPrieto, M., Lovett, M. C., Norman, M. K. (2010). *How learning works: Seven research-based principles for smart teaching.* Jossey-Bass: San Francisco, CA.

Association of American Universities (AAU) (2013). AAU *framework for systemic change in undergraduate STEM teaching and learning.* http://www.aau.edu/policy/article.aspx?id=12588, retrieved 3/20/2014.

Austin, A. E. (2011). *Promoting evidence-based change in undergraduate science education.* Paper commissioned by the Board on Science Education of the National Academies, National Research Council. Washington, D.C.: The National Academies.

Bay View Alliance (BVA), http://bayviewalliance.org/ Accessed 2/28/2015.

DLRC, 2011: https://stemedhub.org/groups/transforminginstitutions

DLRC, 2014: https://stemedhub.org/groups/transforminginstitutions

Fairweather, J. (2005). Beyond the rhetoric: Trends in the relative value of teaching and research in faculty salaries. *Journal of Higher Education, 76,* 401–422.

Fairweather, J., & Beach, A. (2002). Variation in faculty work within research universities:

Implications for state and institutional policy. *Review of Higher Education, 26,* 97–115.

Fry, C. L. (2014). *Achieving systemic change: A sourcebook for advancing and funding undergraduate STEM education.* The Coalition for Reform of Undergraduate STEM Education: Washington, DC.

Henderson, B. C., Beach, A., and Finkelstein, N. (2010). Facilitating change in undergraduate STEM instructional practices: An analytic review of the literature. *Journal of Research in Science Teaching, 48*(8), 952–984.

JGRC, 2011, Jain Stories, http://www.jainworld.com/education/stories25.asp, retrieved 2/26/2015.

Kezar, A. J. (2011). *Understanding and facilitating organizational change in the 21st century.* San Francisco, CA: Jossey-Bass.

Kober, N. (2015). *Reaching students: What research says about effective instruction in undergraduate science and engineering.* Washington, DC: The National Academies Press.

Kuh, G. D. (2008). *High-impact educational practices: What they are, who has access to them, and why they matter.* Washington, DC: AAC&U.

Kuh, G., Kinzie, J., Schuh, J., and Witt, E. (2005). *Student success in college: Creating conditions that matter.* Washington, DC: Association for the Study of Higher Education.

Labov, J. B., Singer, S. R., George, M. D., Schweingruber, H. A., and Hilton, M. L. (2009). Effective practices in undergraduate STEM education, part 1: Examining the evidence. *CBE—Life Sci. Educ., 8*, 157–161.

Lemke, J. L., and Sabelli, N. H. (2008). Complex systems and educational change: Towards a new research agenda. *Educational Philosophy & Theory, 40*(1), 118–129. doi:10.1111/j.1469-5812.2007.00401.x

National Academies: Institute of Medicine, National Academy of Sciences, and National Academy of Engineering. (2007). *Rising above the gathering storm: Energizing and employing America for a brighter economic future.* Washington, DC: The National Academies Press.

National Research Council. (1999). *How people learn: Bridging research and practice.* Washington, DC: The National Academies Press.

National Research Council. (2011). *Promising practices in undergraduate science, technology, engineering and mathematics education: Summary of two workshops.* N. Nielsen, Rapporteur. Board on Science Education, Division of Behavioral and Social Sciences and Education. Washington, DC: National Academies Press.

National Research Council. (2012a). *Discipline-based education research: Understanding and improving learning in undergraduate science and engineering.* S. R. Singer, N. R. Nielsen and H. A. Schweingruber, Eds. Board on Science Education, Division of Behavioral and Social Sciences and Education. Washington, DC: National Academies Press.

National Science Foundation (NSF). List of awards. http://www.nsf.gov/award search/. Accessed 2/20/2015.

National Science Foundation (NSF). NSF Budget Requests to Congress and Annual Appropriations. http://www.nsf.gov/about/budget/. Accessed 2/20/2015.

Maton, K. I., F. A. Hrabowski, M. Özdemir and H. Wimms. (2008). Enhancing representation, retention, and achievement of minority students in higher education: A social transformation theory of change. In *Toward Positive Youth Development*, M. Shinn and H. Yoshikawa, (Eds.). Oxford University Press.

President's Council of Advisors on Science and Technology (PCAST). (2012). *Engage to excel: Producing one million additional college graduates with degrees in science, technology, engineering, and mathematics*. Washington, DC: PCAST. http://www.whitehouse.gov/sites/default/files/microsites/ostp/pcast-executive-report-final_2-13-12.pdf. Accessed 10/20/2014.

Scileppi, J. A. (1988). *A systems view of education: A model for change*. Lanham, MD: University Press of America.

Sorcinelli, M. D. (2014). *Evidence-based teaching: What we know and how to promote it on your campus*. Keynote address at the New England Student Success Conference. Amherst, MA.

Svinicki, M. D. (2004). *Learning and motivation in the postsecondary classroom*. San Francisco, CA: Anker (now Jossey-Bass).

Weimer, M. (2013). *Learner-centered teaching: Five key changes to practice*. San Francisco, CA: Jossey-Bass.

ABOUT THE AUTHOR

Gabriela C. Weaver is the Vice Provost for Faculty Development and Director of the UMass Institute for Teaching Excellence and Faculty Development and Professor in the Department of Chemistry at the University of Massachusetts, Amherst in Amherst, MA.

2

Transforming Undergraduate STEM Education: Responding to Opportunities, Needs and Pressures

Martin Storksdieck

Comparative international student assessments like the Trends in International Mathematics and Science Study (TIMSS) and the Program for International Student Assessment (PISA) have shown consistently over the last 15 to 20 years that U.S. secondary students perform relatively weak academically compared to students in other developed countries, leading to concerns that the U.S. might lose its economic competitiveness in the long run (IOM, NAS, NAE 2007; NRC 2010). Moreover, the National Assessment of Educational Progress (NAEP), a test of students' performance across all 50 states conducted by the U.S. Department of Education's National Center for Education Statistics regularly reveals low performance as well. For instance, less than a third of eighth-grade students performed at or above the "proficiency" level in science in 2009 and 2011[1]. In mathematics in 2013, only 42% of fourth graders, 36% of eighth graders, and 26% of 12th graders[2] performed at or above "proficiency" level. All of these tests reveal major achievement gaps based on socio-economic status, parental education, and race/ethnicity, indicating an overall failure of elementary and secondary schooling to compensate systematically for broader societal inequities and calling into question basic notions of fairness. While these trends have long driven education policy, a 2010 report by the Presidents' Council of Advisors on Science and Technology (PCAST) on STEM education, followed by a National Research Council report on K–12 science education (NRC, 2012) that served as the guiding framework for the Next Generation Science Standards (NRC, 2013a) and two reports by the NRC on improving STEM education (NRC, 2011a; 2013b), have provided major momentum for significant changes in elementary and secondary STEM education in the U.S.

While the K–12 U.S. education system has long been portrayed as being in a state of crisis and in apparent need of widespread improvement, post-secondary education in the United States overall has long been considered a major

---

1. http://www.nationsreportcard.gov/science_2011/g8_nat.aspx?tab_id=tab2&subtab_id=Tab_1#chart or http://goo.gl/dvCJqf
2. http://www.nationsreportcard.gov/math_2013/

point of national pride and a system of international envy. However, recent high-visibility reports and articles in national newspapers are putting a spotlight on problems in post-secondary education in general, and undergraduate STEM education in particular. They suggest an urgent need for major reform that parallels that at the elementary and secondary level. Ultimately, insights on how people learn (NRC, 1999) are beginning to shape expectations for how we should teach, not only in K–12, but also in higher education. A 2012 report by the White House (PCAST, 2012) linked poor STEM education at the undergraduate level to a lack of STEM graduates overall. The report concludes that sub-par STEM education in the nation's colleges and universities will indirectly limit future economic growth and competitiveness of the U.S. economy.

A series of reports by the National Research Council between 2011 and 2015 focused on undergraduate STEM education in community colleges (NRC, 2012b) and introductory courses of baccalaureate granting institutions (NRC, 2011a; 2012c; 2015a); the reports found evidence-based practices that support student learning and retention to be widely missing. A separate report indicated a need to expand participation of underserved minorities in STEM at the college level (National Academy of Sciences, National Academy of Engineering, and Institute of Medicine, 2011). The above-mentioned reports summarized the tremendous amount of scholarship that has emerged over the last 20 years (Seymour & Hewitt, 1997), as did a seminal report that resulted from a joint initiative by the American Association for the Advancement of Science and the National Science Foundation, entitled *Vision and Change in Undergraduate Biology Education* (American Association for the Advancement of Science, 2011). A more recent report spearheaded by the Association of American Colleges and Universities (The Coalition for Reform of Undergraduate STEM Education, 2014) echoes these ideas. Research summarized in these and many other publications not only demonstrates an urgent need for improving STEM education at the undergraduate level, but many also provide guidance on how to achieve transformational change.

And transformational change is urgently needed, if only to respond to a new crop of students who will soon learn science, engineering and mathematics differently from previous generations of college-going students. The Next Generation Science Standards, Common Core Standards in Mathematics and English Language Arts, and the College Board's Advanced Placement Redesign all stress that learning should be facilitated through the practices in the relevant disciplines. This has implications for teaching itself, and for the expectations that students might have for what it means to receive a quality education. Hence, ongoing reforms in K–12 education might soon begin to influence graduating high school students, and could put pressure on two- and four-year institutions to provide more engaging STEM learning opportunities than is common today for incoming students. The cultural expectation that equates university education with anonymous large-scale lecture classes in which an instructor or professor acts as *sage on the stage* may soon be considered a quaint relic of the past (Deslauriers, L., Schelew, E., & Wieman, C., 2011).

Unfortunately, institutions of higher learning, maybe more so than most other organizations, are complex and somewhat resistant to change; neither top-down, nor bottom-up strategies alone are promising to be effective (Austin, 2012; Kezar, 2011). Additionally, higher education institutions today are operating in a national context that simultaneously calls for—yet puts strain on—their ability to provide improved STEM education: a mounting student debt crisis that is tied to decreases in public support for higher education; an associated threat to the business model in higher education, spurred by potentially disruptive innovations in educational technologies making online education increasingly possible and acceptable; a changing student population that requires renewed focus on so-called co-curricular services to help with retention, persistence and (on-time) graduation; and for-profit colleges that seem to speak to non-traditional student needs (even if they may not meet them in the end).

THE CONTEXT OF STEM REFORM IN HIGHER EDUCATION

The business model for many public institutions of higher learning is becoming unsustainable. Declining funding at the state level leads to associated unsustainable increases in tuition, making higher education increasingly unaffordable for coming generations, or leading to reduction in services at a time when increasingly more students are expected to attend college. At the same time federal funding for research does not keep up with the ever-increasing research enterprise at colleges and universities; returned overhead through research dollars will not provide additional income to universities to make up for dwindling state support. Tuition money is filling the gap, making post-secondary education in the U.S. one of the most expensive in the world for the student (OECD, 2014), resulting in the skyrocketing of student debt over the last decade to dangerous and unprecedented heights (U.S. Department of Education, 2013). While a "good" post-secondary education remains a smart investment for an individual, the benefit-cost analysis for many two- or four-year degrees might begin to look less attractive, particularly when students graduate into difficult economic conditions or are forced into jobs with fewer options to advance over time in order to service their high residual debt. Median family income for those with bachelor degrees for 2013 was almost $80,000, compared to associate degrees with

$56,000.[3] Even attending some college leads to a 20% income premium over stopping at a high school degree ($49,700 versus $40,700). Nonetheless, the immense costs of a college education in the U.S. to the individual (compared with most other countries in the world, including much of Europe) is seen as a "barrier to entry" into higher education, and is now spawning various initiatives to reduce college costs for students while improving the "quality of service", i.e., the effectiveness of the education that students receive. This is putting tremendous pressure on colleges and universities to rethink how they deliver an education, causing universities to struggle with the expectation to simultaneously improve education and lower cost. It is not clear right now whether all U.S. institutions of higher learning will succeed in this challenging task.

But quality versus cost is not the only balancing act for research-intensive colleges and universities. The evidence for student-centered, cognitively engaging undergraduate STEM teaching approaches is overwhelming and it is now less a question of whether, but when, they will become the norm rather than the exception. These approaches include, among others, using interactive and engaging techniques in large introductory courses and labs; providing students with authentic research and service experiences in freshman and sophomore years; using adapted mathematics and literacy support for struggling students; creating so-called "wrap-around" services known to support retention and student success for non-traditional, minority and first-generation college goers; and improving mentoring and coaching for students. This host of measures known to improve student success creates a conflict of priorities within those institutions that also want to (simultaneously) improve their research output. The current financial model of tenure-track faculty who use research funds to buy out teaching obligations and are replaced with contingent faculty or full-time instructors, with the overall trend toward a smaller fraction of courses being taught by tenure-track faculty members, is challenging the overall academic model of the research university. It is creating an unhealthy two-tier system in which obligations for teaching and student support are seen as secondary and lower-ranking compared to the university business of conducting research through external funding.

A recent public fascination with massive open online courses (MOOCs) has focused attention on new learning technologies and their promise for changing the way instruction occurs. New education technologies provide opportunities for individualized and adaptive learning, and if used prudently might benefit student success across the board. They can enhance on-campus courses by providing multiple modes of engaging students with the content, such as through

---

3. http://www.census.gov/hhes/www/income/data/historical/household/2013/h13.xls

blended models of in-class and online elements. Digitally-facilitated education is beginning to provide new models for how higher education might be delivered and how degrees may be granted in the near future. This technology-enhanced education can be delivered systematically as online degree programs, supplement existing programs with MOOC-format courses, or serve as part of a rapidly changing culture in which online classroom management systems like Canvass or Blackboard become the norm. Patchwork degrees with MOOCs in French Literature from Harvard, inexpensive, in-person seminars on writing from a local community college and a chemistry lab course from a local land-grant university, might become the new norm. Degrees might be granted based on competencies and portfolios by institutions that may not yet exist, at a cost that is far below the current norm. The discussion about educational technology application in higher education, however, oscillates between excitement about its potential, and concern about a two-tier system between a high-quality, high-touch, in-person education for those who can afford it, and a low-cost online and blended education based primarily on self-study for the rest, with serious questions about their equivalency.

Whether delivered in-person, in blended environments or online, there is new appreciation for the value of two-year associate degrees and post-secondary certificates as attractive alternatives to a baccalaureate degree, particularly for the many so-called "middle-skilled" jobs in STEM and health-related fields (professions that require postsecondary qualifications, but not a BS or BA degree). The current movement towards strengthening degree programs and post-secondary certificates from community and junior colleges, particularly in STEM fields, will take "business" away from four-year institutions. Already, transfer students from two-year institutions comprise more than 40% of the baccalaureate-bound students and community or junior colleges are delivering the education cheaper and in more intimate settings than universities. For-profit providers of post-secondary certificates and associate degrees had been gaining popularity in the last decade, despite concerns and public debates about quality, value and cost, and doubtful outcomes for participating students. Nonetheless, for-profits have reached an increasingly higher number of students, particularly from underserved minorities, veterans, and other nontraditional students. The reasons for their success in attracting students is based, among others, on targeted marketing and flexible ways to fit educational offerings into the schedule of working adults; this may provide lessons in how to address those students' educational needs that the more established public and private nonprofit colleges and universities could also embrace (Kinser, 2013). In fact, the increasing supply of online bachelor's degrees from established research universities might provide educational options for students with limited means and a high need for flexibility.

Taken together, these pressures and trends require much-needed change in colleges and universities. Current attempts by many of them to "grow" their way out of the impending financial, educational and technological crisis by attracting more out-of-state or international students who pay top tuition dollars, or by expanding the research enterprise in order to attract extramural (mostly federal) funding will quickly reach a limit of diminishing returns. Federal research funding is currently not keeping pace and, in fact, might even shrink relative to inflation in the near future. Another strategy, that of attracting international students, which taps into potentially endless demand, comes at yet to be determined costs since the universities will have to respond to the implicit promise of a superior tertiary education with costly delivery. Something has to give.

LOOKING TOWARD SOLUTIONS

Complex problems require complex and multi-pronged solutions. A host of national institutions that focus on the quality of higher education, from the National Academy of Sciences to the American Association for the Advancement of Science, and from all major university associations to PCAST, are beginning to not only write comprehensive reports about the student success dilemma, but are beginning to build on initial efforts to address it through structural change projects, supported with funding from the National Science Foundation, the National Institutes of Health, and some forward-looking private foundations, such as the Helmsley Trust, the Alfred P. Sloan Foundation, or the Howard Hughes Medical Institute. Beyond that, there are now hundreds of efforts around the country, in many colleges and universities, to address the quality of classroom-based instruction and the nature of laboratory courses, or to provide research experiences for undergraduate students, or to address more broadly questions of completion, retention, persistence and affordability. But individual, mostly course-level or small-scale efforts have not resulted in broad and sustainable approaches that fully address the challenges described above.

The chapters in this book provide a multitude of perspectives on enacting and supporting change at large scale and with lasting results. It is clear from this collection of narratives that there are many dimensions to consider for this level of transformation. Most recently, an impatient political system that seeks to protect students is demanding publicly available, simple indicators for college success, implying the existence of fast-acting simple solutions. If institutions of higher learning do not begin to address concerns about the quality of undergraduate (STEM) education, and embrace potential solutions actively, prudently, urgently and based on evidence, solutions with the potential for questionable results may be imposed externally, just as they were to elementary and secondary education.

REFERENCES

American Association for the Advancement of Science. (2011). *Vision and change in undergraduate biology education: A call to action.* Washington, DC: American Association for the Advancement of Science. Retrieved from http://vision andchange.org/finalreport.

Austin, A. (2011). *Promoting evidence-based change in undergraduate science education.* Paper Commissioned by the National Research Council's Board on Science Education. Retrieved from http://sites. nationalacademies.org/DBASSE /BOSE/DBASSE071087.

Deslauriers, L., Schelew, E., & Wieman, C. (2011). Improved learning in a large-enrollment physics class. *Science,* Vol. 332 no. 6031 pp. 862–864.

Institute of Medicine, National Academy of Sciences, and National Academy of Engineering. (2007). *Rising above the gathering storm: Energizing and employing America for a brighter economic future.* Washington, DC: The National Academies Press.

Kezar, A. (2011). What is the best way to achieve broader reach of improved practices in higher education? *Innovative Higher Education,* 36 (4): 235–247.

Kinser, K. (2013). *For-profit pathways into STEM.* Retrieved from http://sites.nation alacademies.org/cs/groups/dbassesite/documents/webpage/dbasse_088833 .pdf.

National Academy of Sciences, National Academy of Engineering, and Institute of Medicine. (2011). *Expanding underrepresented minority participation: America's science and technology talent at the crossroads.* Washington, DC: The National Academies Press.

National Research Council. (1999). *How people learn: Brain, mind, experience, and school.* Washington, DC: The National Academies Press.

National Research Council. (2005). *America's lab report: Investigations in high school science.* Washington, DC: The National Academies Press.

National Research Council. (2009). *Learning science in informal environments: People, places, and pursuits.* Washington, DC: The National Academies Press.

National Research Council. (2010). *Rising above the gathering storm, revisited: Rapidly approaching category 5.* Washington, DC: The National Academies Press.

National Research Council. (2011a). *Successful K–12 STEM education: Identifying effective approaches in science, technology, engineering, and mathematics.* Washington, DC: The National Academies Press.

National Research Council. (2011b). *Promising practices in undergraduate science, technology, engineering, and mathematics education: Summary of two workshops.* Washington, DC: National Academies Press.

National Research Council. (2012a). *A framework for K–12 science education: Practices, crosscutting concepts, and core ideas.* Washington, DC: The National Academies Press.

National Research Council. (2012b). *Community colleges in the evolving STEM education landscape: Summary of a summit.* Washington, DC: The National Academies Press.

National Research Council. (2012c). *Discipline-based education research: Understanding and improving learning in undergraduate science and engineering.* Washington, DC: National Academies Press.

National Research Council. (2013a). *Next Generation Science Standards: For states, by states.* Washington, DC: The National Academies Press.

National Research Council. (2013b). *Monitoring progress toward successful K–12 STEM education: A nation advancing?* Washington, DC: The National Academies Press.

National Research Council. (2015a). *Reaching students: What research says about effective instruction in undergraduate science and engineering.* Washington, DC: The National Academies Press.

National Research Council. (2015b). *Guide to implementing the Next Generation Science Standards.* Washington, DC: The National Academies Press.

OECD (2014). *Education at a glance 2014: OECD indicators.* OECD Publishing. Retrieved from http://dx.doi.org/10.1787/eag-2014-en.

President's Council of Advisors on Science and Technology. (2010). *Prepare and inspire: K–12 science, technology, engineering, and math (STEM) education for America's future.* Washington, DC: The White House. Retrieved from https://www.whitehouse.gov/sites/default/files/microsites/ostp/pcast-stem-ed-final.pdf

President's Council of Advisors on Science and Technology. (2012). *Engage to excel: Producing one million additional college graduates with degrees in science, technology, engineering, and mathematics.* Washington, DC: The White House. Retrieved from http://www.whitehouse.gov/sites/default/files/microsites/ostp/pcast-engage-to-excel-final2-25-12.pdf.

Seymour, E. & Hewitt, N.M. (1997). *Talking about leaving: Why undergraduates leave the sciences.* Boulder, CO: Westview Press.

The Coalition for Reform of Undergraduate STEM Education. (2014). *Achieving systemic change: A sourcebook for advancing and funding undergraduate STEM education.* Fry, C.L. (ed.). Washington, DC: Association of American Colleges and Universities. https://www.aacu.org/sites/default/files/files/publications/E-PKALSourcebook.pdf

U.S. Department of Education. (2013). Degrees of debt, student borrowing and loan repayment of bachelor's degree recipients one year after graduating: 1994, 2001, and 2009. *Stats in Brief, NCES* 2014-011. Retrieved from http://nces.ed.gov/pubs2014/2014011.pdf.

ABOUT THE AUTHOR

Martin Storksdieck is the Director of the Center for Research on Lifelong STEM Learning and a professor in the College of Education and the School of Public Policy at Oregon State University in Corvallis, Oregon.

SECTION A
Theories and Models of Institutional Transformation

This section begins by placing the reform of undergraduate STEM instruction in the larger context of desired outcomes across the levels of education, and emphasizes the importance of approaching the work of transformation with awareness of how particular interventions fit within a knowledge base and contribute to it. Kezar and Holcombe review understandings from a broad literature on organizational change, as they apply specifically to higher education. Miller and Fairweather, and Elrod and Kezar present two initiatives that use frameworks as a tool for supporting institutional level change across a network of participating institutions. The first chapter begins with a logic model and invites participating institutions to respond with local initiatives within the model. The second chapter, grounded in well-known principles of organizational change, invites the participating institutions to develop their change model based on perceived needs for change.

1

The Reform of Undergraduate Science, Technology, Engineering, and Mathematics Education in Context: Preparing Tomorrow's STEM Professionals and Educating a STEM- Savvy Public

Joan Ferrini-Mundy, Layne Scherer, and Susan Rundell Singer[1]

INTRODUCTION

The myriad calls for the transformation and improvement of our nation's undergraduate science, technology, engineering and mathematics (STEM) education enterprise are situated within a broader set of concerns about the preparation of the scientific workforce and the education of a public that will use and support science. We wish to focus on two main drivers for the current priority on the transformation of undergraduate STEM education and describe the contexts and approaches that guide the National Science Foundation's (NSF) investment priorities in this area.

The first driver is the demand for a diverse STEM professional workforce prepared to advance the frontiers of science and engineering given the increasing need for STEM knowledge to address the most pressing challenges faced by society. The second is the need to consider the undergraduate STEM experience as an opportunity to provide students foundational skills and knowledge required by the evolving role of technology and data across the spectrum of personal and society-level decisions including health, food, energy, the environment, and finances. Both of these drivers recognize the vital role that undergraduate STEM education has on developing not only the experts that will expand the frontiers of discovery, but also on educating a STEM-literate public equipped with the skills they need to be informed consumers, and increasingly, generators, of data.

In the summary of the 1944 landmark report *Science the Endless Frontier*, American engineer Vannevar Bush proposed to President Roosevelt the establishment of the basic structures that would later become NSF:

1. Any opinion, findings and conclusions or recommendations expressed in this material are those of the authors and do not necessarily reflect the views of the National Science Foundation.

Basic scientific research is scientific capital . . . How do we increase this scientific capital? First, we must have plenty of men and women trained in science, for upon them depends both the creation of new knowledge and its application to practical purposes.

The subsequent National Science Foundation Act of 1950 affirms that perspective in authorizing a federal agency to both support research in science and to strengthen education.

Our work in the Directorate for Education and Human Resources (EHR) at NSF attends to both goals of building the workforce, in addition to public understanding, and we recognize the unique circumstances of the NSF in the full integration of science and education, both organizationally and conceptually.

STEM UNDERGRADUATES, STEM CAREERS, AND PUBLIC UNDERSTANDING OF STEM

Features of U.S. undergraduate STEM education provide context for understanding the landscape in which NSF investments are planned and executed. In 2010, 70% of STEM degree recipients were working in fields closely aligned with their degree and 24% were working in a related field, according to the National Science Board (NSB, 2014). This contrasts with 35% of degree recipients in other fields working in jobs closely aligned with their college major and 32% working in fields somewhat related to their degree. The tighter correspondence with employment underscores the value of high quality STEM degree preparation. The projected demand for STEM undergraduates remains crucial as the U.S. Bureau of Labor Statistics (2014) anticipates that employment in STEM-related occupations will grow by more than 9 million between 2012 and 2022. Producing more high-quality scientists and engineers, with the conceptual understanding and adaptability to succeed in the rapidly changing world of work, is a national need. Given the tight link between degree and employment, aligning collegiate learning and learning environments with workforce skills, including intra- and interpersonal-skills, is a promising practice in STEM education.

A greater focus on undergraduate STEM education can have impacts that reach far beyond the students who earn bachelor's degrees in science and engineering. Within the workplace, an increasing number of occupations require some STEM skills. Jonathan Rothwell (2013) mined the Department of Labor's Occupational Information Network Data Collection Program (O*NET) to identify jobs that require STEM knowledge, though they may not be classified as science and engineering using the Bureau of Labor Statistic's definition. Rothwell found that 20% of all jobs in the U.S. require some STEM knowledge and half of these jobs require less than a bachelor's degree. Individuals in STEM and STEM-related jobs requiring less than a bachelor's degree enjoy a 10% pay differential compared to jobs with similar education requirements (14% differential for those requiring a bachelor's degree). According to Rothwell, metropolitan areas with greater numbers of jobs requiring some STEM knowledge have stronger economies with less unemployment and a greater share of exports as a function of GDP. Growth in STEM-related fields is illustrated by a 167% increase in associate's degrees granted in health care fields from 2002 to 2012 (NSB, 2014).

High-quality STEM education is also relevant for college graduates who will not utilize STEM skills in the work place. According to the National Science Board (2014), only 8% of all associate's degrees and 32.6% of all bachelor's degrees granted in 2012 were in traditional science and engineering fields, excluding health care fields. For the majority of college students, a required general education quantitative literacy or science course may be the last formal study of the subject for such students. While these students will go on to enter a range of career pathways with varying direct application of STEM skills, they will also go on to be parents, consumers, community leaders, and school board members. Currently, there is limited formal research available to describe the uses of STEM knowledge in everyday life, but there is an increasing body of knowledge about how to best engage the public in STEM. One thing we do know is that the U.S. public seems interested in science. Recent results from the National Science Board (2014) indicate that a majority of Americans say they are "interested in 'new scientific discoveries' . . . that the benefits of science outweigh the potential harms." Americans say that they "hold positive views of scientists and engineers," and the public's level of factual knowledge about science is comparable generally to levels in Europe. Citizen science activities, for example, are becoming more accessible to a wider range of individuals and groups and provide interesting opportunities for intermittent engagement with science over the life span (Bonney et al., 2014). Remembering that undergraduate STEM education may be the final encounter with these disciplines in a formal educational setting for the majority of associate's and bachelor's degree earners, it may be important to better understand how those experiences contribute best to equipping people with the foundational knowledge, skills, and understanding of the nature of science to inform and enrich their daily lives.

KEY CONTEXTS FOR NSF INVESTMENT IN UNDERGRADUATE EDUCATION

Recently, the federal government has increased its level of attention to STEM education with undergraduate education as one of the key areas to leverage funding across agencies for greater impact. A significant milestone for improving government investment in undergraduate education came with the Committee on STEM Education's (CoSTEM) publication of the *Federal Science, Technology, Engineering, and Mathematics (STEM) Education 5-Year Strategic Plan* (2013). This report, requested by Congress in the America Competes Reauthorization Act of 2010, identifies "five priority STEM education investment areas where a coordinated Federal strategy can be developed. . . ." (CoSTEM, 2013, p. iii). One of the priority areas is undergraduate education, where the goal is to "Enhance the STEM experience of undergraduate students. Graduate one million additional students with degrees in STEM fields over the next 10 years" (CoSTEM, 2013, p. 10). The four strategies proposed for achieving the undergraduate goal are:

- implement evidence-based instructional practices and innovations to improve undergraduate STEM learning and retention;
- focus on STEM education at two-year colleges and build bridges to four-year programs;
- expand partnerships outside of higher education to provide early research experiences; and
- address the gatekeeper issue of beginning college mathematics.

Given that fourteen agencies are working together to build coherence and coverage across the federal investments, each agency's investments are situated within this larger set. In particular, NSF plays a distinctive role in funding basic research to advance the progress of science and build the prosperity of the nation, while also funding a robust portfolio of investments in education. As the only federal agency with a dual mandate in research and education, NSF also provides leadership in integrating frontier science with cutting-edge education research.

Additionally, one of the other priority goals focuses on how to better serve students from groups that have been historically underrepresented in STEM: "Increase the number of underrepresented minorities who graduate college with STEM degrees in the next 10 years and improve women's participation in areas of STEM where they are significantly underrepresented" (CoSTEM, 2013, p. 11). These two goals, which coincide with the NSF's aim to build a diverse STEM workforce through improved undergraduate experiences, indicate the growing demand for a broader conversation to ensure federal funds support effective programs to improve outcomes for students.

In the FY 2014 NSF President's Budget Request, NSF introduced plans for a coherent undergraduate STEM education investment by announcing "Catalyzing Advances in Undergraduate STEM Education" (renamed "Improving Undergraduate STEM Education" subsequently). The goal is "to provide coherence across all NSF undergraduate education programs to maximize the effectiveness of NSF learning experiences of all undergraduates" (NSF, 2013).

EHR also has a framework to guide investments across this directorate. All investments are now identified within one of these themes: improving learning and learning environments; broadening participation and capacity; and building a STEM professional workforce (NSF Federal Advisory Committee for Education and Human Resources, 2014). These themes provide different lenses to view the portfolio of our work, providing a schema to understand how the specific outcomes targeted by each project and program build into a complementary and comprehensive group.

EHR's efforts to enhance its undergraduate investment are guided by all of these contexts: the CoSTEM plan, the agency-wide undergraduate STEM education framework, and the EHR investment themes. At the time of this writing, there are seven distinct EHR programs[2] that provide funding for the improvement of undergraduate education and are categorized as undergraduate programs. There are several other programs that are closely related to, or dependent on, the improvement of undergraduate STEM education, including programs supporting K–12 education and graduate education. These programs recognize the movement of students through and across the educational ecosystem and that the benefits students gain at one level may improve their abilities further downstream.

DEVELOPING AN INVESTMENT STRATEGY

The framework below has been a helpful guide for clarifying the different contributions to research and development, as well as human capital for our programs. For example, the *Improving Undergraduate STEM Education (IUSE): EHR* program (http://www.nsf.gov/funding/pgm_summ.jsp?pims_id=505082) encompasses design research, small-scale implementation, and in some cases, scaling. The Scholarships for Science, Technology, Engineering, and Mathematics (S-STEM) program (http://www.nsf.gov/funding/pgm_summ.jsp?pims_id =5257) builds capacity by providing scholarships.

---

2. Improving Undergraduate STEM Education (IUSE): Education and Human Resources; Advanced Technological Education (ATE); Scholarship in Science, Technology, Engineering, and Mathematics (S-STEM); Historically Black Colleges and Universities Undergraduate Program (HBCU-UP); Louis Stokes Alliances for Minority Participation (LSAMP); Tribal Colleges and Universities Program (TCUP); and the Robert Noyce Teacher Scholarship Program. See Summary Table 13—*National Science Foundation CoSTEM Inventory and Postdoctoral Fellowship Programs By Level of Education, Request to Congress* in NSF FY2016 Budget.

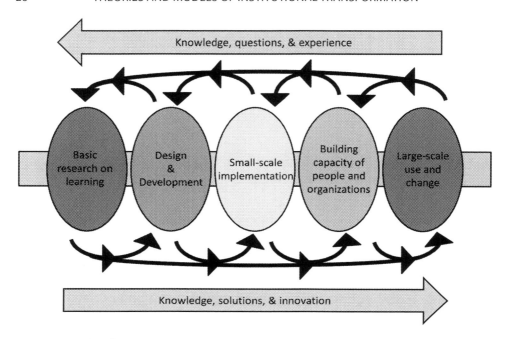

FIGURE 1. Cycle of Innovation

EHR continually reviews the fit of our own programs with one another, as well as within the context of the other education-focused programs at NSF. For example, IUSE explicitly coordinates investments in undergraduate education across the NSF to leverage collective impact. Regular Committee of Visitors reviews, ongoing external evaluations of program impact, new reports from the National Academies and various professional societies, and findings from education research all are considered in the design and development of programs and funding approaches. In considering programs, we ask such questions as:

- What are the main goals that this program will address?
- Do these goals align with documented national challenges in education?
- How do the goals of this program fit with, and complement or leverage, existing programs in EHR and in NSF?
- Will the proposed program mechanisms bring in a body of work that will address the national need?
- How does existing research support the approach we hope to implement and what data systems exist currently to provide information on the issue?

- What body of knowledge will be enhanced through these investments?
- What can we expect to be the national impact on STEM education?

Increasingly across NSF, planning tools such as logic models are used at the program level and are encouraged at the project level (W.K. Kellogg Foundation 2004). The NSF FY 2016 Budget request sets forth a goal for FY 2016 that "NSF will have incorporated logic models/theory of change language in the rationale for all new programs" (NSF, 2015, Performance, p. 44). There are various approaches to logic modeling, but in some form, all call for articulating the following core components: the overall goal of a program; assumptions and contexts to consider; the inputs (e.g. available resources, human capital) the activities or program components that will be implemented; and a series of short-term, intermediate, and long-term outputs, outcomes, and impacts.

Inputs include the nation's institutions of higher education, the set of funding programs, recommendations from the National Academies, the President's Council of Advisors on Science and Technology, and professional societies, and growing bodies of research to guide improvement. The set of programs calls, collectively, for *activities* in all components of the cycle in Figure 1: research, design and development of instructional approaches; implementation and testing of innovative interventions; capacity-building through faculty development and student support; and large-scale efforts, for example using technology.

Short-term *outcomes* and *outputs* include a diverse program portfolio, a growing evidence base about high quality instructional practices, assessment tools, faculty and graduate students who are implementing improved approaches, and increased numbers of students from groups that have been underrepresented in selecting STEM fields. Longer-term *outcomes* include not only wide use of instruction approaches, but measurable change and improvement in student learning, retention of students from underrepresented groups in STEM majors, numbers of students completing STEM degrees in the transformed institutions, transition to and success in graduate study, and ultimately successful placement in and success in the STEM workforce, along with a science-literate public.

MAKING IMPROVEMENTS AT SCALE

The NSF has a pivotal role to play in the improvement of STEM education; however, given the scope of the U.S. education system, NSF must deploy resources strategically to achieve the greatest impact. Overall, there is approximately $1.1 trillion invested in U.S. education annually. Of that $1.1 trillion, 94% of the investment is spent on the nation's state and local K–12 and postsecondary systems. Most of the remaining 6% is spent by the U.S. Department of Education on the K–12 system. Only 0.3% of all federal investment is directed at STEM education, and only a small portion of this is NSF's undergraduate investment (Federal Inventory of STEM Education Fast-Track Action Committee, 2011). Thus NSF investments will not directly reach all institutions and all STEM students, so having strategies in mind for "scaling" or propagating is a critical part of our activity. We discuss four here.

Scaling Through Research

In Figure 1, the feedback and feed-forward arrows are particularly important parts of the system of investment. In order to move, for instance, to large-scale implementation of well-tested ideas, there will be cycles and iterations where learnings from one line of work are influencing and informing the next, and then the challenges that come up in that phase result in feedback to inform a next round of development. Elements of this system are captured in the *Common Guidelines for Education Research and Development,* a report produced jointly by NSF's EHR and the U.S. Department of Education's Institute of Education Sciences (IES). Six types of research are described, and in each description it is clear how the evidence and findings of each type are related to the other types (IES & NSF, 2011, p. 9). For instance, *foundational research* will produce basic understandings that will "influence and inform research and development in different contexts." *Design and development research* (much NSF funding is in this category) "develops solutions for achieving a learning goal, and such studies themselves have internal iterations." Successful design and development projects, where interventions are well tested and improved in a given setting, can lead to the more extensive efficacy and effectiveness studies that would provide evidence that an approach is ready for wide scaling. Of course with technological approaches that can be used widely immediately after (or even during) design, the "stages" suggested in Figure 1 may not progress sequentially.

Formal methodologies for synthesizing research studies can provide robust evidence to justify scaling. A recent example is the meta-analysis conducted by Freeman, et al. (2014), that synthesized 225 studies of instructional approaches to promote active learning. The study found that, "average examination scores improved by about 6% in active learning sections, and that students in classes with traditional lecturing were 1.5 times more likely to fail than were students in classes with active learning" (Freeman, 2014). The study also points to class size as a factor as the greatest effects were seen in small classes ($n \leq 50$). Critical to meta-analysis is sound design and thorough reporting of the source studies.

Another way in which research is related to scale is in building a strong record of learning through implementation research (Fishman, Penuel, Allen,

Cheng, & Sabelli, 2013), or improvement research (Bryk, Gomez, & Grunow, 2011). These approaches bring practitioners and researchers together throughout the process for iterative design, improvement, and assessment cycles.

Scaling Through Commercialization and Wide Use

Scaling of effective ideas can occur through commercialization and passing along the responsibility for widespread marketing and dissemination to the private sector. In past decades, EHR has funded curriculum projects that required the development teams to partner with a potential publisher at the beginning of the project. The Instructional Materials Development (IMD) program, which ran from 1990–2006, required the applicants to include a robust plan for dissemination in their proposal, complete with a timeline for securing a publisher, projected sales, and income.

The NSF Innovation Corps (I-Corps™) program supports a set of activities and programs that prepare scientists and engineers to extend their focus beyond the laboratory and broaden the impact of select, NSF-funded, basic-research projects, providing another vehicle for scaling. We have launched I-Corps for Learning, which encourages proposals that take discoveries and promising practices from education research and development and promote opportunities for widespread adoption, adaptation, and utilization. (http://www.nsf.gov /pubs/2015/nsf15050/nsf15050.pdf)

Scaling Through Partnering

Partnering, with institutions from other sectors or with other government agencies, is a good way of extending the reach of successful interventions and programs. Two recent examples in NSF's undergraduate portfolio demonstrate this. In FY 2012, NSF announced a cooperative activity among NSF, Intel Foundation, and GE Foundation to stimulate comprehensive action at universities and colleges to help increase the annual number of new B.S. graduates in engineering and computer science by 10,000. Proposals for support of projects were submitted under a special funding focus (Graduate 10K+) within the NSF Science, Technology, Engineering, and Mathematics Talent Expansion Program (STEP, see http://www.nsf.gov/funding/pgm_summ.jsp?pims_id=5488; http:// www.nsf.gov/pubs/2012/nsf12108/nsf12108.jsp). The partners from the private sector added funds to NSF's STEP program and thus leveraged reach to a broader set of students with NSF's investment platform as a basis.

The Advanced Technological Education (ATE) program is an interesting example of an undergraduate partnership with another government agency. ATE, with a budget of $66M in FY 2015, has built expertise over two decades, and both informs and leverages a shorter-term infusion of funds by the U.S.

Department of Labor in the Trade Adjustment Assistance Community College and Career Training Grant Program (TAACCT), that invested almost $2 billion from 2010 to 2015 to support displaced adults in developing employable skills. Several ATE Centers have combined their collective expertise to provide technical assistance to TAACCT awardees, through webinars and other activities with the effort coordinated by the National Convergence Technology Center (http://www.connectedtech.org/).

Using Alliances and Networks to Spread Best Practices

One role that the NSF currently plays and can continue to advance is to help connect researchers and practitioners across networks to encourage the sharing of best practices, innovative theories, and wisdom of experience. One way that such scaling happens is when large numbers of individual institutions within classes of institutions are part of an NSF program. For example, the EHR-based Historically Black Colleges and Universities Undergraduate Program (HBCU-UP), and Tribal Colleges and Universities Program (TCUP) are interesting because they can reach most of the institutions in the eligible cohorts, so the "scaling" approach includes reaching all institutions. In the 16 years of HBCU-UP, 84 of the 105 HBCUs have had support, and in the 15 years of the TCUP program 45 different institutions have received TCUP funding, encompassing nearly all TCUs. Thus, best practices spread through networks of institutions that have similar goals and contexts.

Alliances of institutions also provide powerful interconnections and networks. Within the Louis Stokes Alliances for Minority Participation (LSAMP), an EHR program now in its 24th year, the 49 distinct alliances supported have included more than 700 institutions. Colleges and universities with different contexts and student populations work together to increase capacity, including student support, to improve institutional effectiveness for bringing more students from underrepresented groups into science.

A number of NSF's undergraduate programs have or are seeking proposals for resource networks and centers, organizations designed to serve both as repositories of best practices and to push ideas to the field, and to catalyze and facilitate interactions. Examples include STEM Central (https://stem-central .net), which originated as an NSF-funded center focused on connecting principal investigators involved in the STEM Talent Expansion (STEP) program and ATE Central (https://atecentral.net/) which supports the highly collaborative ATE community. EHR's Division of Human Resource Development is funding a set of design projects that will lead to the development of a proposal for a Broadening Participation in STEM Resource Network (BPS-Resource Network).

LOOKING FORWARD

In the transformation of undergraduate STEM education, there is a critical role for scientists and engineers who, with their deep disciplinary knowledge, can offer leadership, especially designing how to incorporate attention to areas and practices that are emerging in science and engineering. For instance, as the practice of science across all fields is becoming data intensive—how might this be reflected in the undergraduate education in STEM in the future? The NSF has as a priority goal in the current strategic plan: "Improve the nation's capacity in data science by investing in the development of human capital and infrastructure" (NSF, 2014). For undergraduate education, this raises questions around the curricular topics and approaches for engaging students, and points to the need for research on conceptual understanding and learning progressions in data science.

The integration of data science into undergraduate STEM curricula would not only serve graduating students entering the job market with valuable skills, but could also increase the number of students drawn to STEM fields. At the 2014 Conference Board of Mathematical Sciences Forum on *The First Two Years of College Math: Building Student Success,* Time Warner data manager Sabrina Schmidt reflected on her experience as a mathematics major at Vassar College. In her session called "What I Wished I Had Learned," Schmidt discussed how she wished she had "learned more about math's crucial and expanding contributions in today's business landscape" and commented:

> If students were more aware of which areas of math are prominently used in the applications that most interest them, I believe they would be more apt to take some introductory courses in statistics and computer science in their first two years of college (Schmidt, 2014).

Schmidt's suggestions lay bare the critical need to show students the practical applications of STEM concepts as a way to spark interest.

Biology is an example of a field that has made a very sharp transition from a descriptive science to a data-intensive science, presenting both opportunity and challenge for the preparation of future biologists. Over an eight-year period, NSF in partnership with the Howard Hughes Medical Institute, the National Institutes of Health, and the U.S. Department of Agriculture has supported a broad community initiative, *Vision and Change in Undergraduate Biology Education* (http://visionandchange.org) that has rethought both the mathematical and otherwise interdisciplinary nature of biology education. Most recently, in collaboration with the Directorate for Mathematics and Physical Sciences and the Directorate for Computer and Information Science and Engineering,

EHR funded the Quantitative Undergraduate Biology Education and Synthesis (QUBES) project (http://qubeshub.org). QUBES is an alliance of mathematical and life science organizations and universities to develop resources and professional development needed to grow the next generation of quantitative biologists who can innovate in the world of data science. QUBES is designed through both scale and partnerships to propagate effective practice in data science applications for biologists.

Another key consideration is the growing interdisciplinary nature of science. At the University of California, Los Angeles, a freshman seminar titled "Diversity and Complexity: Introduction to Modeling Complex Systems" will be taught jointly by a mathematician (Mark Green) and a sociologist (Jacob Foster). New approaches to the design of courses that demonstrate interdisciplinarity may be very engaging for students. The National Science Foundation Research Traineeship program (NRT) has a focus on building interdisciplinary programs for graduate students in STEM, and thus raises the interesting question of what might be the key elements in the undergraduate preparation for such programs.

It is indeed an exciting time in STEM education, with innovation and change at the undergraduate level providing great opportunities for new generations of scientists and for a literate public. We in the Directorate for Education and Human Resources are pleased to be a part of this transformation.

REFERENCES

Bonney, R., Shirk, J. L., Phillips, T. B., Wiggins, A., Ballard, H. L., Miller-Rushing, A. J., & Parrish, J. K. (2014). Next steps for citizen science. *Science,* 343, 1436–1437.

Bryk, A. S., Gomez, L. M., and Grunow, A. (2011). Getting ideas into action: Building networked improvement communities in education. In M. T. Hallinan (Ed.), *Frontiers in Sociology of Education* (pp. 127–162).

Bush, V. (1945). *Science: The endless frontier* (Washington, D.C.). Retrieved from http://www.nsf.gov/about/history/vbush1945.htm#transmittal

Fishman, B. J., Penuel, W. R., Allen, A. R., Cheng, B. H., & Sabelli, N. (2013). Design-based implementation research: An emerging model for transforming the relationship of research and practice. *National Society for the Study of Education*, 112, 136–156.

Freeman, S., Eddy S. L., McDonough, M., Smith, M. K., Okoroafor, N., Jordt, H., & Wenderoth, M. P. (2014). Active learning increases student performance in science, engineering, and mathematics. *Proceedings of the National Academy of Sciences of the United States of America.* 111, 8410–8415. Retrieved from http://www.pnas.org/content/111/23/8410.full

Institute of Education Statistics, United States Department of Education, & the National Science Foundation. (2011). *Common guidelines for education research and development*. Retrieved from http://ies.ed.gov/pdf/CommonGuidelines.pdf

National Science and Technology Council, Committee on STEM Education, Federal Inventory of STEM Education Fast-Track Action Committee. (2011). *The federal science, technology, engineering, and mathematics (STEM) education portfolio*. Retrieved from http://www.whitehouse.gov/sites/default/files/microsites/ostp /costem__federal_stem_education_portfolio_report.pdf

National Science and Technology Council, Committee on STEM Education. (2013). *The federal science, technology, engineering, and mathematics (STEM) 5-year strategic plan*. Retrieved from http://www.whitehouse.gov/sites/default/files/microsites/ostp/ stem_stratplan_2013.pdf

National Science Board (2014). Science and technology: Public attitudes and understanding. In *Science and Engineering Indicators 2014*. Retrieved from http://www.nsf.gov/statistics/seind14/index.cfm/chapter-7

National Science Foundation Act of 1950, 42 U.S.C. §1861.

National Science Foundation. (2013). *FY 2014 Budget Request to Congress*. Retrieved from http://www.nsf.gov/about/budget/fy2014/pdf/fy2014budget.pdf

National Science Foundation. (2014). *National Science Foundation: Investing in science, engineering, and education for the nation's future, strategic plan for 2014–2018*. Retrieved from http://www.nsf.gov/pubs/2014/nsf14043/nsf14043.pdf

National Science Foundation. (2015). *FY 2016 Budget Request to Congress*. Retrieved from http://www.nsf.gov/about/budget/fy2016/pdf/fy2016budget.pdf

National Science Foundation Federal Advisory Committee for Education and Human Resources. (2013). *Strategic re-envisioning for the Education and Human Resources Directorate: A report to the Directorate for Education and Human Resources, National Science Foundation*. Retrieved from http://www.nsf.gov/ehr /Pubs/AC_ReEnvisioning_Report_Sept_2014_01.pdf

Rothwell, J. (2013). *The hidden STEM economy*. Washington, D.C.: Brookings Institute. Retrieved from http://www.brookings.edu/research/reports/2013/06/10 -stem-economy-rothwell

Schmidt, Sabrina. (2014). *What I wish I had learned in the first two years of college mathematics*. Retrieved from www.cbmsweb.org/Forum5/Schmidt.docx

United States Department of Education, Institute of Education Statistics, National Center for Education Statistics. (2009). Table 3.5. Relatedness of postbaccalaureate job and bachelor's degree major: Among 2007–08 first-time bachelor's degree recipients who were employed, percentage who did not consider their current job related to their undergraduate major and percentage distribution of reasons for working outside their bachelor's degree field, by selected individual and institutional characteristics: 2009. *2008/09 Baccalaureate and Beyond Longitudinal Study*. Retrieved from http://nces.ed.gov/datalab/tableslibrary/view table.aspx?tableid=8856

United States Department of Education, Institute of Education Statistics, National Center for Education Statistics. (2012). Table 2. Employment and enrollment: Percentage distribution of 2007–08 bachelor's degree recipients' employment and additional postsecondary enrollment status, by demographic and enrollment characteristics: 2012. *2008/12 Baccalaureate and Beyond Longitudinal Study.* Retrieved from http://nces.ed.gov/datalab/tableslibrary/viewtable.aspx?tableid=9501

United States Department of Education, Institute of Education Statistics, National Center for Education Statistics. (2012). Table 2.2 Major field of study: Percentage of undergraduates majoring in STEM fields and percentage distribution of undergraduates' major field of study, by selected institution and student characteristics: 2011–12. *2011–12 National Postsecondary Student Aid Study.* Retrieved from http://nces.ed.gov/datalab/tableslibrary/viewtable.aspx?tableid=9538

U.S. Department of Labor, Bureau of Labor Statistics. (2014, March). *STEM 101: Intro to tomorrow's jobs.* Retrieved from http://www.bls.gov/careeroutlook/2014/spring/ art01.pdf

W.K. Kellogg Foundation. (2004). *Logic Model Development Guide.* Retrieved from http://www.wkkf.org/~/media/pdfs/logicmodel.pdf

ABOUT THE AUTHORS

Joan Ferrini-Mundy is the Assistant Director, Directorate for Education and Human Resources at the National Science Foundation in Arlington, Virginia.

Layne Scherer is a Science Assistant, Directorate for Education and Human Resources at the National Science Foundation in Arlington, Virginia.

Susan Rundell Singer is the Division Director, Division of Undergraduate Education, Directorate for Education and Human Resources at the National Science Foundation in Arlington, Virginia.

2

Institutional Transformation in STEM: Insights from Change Research and the Keck-PKAL Project

Adrianna Kezar and Elizabeth Holcombe

THE NEED FOR AN INSTITUTIONAL FOCUS FOR CHANGE

Over the last several decades, policy makers, business leaders, and other higher education stakeholders have repeatedly called attention to problematic trends in science, technology, engineering, and math (STEM) education in colleges and universities. While many colleges have begun making changes to the way they teach and support STEM students, these change efforts have almost always occurred at the departmental level (Austin, 2011, Fairweather, 2008; Henderson, Beach, & Finkelstein, 2011). Few have reached the institutional level of entire programs or colleges in the STEM disciplines. There is growing recognition that reform in STEM is an institutional imperative, rather than only a departmental one (Austin, 2011; Fairweather, 2008; Henderson et al., 2011; Kezar, 2011). For example, institution-wide implementation of high-impact practices (HIPs) has shown to dramatically improve the graduation rates of under-represented minority (URM) students (Kuh & O'Donnell, 2013). Student advising, faculty professional development, student research mentoring, academic support programs, clear STEM-focused institutional articulation agreements, external partnerships with business and industry related to internships, among other critical areas, are often overlooked within reform efforts, yet have been identified as central to STEM student success (Fairweather, 2008; Kuh, 2008). The Meyerhoff Scholars Program at the University of Maryland, Baltimore County, epitomizes this type of institution-wide effort and combines specific academic, social and research support interventions that have resulted in dramatic improvements in graduation of minority STEM students (Maton, Hrabowski, & Schmitt, 2000).

Most prior initiatives and reports about STEM reform have been aimed at altering individual faculty or departmental activities (Henderson et al., 2011; Fairweather, 2008; Austin, 2011). There is little research that has helped leaders to understand various interventions that might be implemented that extend beyond departments and create an institutional vision for STEM reform. In addition, earlier efforts have not addressed the policies and practices at the institutional level that often hinder reforms but can be leveraged to enable greater changes. There is research about the institutional change process that can help campus leaders in STEM undertake this type of systemic reform (Association of American Universities, 2013; Beach, Henderson, & Finkelstein, 2012). This chapter will use the new Keck-Project Kaleidoscope (Keck-PKAL) change model to examine institutional reform efforts in STEM education. We will frame this discussion with a review of three important theoretical concepts from the institutional change literature that are helpful for understanding this institutional change process. After reviewing organizational culture, sense-making and multi-frame leadership, we will conclude with a brief discussion of how this systemic and comprehensive approach to STEM reform requires leadership at all levels of the institution, from department faculty to student affairs professionals, to deans, provosts and presidents. Culture changes much more readily when institutional actors are aligned (Kezar, 2013).

WAYS TO SUPPORT INSTITUTIONAL CHANGE— KECK-PKAL PLANNING GUIDE

One resource for helping to navigate the challenge of creating institutional-level change is the new Keck-Project Kaleidoscope (Keck-PKAL) change model, also reviewed in another chapter in this volume in greater detail. Sponsored by the W.M. Keck Foundation and AAC&U's Project Kaleidoscope, this initiative built a comprehensive institutional model to help campuses implement evidence-based STEM reforms. The most important aspect of this model is that it takes a systems approach to change. Previous efforts at STEM reform largely ignore the broader ecosystem of areas that need to be changed to support student learning and the ecosystem of structures, policies and practices that need to be adjusted to maintain and sustain change, such as tenure and promotion policies and institutional culture. The Keck-PKAL model begins by establishing a vision and goals for the change project and then guides campus teams through an analysis phase to gather data and collect information about the current STEM learning and student success landscape. This analysis leads to the identification of specific campus challenges defined by the data and couched in the context, mission, and priorities of the campus. These challenges establish the outcomes of the change project and lead teams to choose, implement and evaluate specific strategies that will address the challenges and improve STEM student learning and success. Any change process is dynamic and nonlinear so this model is described as a flow, much like a river, where there are multiple points of entry (and exit!), as well as obstacles that might be encountered along the way that create eddies in the flow.

The Keck-PKAL model is based on practices of organizational learning (Senge, 1990; Bauman, 2005; Kezar, 2005). Within this approach, information gathering and data analysis play a central role in helping individuals identify directions and appropriate interventions for making strategic forward progress. Participants in organizational learning processes also require foreground reflection and dialogue—often based on the data, but also to question typical assumptions. Faculty and staff involved in organizational learning work in non-hierarchical teams to develop this learning and to consider innovative approaches. This means having campus teams look at data related to student success in order to determine the specific challenges and problems, and to orient themselves toward a vision for change. An organizational learning model focuses on learning throughout the change process.

Reflection is another hallmark of an organizational learning process. The teams in the Keck-PKAL project were asked to reflect at each stage and to correct for errors and identify problems that inherently emerge with any change process. Therefore the teams, through reflection, were able to realize when they did not have adequate buy-in to initiate a change process, or when the vision was too top-down or fragmented, or when politics were emerging that might sidetrack their efforts, or why they needed measurement of results to ensure future support of the initiative. Organizational learning change processes utilize data and information to help guide and make choices, but may also make use of outside facilitators (both consultants as well as project leaders) to help campus teams carry out required reflection on their process in order to adapt along the way. The Keck-PKAL guide itself has many questions built in that help facilitate this type of learning and reflection.

The Keck-PKAL model focuses on facilitating organizational learning, but it also incorporates key ideas from other research on change, such as the need to address politics, develop buy-in, understand the power of organizational culture, and help campus leaders unearth underlying assumptions and values that might create resistance to change (Kezar, 2013). The Keck-PKAL model also includes some of the practices frequently included in strategic planning, a typical approach to change in colleges and universities. These practices include vision-setting, identifying benchmarks, and conducting a landscape analysis. It should be noted that many of these processes—organizational learning, addressing politics, unearthing cultural assumptions, and vision-setting—were extremely hard for STEM leaders in the project to embrace, especially those that are often messy and non-linear. Strategic planning approaches that are linear and less messy were often preferred by the leaders we worked with, which suggests that teams are not naturally inclined to use the strategies that work to create change. In the next section, we explore why change is so challenging in order to demonstrate the need for the complex framework that we offer as part of the KECK-PKAL model.

ORGANIZATIONAL CULTURE

Most STEM reform efforts require a change in the culture of the institution, although few initiatives to date have focused on cultural change. Organizational culture, though, is largely tacit, driven by underlying values and assumptions (Schein, 1985). Because assumptions and values are tacit, institutional actors are not conscious of them and typically act more on impulse and routine. When staff and faculty operate from routines, change can be challenging. Imagine trying to have STEM faculty move from lectures to active learning. Their underlying belief is that good teaching involves delivery of content. Asking them to move to a mode where they do not deliver content violates their unarticulated beliefs about good teaching. Cultural theories of change emphasize the need to analyze and be cognizant of these underlying systems of meaning, assumptions, and values; while often not directly articulated, they can nonetheless shape institutional operations and prevent or facilitate change. Change within an organization entails the alteration of deeply embedded values, beliefs, myths, and rituals (Schein, 1985; Shaw & Lee, 1997). This approach also provides more understanding of why data use and reflection were so important in the Keck-PKAL project for helping faculty and staff make sense of the conflicts between their assumptions and reality. We next describe a few concepts that will help leaders to identify key areas of culture that might impact their change process.

History and traditions are important elements of organizational culture, as they represent the collective experience of change processes over time (Kezar, 2013). Understanding how individuals and groups reacted to earlier efforts, as well as barriers that emerged and values that surfaced, is critical to a change agent's success. Institutional history and traditions have been found to strongly influence change processes and the way people make sense of a processed innovation like new approaches to teaching (Kezar, 2001). Change agents embarking upon a STEM reform project must carefully analyze the history of STEM reforms across their campus. Previous failed change efforts can derail a change process, and political problems that emerged should be used as a cautionary tale. Successful change efforts can be explored and learned from. It is also important to distinguish newer initiatives from past ones so that long-time faculty do not automatically associate a new effort with past ones, particularly if there were concerns.

Perhaps the most important drivers of and barriers to change are an organization's values and underlying assumptions. Values can be elusive because they are sometimes unarticulated and other times aspirational or espoused. Values guide behavior, but often in a way that happens unconsciously. Behaviors of people on campuses reflect a system of values, as do language, artifacts (e.g., a policy), and symbols (e.g., an image); in fact, sometimes these are a better reflection of values (Schein, 1985). For example, if student affairs staff describe a wall between them and faculty members, this language suggests collaboration will be difficult when undertaking a broad, institution-wide STEM reform effort. It is important, too, to recognize that espoused values are not always the true values of an organization. In fact, espoused values, examples of which are found in campus mission and values statements, are often aspirational—they reflect what a campus would like to be. Espoused and aspirational values can be potentially significant levers for change because they represent specific areas where stakeholders across the campus might be willing to invest resources and effort to achieve goals.

A key finding from research is that change agents are more successful when they align their strategies with institutional culture (Kezar & Eckel, 2002a). For example, on a very decentralized campus, trying to develop a universal policy on instructional practice for all STEM divisions and units might be met with great resistance. However, change agents can work from the bottom up within their different units to advance policies that are supportive of better STEM teaching campus-wide. By working within the institutional culture, which is decentralized, the change agents will experience more success and support than is likely to come from pushing for universal policy at the campus level alone. Kezar and Eckel (2002a) demonstrated how savvy change agents conduct a cultural assessment and align strategies with the institutional culture. This is a further resource for those interested in understanding the dynamics of working within their institutional culture.

SENSEMAKING

Sensemaking is a theory that helps explain ways that people can change, especially in relationship to taken-for-granted notions like "good teaching" (Kezar, 2013). Sensemaking processes help people to unearth and examine their underlying assumptions about any concept and therefore make it more open to alteration (Kezar & Eckel, 2002b). As a change strategy, sensemaking can address the deep seated cultural changes that we are often confronted with in higher education. Like the process of organizational learning described above under the Keck-PKAL guide, sensemaking highlights the role of learning and development with regard to change (Kezar, 2005; Kezar, 2012). Studies of resistance to change illustrate that people are often not resisting a change because they disagree with it, but because they do not truly understand its nature or how they might integrate it into their work and role (Weick, 1995; Eckel & Kezar, 2003a, 2003b; Gioia & Thomas, 1996). People may hold unconscious views that shape their worldview that themselves prevent change among individuals. For example, STEM faculty may not be conscious of their views about intelligence and learning that may cause them to believe that many students do not have the aptitude for STEM majors. These unconscious beliefs may prevent them from engaging in changing their instructional practices in a meaningful way.

Part of the difficulty of creating change is realizing that people are interpreting their environment very differently from one another (Cameron & Quinn, 1988). Therefore, the focus of change strategies within sensemaking is how leaders can shape individuals' thinking within the change process through framing and interpretation, and how individuals within an organization interpret and make sense of change (Chaffee, 1983; Harris, 1996; Kenny, 2006). Sensemaking is about changing mindsets, which in turn alters behaviors, priorities, values, and commitments (Eckel & Kezar, 2003a). Studies demonstrate that sensemaking is facilitated by change agents who create vehicles for social interaction, help introduce new ideas into the organization, provide opportunities for social connection, and effectively use language and communication to help facilitate people's evolving thinking (Gioia & Thomas, 1996; Thomas, Clark, & Gioia, 1993; Weick, 1995). Examples of sensemaking strategies that could guide institutional STEM reform efforts include on-going campus brown bag or speaker series focused on key STEM reform topics like concept inventories; using assessment data to support student learning; professional development; creating concept or white papers translating national STEM reports, such as Bio2010, into focused ideas for one's specific campus; and cross-campus teams (with both academic and student affairs) that work together regularly. Yet, our research suggests that sensemaking, while important, cannot be used in isolation. Leaders need to invoke multi-frame leadership in order to foster truly meaningful and lasting changes at the institutional level.

MULTI-FRAME LEADERSHIP

Bolman and Deal (1997) provide one of the most comprehensive overviews of organizational theory and its implications for change. Their book, *Reframing Organizations (1997)*, synthesizes thousands of studies about organizational behavior and theory and describes four major frames (or schools of thought) that help to explain how organizations operate: structural, human resource, political, and symbolic. Frames are important because they help leaders to clarify and negotiate a particular issue within an organizational context. In addition to helping understand how organizations operate, the four frames have important lessons for anyone undertaking a change initiative. Each frame offers a unique view of both barriers to change and essential strategies that support change. We will provide a brief overview of each frame and its assumptions about change before discussing ways in which the frames can be integrated to drive the more holistic approach to change that is necessary for institutional reform in STEM.

Structural Frame

The structural frame is perhaps the most commonly used framework among leaders and the most familiar to those in the general public. The structural frame is often epitomized by the notion of the organizational chart, where people understand how the organization functions through a definition of a variety of roles and the relationships among those roles. Key assumptions underlying the structural frame include the overarching role of structure in defining purpose and solving problems, the importance of goals and objectives, a focus on efficiency and specialization, operation through coordination and control, the necessity of matching structure to circumstance, and the primacy of rationality in decision-making and interactions. Barriers to change when examined through the structural frame include lack of clarity or direction surrounding new roles and responsibilities, no goals or priorities being set, and confusion about authority and decision-making. An example of this barrier is that of an institution-wide STEM reform effort that will broaden and shift views of who is responsible for STEM education on campus; when instruction, peer support, and co-curricular and extracurricular programming are all crucial pieces of a holistic STEM-reform effort, then responsibilities can become diffuse and unclear. Facilitators of change include establishing goals for STEM reform, such as targets for increased graduation, reviewing data on student success, and reviewing policies that may inhibit good teaching, such as allocation of classroom space or scheduling of classes. Bolman and Deal (1997) emphasize the importance of formalizing new structures and policies and communicating changes around roles and expectations when thinking about change through the structural frame.

Human Resource Frame

The human resource frame emphasizes the human subsystem of the organization. This frame focuses on the motivations, needs, commitment, training, hiring and socialization of people within an organization and how this impacts organizational functioning. The assumptions underlying the human resource frame include the crucial role of organizations in serving human needs, the importance of fit between individuals and organizations, and the value in rewarding and meaningful mutually beneficial relationships between people and organizations. Barriers that emerge when employing the human resource frame to think about change include anxiety or uncertainty around new strategies or policies, loss of morale, and feelings of incompetence in the face of new approaches to work that may not be well-understood. Including STEM faculty and staff in planning meetings, encouraging participation among a diverse group of stakeholders throughout each stage of the change process, and providing effective training to promote new knowledge and skills were all effective strategies in the Keck-PKAL project that align with the human resource frame.

Political Frame

Bolman and Deal's (1997) research identified that many people, particularly educators and women, downplayed the political frame for understanding organizational challenges and developing solutions. Politics often has a negative image and is associated with images of ambitious people climbing to the top, willing to engage in unscrupulous activities in order to move their agenda forward. This view of politics is very limited, however. Instead, the political frame can help leaders to understand the important ways that they can build an agenda or common vision for change, mobilize people, use persuasion to influence others, identify sources of power and use them to leverage change, and utilize the power of networks in order to create organizational direction and change. The political frame also helps many conflict-averse leaders to see the value in conflict because it demonstrates where people have competing interests and where negotiation and solutions can be identified. The political frame also challenges leaders with a highly rational approach to their work to think about other conditions that are shaping organizational behavior, such as diverse interests or beliefs. Barriers to change when examined through the political frame include feelings of disempowerment among stakeholders, as well as conflicts that go unaddressed and either simmer under the surface or explode into intense battles leading to widespread loss of support. In the Keck-PKAL project, campuses experienced political problems when they rushed ahead with implementation of reform despite lack of buy-in or with only a small group of faculty supporters; these campuses found strong feelings of disempowerment and resistance to change among those who were not included, and were forced to begin the process again. Campuses that focused on forming broad coalitions of support were much more successful in implementing change.

Symbolic Frame

Perhaps the most underutilized frame is the symbolic perspective of the organization. People inherently need meaning and the symbolic frame helps to provide avenues for people to establish meaning through their work. This frame also demonstrates that mission and vision are important for providing a sense of purpose for faculty and staff engaging in institutional STEM reform; mission and vision create a common language that everyone understands and an image of the future towards which stakeholders can strive. Overall, the symbolic subsystem of organizations sheds light on the values that undergird activities, practices, and policies that typically go unnoticed. Bolman and Deal (1997) point out how the symbolic frame, more so than any others, moves leaders beyond thinking in a purely strategic or highly rational manner and highlights the importance of faith, purpose, emotions, values and spirit for organizational functioning. Assumptions of the symbolic frame include the critical importance of meaning and the multitude of meanings that different actors construct for the same event, as well as the role of culture in uniting people around shared beliefs and values. The symbolic frame and its underlying values are exemplified in campus symbols, rituals, stories, and, as noted above, the mission and vision. Barriers to change when utilizing the symbolic frame include a sense of loss of meaning or shared purpose when making major changes, as well as a tendency to cling tenaciously to past ways of operating. For example, major efforts to change the way STEM students are taught and supported could cause stakeholders to undergo a crisis of meaning or purpose—if past ways were not effective, how might faculty and staff feel about their work and their role? Strategies to overcome such symbolic challenges include explicit recognition of the past, celebrations of future change, and creation of transition rituals, kickoff ceremonies, or new symbols associated with the STEM reform project. Leaders using a symbolic frame help craft a narrative for why change is needed and use common values, such as improving student success, as ways to bolster buy-in.

Multi-Frame Thinking: Pulling the Frames Together

Bolman and Deal's (1997) empirical research suggests that leaders are more successful and effective when they use multi-frame thinking for conceptualizing issues within organizations. Their research also suggests that leaders generally tend to use a single or a couple of frames in order to understand and analyze issues within an organization. Like other leaders, STEM faculty are often more comfortable using a solely structural framework to lead. The complexity of STEM reform at the institutional level, however, requires multi-frame leadership.

Institutional transformation is inherently a process that involves politics, relationship-building, and symbolic or visionary elements—multi-frame leadership.

In our study, we saw campuses begin to shift from a focus on just implementing a new pedagogical approach or support program to a broader, multi-frame approach to the change process. Campus leaders leveraged the power of the symbolic frame by creating a shared vision for change in STEM education. They attended to the human resource frame by planning meaningful professional development opportunities for STEM faculty and staff. They considered political issues by working to get buy-in from key actors across campus, as well as structural issues by reviewing data to understand problems in student success. This shift to a multi-frame approach did not happen spontaneously; rather, it often happened as a result of campus teams seeking out leadership training through groups like Project Kaleidoscope, or being mentored by someone inside or outside their organization. Adopting a multi-frame approach to leadership is not easy or automatic, but it is critical for creating lasting change in STEM education at the institutional level.

MULTI-LEVEL LEADERSHIP

Research on change in colleges and universities demonstrates that systemic change occurs when leaders across an institution work in concert toward a solution (Kezar, 2014). STEM reform at the institutional level is unlikely without leadership capacity built at multiple levels and harnessed as changes are implemented. Senior leaders have long been documented to be significant players in implementing changes, but are generally not brought into STEM reform efforts (Henderson et al., 2011; Kezar, 2013). Senior leaders, such as provosts, can shape and change incentives and rewards and can create more robust data systems to enhance data driven decision-making and foster organizational learning. Deans, provosts, and presidents are needed to examine policies around tuition, articulation, and course credit. Pressure from external players, such as accrediting organizations, legislative bodies, government agencies, and business and industry leaders, has also been well-documented; senior administrators are also positioned to use these as a lever for STEM reform (Eckel & Kezar, 2003).

Middle-level leaders, such as department chairs, are important for sensemaking; they might create a departmental learning community on active learning, have discussions about learning goals and competencies, support instructional and curricular changes, or help faculty in obtaining professional development to conduct assessment. Department chair leadership programs have been shown to be instrumental to other types of STEM changes, such as getting more women and underrepresented minorities into STEM disciplines as faculty through ADVANCE grants (Rosser, 2009). Individual campuses are increasingly offering department chair training because they recognize that these individuals are critical to policy implementation; however, recent studies also demonstrate their important role in change. Grassroots faculty leaders are also needed to motivate their colleagues and obtain buy-in for change, and advisors in student affairs are needed to help students obtain support or relevant career and academic information. Faculty and staff leaders often pilot or test new programs and interventions and conduct assessment of these programs. Leaders at all levels are critical for altering the culture by reshaping values and redefining what is considered normative.

CONCLUSION

Our hope in reviewing the literature on change and describing how these principles are embedded in one recent initiative—the Keck-PKAL project—is that other STEM leaders will understand that a systems approach is needed for change in STEM. Furthermore, cultural change is necessary for meaningful institutional STEM reform. Creating cultural change on campuses will require a complex and multi-faceted approach and a strong understanding of institutional culture, as well as attention to sensemaking strategies, organizational learning, multi-frame thinking, and multi-level leadership.

REFERENCES

Anderson, W. A., Banerjee, U., Drennan, C. L., Elgin, S. C. R, Epstein, I. R., Handelsman, J., Hatfull, G. F., Losick, R., O'Dowd, D. K., Olivera, B. M., Strobel, S. A., Walker, G. C., & Warner., I. M. (2011). Changing the culture of science education at research universities. *Science, 331,* 152–153. http://www.physics.emory.edu/~weeks/journal/anderson-sci11.pdf

Association of American Universities. (2013). *Framework for systemic change in undergraduate STEM teaching and learning.* http://www.aau.edu/WorkArea/DownloadAsset.aspx?id=14357

Austin, A. E. (2011). *Promoting evidence-based change in undergraduate science education.* Paper commissioned by the Board on Science Education of the National Academies National Research Council. Washington, DC: The National Academies. http://sites.nationalacademies.org/DBASSE/BOSE/DBASSE_071087#.UdIxQvm1F8E

Bauman, G. L. (2005). Promoting organizational learning in higher education to achieve equity in educational outcomes. In A. Kezar (Ed.), *Organizational Learning in Higher Education* (Vol. 131). San Francisco: Jossey-Bass.

Beach, A. L, Henderson, C. & Finkelstein, N. (2012). Facilitating change in undergraduate STEM education. *Change: The Magazine of Higher Learning*, 44:6, 52–59.

Bolman, L. & Deal, T. (1997). *Reframing organizations*. San Francisco: Jossey-Bass.

Cameron, K. S., & Quinn, R. E. (1988). Organizational paradox and transformation. In K. Cameron & R. Quinn (Eds.), *Paradox and transformation*. New York: Bellinger.

Chaffee, E. E. (1983). *Rational decisionmaking in higher education*. Boulder, CO: National Center for Higher Education Management Systems.

Eckel, P. & Kezar, A. (2003a). Strategies for making new institutional sense: Key ingredients to higher education transformation. *Higher Education Policy, 16*(1), 39–53.

Eckel, P. & Kezar, A. (2003b). *Taking the reins: Institutional transformation in higher education*. Phoenix, AZ: ACE-ORYX Press.

Fairweather, J. (2008). *Linking evidence and promising practices in science, technology, engineering, and mathematics (STEM) undergraduate education: A status report for the National Academies National Research Council Board on Science Education*. Commissioned Paper for the National Academies Workshop: Evidence on Promising Practices in Undergraduate Science, Technology, Engineering, and Mathematics (STEM) Education. http://www.nsf.gov/attachments/117803/public/Xc--Linking_Evidence--Fairweather.pdf

Gioia, D. A. & Thomas, J. B. (1996). Identity, image, and issue interpretation: Sensemaking during strategic change in academia. *Administrative Science Quarterly, 41*, 370–403.

Harris, S. G. (1996). Organizational culture and individual sensemaking: A schema-based perspective. In J.R. Meindl, C. Stubbart, & J.F. Poroc (Eds.), *Cognition in groups and organizations*. London: Sage.

Henderson, C., Beach, A., & Finkelstein, N. (2011). Facilitating change in undergraduate STEM instructional practices: An analytic review of the literature. *Journal of Research in Science Teaching, 48*(8), 952–984.

Kenny, J. (2006). Strategy and the learning organization: A maturity model for the formation of strategy. *Learning Organization, 13*(4), 353–368.

Kezar, A. (2001). *Understanding and facilitating organizational change in the 21st century: Recent research and conceptualizations*. ASHE-ERIC Higher Education Report, 28 (4). San Francisco: Jossey-Bass.

Kezar, A. (2005). What campuses need to know about organizational learning and the learning organization. *New Directions for Higher Education, 2005*(131), 7–22.

Kezar, A. (2011). What is the best way to achieve reach of improved practices in education? Innovative Higher Education, 36(11), 235–249.

Kezar, A. (2012). Understanding sensemaking in transformational change processes from the bottom up. *Studies in Higher Education, 65*, 761–780.

Kezar, A. (2013). *How colleges change.* New York: Routledge.

Kezar, A., & Eckel, P. (2002a). The effect of institutional culture on change strategies in higher education: Universal principles or culturally responsive concepts? *The Journal of Higher Education, 73*(4), 435–460.

Kezar, A., & Eckel, P. (2002b). Examining the institutional transformation process: The importance of sensemaking, inter-related strategies and balance. *Research in Higher Education, 43*(4), 295–328.

Kuh, G. D. (2008). Excerpt from *High-impact educational practices: What they are, who has access to them, and why they matter.* Washington, D.C.: Association of American Colleges and Universities.

Kuh, G. D. & O'Donnell, K. (2013). *Ensuring quality and taking high-impact practices to scale.* Washington, DC: American Association of Colleges and Universities.

Maton, K. I., Hrabowski, F. A. & Schmitt, C. L. (2000). African American college students excelling in the sciences: College and postcollege outcomes in the Meyerhoff Scholars Program. *Journal of Research in Science Teaching, 37*(7), 629–654.

Schein, E. H. (1985). *Organizational culture and leadership.* San Francisco: Jossey-Bass.

Rosser, S. V. (2009). Creating a new breed of academic leaders from STEM women faculty: NSF's ADVANCE program. In A. Kezar (Ed.), *Rethinking leadership in a complex, multicultural, and global environment* (117–130). Sterling, VA: Stylus.

Senge, P. (1990). *The fifth discipline: The art and practice of the learning organization.* New York, NY: Doubleday.

Shaw, K. A., & Lee, K. E. (1997). Effecting change at Syracuse University: The importance of values, mission, and vision. *Metropolitan Universities: An International Forum, 7*(4), 23–30.

Thomas, J. B., Clark, S. M., & Gioia, D. A. (1993). Strategic sensemaking and organizational performance: Linkages among scanning, interpretation, action, and outcomes. *Academy of Management Journal, 36*(2), 239–270.

Weick, K. E. (1995). *Sensemaking in organizations.* Thousand Oaks, CA: Sage.

ABOUT THE AUTHORS

Adrianna Kezar is a Professor of the Rossier School of Education and Co-Director of the Pullias Center for Higher Education at the University of Southern California in Los Angeles, California.

Elizabeth Holcombe is the Provost's Fellow in Urban Education Policy of the Rossier School of Education at the University of Southern California in Los Angeles, California.

3

The Role of Cultural Change in Large-Scale STEM Reform: The Experience of the AAU Undergraduate STEM Education Initiative

Emily R. Miller and James S. Fairweather

The need for improving undergraduate STEM education has received increased attention and taken on new urgency in recent years. New research on teaching and learning has led to the development of instructional methods that are more engaging and more effective at helping students learn than the long-established model of the expert lecturer transmitting knowledge. These teaching practices, which encourage student engagement and active learning, have been documented and affirmed by recent high-level reports and policy papers (Freeman et al, 2014; Handelsman et al, 2004; Singer et al, 2012; NRC, 2015). There also is robust evidence that active learning pedagogies can positively impact the performance and persistence of underrepresented students (Freeman et al, 2011; Lorenzo et al, 2006; Eddy & Hogan, 2014).

*At the same time, t*he national policy environment is reflecting a more coordinated vision and effort to improve undergraduate STEM education across and within relevant organizations and actors (AAAS, 2011; ACS, 2010; ASEE, 2013; Fry, 2014; NRC, 2009, 2013; NSTC, 2013; PCAST, 2012). We have seen a shift away from isolated directives within individual disciplines (e.g., calculus reform, change in the ABET engineering program accreditation criteria to reflect improvement in student learning) and nationally funded efforts that do not require long-lasting and institutional reforms within colleges and universities (e.g., the funding of individual course and curriculum development projects by the NSF; Fairweather, 2008). A selection of current examples that demonstrate this shift toward larger-scale change include the portfolio of multi-institution reform efforts committed to improving instructional practices funded by The Leona M. and Harry B. Helmsley Charitable Trust, the HHMI sustaining excellence institutional awards aimed at helping improve persistence of students studying STEM disciplines and reinvigorating introductory science courses, and the NSF Improving Undergraduate STEM Education (IUSE) solicitation that has an institutional transformation track.

Despite this movement toward developing and supporting systemic re-form in STEM undergraduate education, a majority of university STEM faculty members who teach undergraduate science and engineering classes have re-mained inattentive to the shifting landscape. Student-centered, evidence-based teaching practices are not yet the norm in most undergraduate STEM educa-tion courses, and the desired magnitude of change in STEM pedagogy has not materialized (Anderson et al, 2011; Dancy & Henderson, 2010; Dancy et al, 2013; Eiseman & Fairweather, 1996; Fairweather & Beach, 2002; Fisher, Zelig-man, & Fairweather, 2005; Henderson et al, 2007, Singer et al, 2012).

A principal reason for the lack of widespread pedagogical reform in STEM is the use of theoretical perspectives whose focus is primarily on individual faculty members and the students in their classrooms (Dancy & Henderson, 2005; Fairweather, 2009). Much of this literature centers on micro-level assess-ments of the classroom, which is crucial to assessing the effect of pedagogy on student learning, but does not address the importance of the dissemination and institutionalization of reforms. It often ignores the larger environment, such as institutional culture, disciplinary and departmental contexts, and the role of philanthropy, governing bodies and accrediting organizations. In addition, a micro-level focus on reform does not take into account the costs and political challenges in scaling up reforms (Fisher, Fairweather, & Amey, 2003). Concern about more macro-level environments requires a change in assessment from looking for benefits and learning outcomes to a more nuanced consideration of factors that facilitate, impede, or influence wide-spread transformation in undergraduate STEM education.

To increase the implementation of instructional strategies shown to be ef-fective requires a model of change, including the roles of research evidence, leadership, resources, faculty workload and rewards, and resources. In this con-text, empirical evidence is only one part of the reform effort. As Fairweather (2008, p.11) has explained, "research evidence of instructional effectiveness is a necessary but not sufficient condition. . . ." for faculty to change their teaching practices. Fairweather suggests that the assumption that "the instructional role can be addressed independently from other aspects of the faculty position, par-ticularly research, and from the larger institutional context" is misguided (2008, p. 3). Given the size and scale of higher education, changing individual faculty members or even isolated departments will have minimal impact. To achieve long-lasting and broadly disseminated educational reforms, efforts must go well beyond this micro-level focus on individual faculty members.

Fairweather (2008) and Austin (1996, 2011) recommend that sustain-able STEM reform requires engaging institutional leaders such as department chairs, deans, and presidents in rethinking institutional structures and culture. Austin (2011, 2014) also recommends that a variety of external stakeholders, such as disciplinary societies, government agencies, and employers, are crucial to long-lasting change. Further, a recent case study of undergraduate STEM reform conducted at the University of Colorado Boulder concludes that top-down and bottom-up reforms alone are inadequate for sustained institutional improvement to undergraduate education; middle-out reforms are also required (Corbo, 2014). Past efforts to increase emphasis on teaching have likely failed because they focused too little on cultivating change at the departmental and college levels. In sum, transforming undergraduate STEM education requires multiple facilitators or "levers" pushing for change that can counterbalance the forces that sustain ineffective instructional practices and address the obstacles inherent in the system in which educational innovations take place (Anderson et al., 2011; Austin, 2011; Beach et al., 2012).

AAU UNDERGRADUATE STEM EDUCATION INITIATIVE

The Association of American Universities (AAU) staff had long recognized that member institutions were vulnerable to criticism about undergraduate science, technology, engineering and mathematics (STEM) teaching, learning, and re-tention. A 2009 survey of AAU member institutions' activities in undergradu-ate STEM education revealed very few that focused on pedagogy inside the classroom.

In an effort to initiate action, AAU staffed gathered data to map major as-sociation and disciplinary society efforts in STEM reform and to identify areas of overlap among various organizations. As a part of this effort, AAU developed a matrix of STEM undergraduate education reform efforts. The association also convened an advisory committee composed of national experts in undergradu-ate STEM teaching and learning. AAU wrote a discussion draft that framed the issue and defined the niche focus of the AAU Undergraduate STEM Educa-tion Initiative ("initiative") (AAU 5-Year Initiative to Improve Undergraduate STEM Education Discussion Draft, October 2011).

In 2011, AAU launched a five-year initiative in partnership with member institutions to improve undergraduate teaching and learning in STEM fields. The overall objective of the initiative is to influence the culture of STEM de-partments at AAU institutions so that faculty members are encouraged and supported to use teaching practices proven by research to be more effective in engaging students in STEM education and in helping students learn.

The goals of AAU's STEM Initiative are to:

1. develop a framework for assessing and improving the quality of STEM teaching and learning;
2. support AAU STEM project sites at a subset of AAU universities to implement the framework, and develop a broader network of AAU universities committed to implementing STEM teaching and learning reforms;
3. explore mechanisms that institutions and departments can use to train, recognize, and reward faculty members who want to improve the quality of their STEM teaching;
4. work with federal research agencies to develop mechanisms for recognizing, rewarding, and promoting efforts to improve undergraduate learning; and
5. develop effective means for sharing information about promising and effective undergraduate STEM education programs, approaches, methods, and pedagogies.

The work currently is supported by grants from The Leona M. and Harry B. Helmsley Charitable Trust and from the National Science Foundation WIDER and IUSE programs. AAU also received funds from Burroughs Wellcome Fund and Research Corporation for Science Advancement to support the development of a broader reform network.

AN ALTERNATIVE PERSPECTIVE ON REFORM

The initiative *from the beginning* has been informed by broader theoretical perspectives about organizational and cultural change in academia and about faculty work and rewards. Theory informs the initiative in two principal ways: (1) "to influence the culture of STEM departments at AAU universities so that they will use sustainable, student-centered, evidence-based, active learning pedagogy in their classes, particularly at the first-year and sophomore levels" and (2) to develop and use a framework for reform based on culture change (AAU, 2011).

With a focus on changing the culture of higher education as its overreaching goal, the initiative took a more systemic view of educational reform within academia, including understanding the wider setting in which educational innovations take place—the department, college, institution, and national discipline.

MAKING IMPROVING UNDERGRADUATE STEM EDUCATION A PRIORITY

The initiative has benefitted from a confluence of various events and environments. The external environment was ripe. We now have available literature to know what types of instructional practices work (Singer et al, 2012; NRC, 2015) and that the primary focus must be on strategies to implement them (PCAST, 2012). Concern about the cost to students and completion rates in STEM, as well as the increasing importance of STEM preparation in the global economy, have made improvement in undergraduate education a priority of the U.S Department of Education, the NSF, and other national associations.

AAU as an organization also has played an important role. When Hunter R. Rawlings III became AAU's president in 2011, he made it clear that this was a priority area for the association. Unique in its history, the AAU has used its position with its member institutions to build an effective network among members to promote reform in undergraduate STEM education. Also crucial has been the role of the leadership in AAU institutions in highlighting the importance of lower-division STEM courses at institutions whose primary focus is graduate education and research.

FRAMEWORK: GENERATING BUY-IN TO A SYSTEMS APPROACH

AAU developed a Framework for Systemic Change to Undergraduate STEM Teaching and Learning to identify the relevant influences on sustainable local STEM reform, including national, institutional, and departmental actors (Figure 1). The framework was developed in collaboration with member universities. This approach combined top-down support for STEM undergraduate reforms in a manner that encouraged local institutions and their faculties to buy into the initiative. The core of AAU's framework is pedagogy: the practices used by faculty members to engage students and guide and support their learning. To successfully enact and institutionalize the use of evidence-based teaching techniques, two layers around this pedagogical core are necessary: scaffolding, or support, for both faculty and students, and larger cultural change to facilitate changing teaching practices. Ultimately, at an institutional level the Framework provides a set of key elements that need to be addressed to bring about sustainable change. The Framework also can take into account the wide variety of settings in which academic work takes place; different strategies will almost certainly be used to achieve improvement in STEM teaching and learning at different institutions. From a multi-institutional perspective, the Framework provides a unifying goal and a commitment to a systems approach to change (Kania and Kramer, 2011).

FIGURE 1. The framework provides a set of key institutional elements that need to be addressed in order to bring about sustainable change to undergraduate STEM teaching and learning. The core of the framework is pedagogy, the practices used by faculty members to teach students and guide and support their learning. Two layers around this pedagogical core are necessary—scaffolding and cultural change—to incubate and then sustain evidence-based teaching.

EXPECTATIONS FOR AAU STEM INITIATIVE PROJECT SITES

AAU selected eight member campuses to serve as project sites. Over three years, each of the eight project sites will implement a major undergraduate STEM education project that incorporates key elements of the Framework. The eight sites were chosen from among 31 AAU universities that submitted concept papers. A number of criteria were considered, such as the degree of department and faculty engagement, institutional commitment, likelihood of sustained organizational change, and commitment to evaluation and assessment. A lesson learned by AAU is that a well-framed request can direct and focus a proposed scope of work and can insist upon the integration of critical elements into proposed projects.

In addition, the process was designed to ensure that project sites represent the diversity of the AAU membership (i.e., public and private universities, large and small enrollments) and that they address a wide range of the elements outlined in the Framework. AAU STEM Project Sites are carrying out projects that focus on overcoming a specific challenge their campus encounters in improving undergraduate STEM teaching and learning. Ultimately, these project sites serve as laboratories to implement the Framework and are the first phase in an effort to encourage broad-based reform of undergraduate teaching practices at AAU research universities and beyond.

AAU UNDERGRADUATE STEM EDUCATION
INITIATIVE PROJECT SITES

Brown University: The project focuses on gateway courses in physics, applied mathematics, chemistry, and engineering by introducing intensive, small group, collaborative problem-solving sessions. Facilitators for the sessions represent all levels of the Brown community, from faculty to postdocs to graduate students to undergraduate students, all collaborating together to improve STEM education. The goals of these problem-solving sessions are to provide a student-centered learning process, a community within and across courses, and a larger context in which students can situation their disciplinary knowledge. Future reform efforts will focus on (a) creating a template for "AAU STEM" courses that will establish a new norm in introductory STEM education at Brown, (b) building a "train-the-trainer" program through Brown's Sheridan Center for Teaching and Learning in which graduate student instructors become the change agents behind the pedagogical strategies used in problem-solving sessions, and (c) providing a forum for past and present "AAU STEM" instructors to engage one another as a community of innovative educators.

Michigan State University: The project focuses on the transformation of the large gateway courses in biological sciences, chemistry, and physics. MSU is bringing together faculty from these disciplines to identify core ideas in the disciplines, crosscutting concepts that span disciplines, and science practices that allow students to use knowledge. Engaging faculty in this process opens up both departmental and interdisciplinary conversations about the goals and emphases of the gateway STEM courses. MSU will develop assessments that emphasize both these core ideas and scientific practices. MSU has also established the STEM Gateway fellowship for faculty who teach these courses, which will further develop the community involved in transformation efforts.

The University of Arizona: The project focuses on the redesign of foundational courses in biology, chemistry, chemical engineering, computer programming, and physics that serve thousands of STEM majors. The curricula and teaching practices in these courses are being critically examined to emphasize core disciplinary ideas, problem-solving abilities, critical thinking, and teamwork, to ensure students develop meaningful understandings and analytical reasoning skills. The University of Arizona is also promoting a change in faculty culture by leading STEM faculty learning communities, teaching talks, and workshops, offering innovative teaching awards, and making changes in promotion and tenure policies toward teaching.

University of California, Davis: The mission of the AAU-UCD partnership is to foster evidence-based, sustainable innovation in STEM instruction. UCD aims to do so through development of cultures of data and evidence around instruction and learning that encourage experimentation, build urgency, and enable change. As part of, and to highlight change, large-scale introductory courses are targeted for evidence-based experimentation that will be thoroughly assessed and results shared within and between STEM departments.

University of Colorado, Boulder: CU Boulder aims to improve undergraduate STEM education by professionalizing educational practice through measurement, assessment, and cultural change. The focus is on department-wide change to achieve more coherent, long-lasting reforms. The project uses a three-layer approach: (1) work with individual faculty and groups of faculty to support transformation of high-impact and high-need courses within individual STEM departments (bottom up); (2) apply targeted approaches to individual departments to stimulate cultural change (middle out); and (3) work with the administration and faculty senate to promote and incentivize the use of evidence-based teaching practices (top down). CU Boulder supports these three layers with infrastructure provided by the AAU and our collaborations with our Office of Informational Technology (OIT) to develop and import technology for better utilizing already existing institutional student data.

The University of North Carolina at Chapel Hill: UNC-CH is creating a support framework to facilitate the implementation of evidence-based teaching practices in large courses that have traditionally been taught by the lecture method. The principal goals are to continue the transition of our large lecture gateway courses in biology, chemistry and physics-astronomy into high-structure, high-engagement learning environments consistent with best practices in science education. To achieve widespread adoption of evidence-based methods among faculty across ranks, UNC-CH is developing and implementing a mentor-apprentice program that facilitates the transfer of effective techniques from mentors with experience in evidence-based practices to apprentices, who are faculty members with less experience in these methods.

University of Pennsylvania: Penn's AAU/STEM Initiative seeks to improve introductory courses in mathematics, chemistry, physics, and bioengineering through teaching practices that foster active learning and enable students to gain a sense of engagement in these disciplines. Penn's initiative supports faculty in developing Structured, Active, In-class Learning (SAIL) and evaluating the impact of that teaching. This includes disseminating of SAIL practices across campus, developing appropriate classroom spaces, and providing support for video creation. As part of this plan, Penn has constituted a faculty board to help departments better understand how students apply materials from introductory courses as they take courses in other departments.

Washington University in St. Louis: The overarching mission is to foster broad-based incorporation of effective active-learning techniques in STEM courses throughout the university. For students, the goals are to increase engagement, achieve long-term retention of information and deeper conceptual understanding, and increase satisfaction and persistence in STEM career paths. For faculty, the aim is to impart and reinforce knowledge of active learning and successful implementation strategies, and to increase the number of faculty who incorporate and evaluate the impact of active learning in their courses. Lastly, at an institutional level, the focus is on evaluating project work and promoting cultural change toward these evidence-based pedagogies.

ENGAGING ALL OF THE AAU MEMBERSHIP

AAU is actively bringing together more campuses from among its members to form an AAU STEM Network. We envision a collaborative network that will help to support and link AAU institutions grappling with similar challenges and barriers in reforming and improving STEM teaching and learning for undergraduate students. The network will provide a forum to facilitate ongoing interaction and exchange of information and ideas between all AAU institutions, as well as cultivate relationships among those leading reform efforts on their own campus.

This effort to form a network is supported by research from Fairweather (2009), Eckel and Kezar (2003), and Kezar (2001) that demonstrates changes in higher education do not come simply from sharing of evidence about best practices, as has often been assumed in current and past STEM initiatives. Relationships and networks more strongly impact changes in ideas and practices (Valente, 1995; Rogers, 2003).

COORDINATING MOMENTUM

AAU works closely with other key organizations to coordinate our activities related to undergraduate STEM reform, including the Association of Public and Land-grant Universities (APLU), Association of American Colleges & Universities (AAC&U) Project Kaleidoscope, the Bayview Alliance (BVA), the American Association for the Advancement of Science (AAAS), the Business Higher Education Forum, the Cottrell Scholars, and other higher education associations and disciplinary societies. Earlier research not only supports the benefits of moving beyond a change model focused on individual faculty, but also supports working with key organizations at multiple levels to reshape norms and incentives. A few broader efforts by larger organizations, such as disciplinary societies or accreditors, exist that demonstrate promise in promoting reform efforts. The Accreditation Board for Engineering and Technology (ABET) engineering program accreditation criteria focused on implementing student outcomes assessment and helped to fuel significant change nationally (Lattuca et al., 2006). NSF funded engineering coalitions (e.g., ECSEL), which lead to widespread change within the funded institutions. As Austin (2011, 2014) has identified, systemic change in higher education will mean garnering support from multiple levels—departments, institutions, and other key groups that drive norms, such as the AAU. If disciplinary societies, philanthropy, national associations, or key external agencies are not engaged, it may signal a lack of alignment in norms and prevent scale of changes (Fry, 2014). Through the initiative, AAU is attempting to convene and align these various norm-driving groups.

EARLY INDICATION OF PROGRESS

AAU is excited by the visible momentum to improve undergraduate STEM teaching and learning at our member campuses. Through AAU STEM Initiative workshops and conferences, the collection of baseline data, individual project site annual reports, campus visits to AAU member institutions, and opportunities to engage with AAU member institutions at national meetings, AAU has gained a deeper appreciation of the project's goals and objectives, implementation and progress. In addition, the information we have gathered from these sources has allowed AAU to begin to assess the effects of the AAU STEM Initiative.

LAUNCHING THE INITIAL SET OF AAU DEMONSTRATION PROJECTS

All eight AAU STEM project sites launched their campus reform efforts in the fall 2013 semester. Each site has also submitted their first interim report, which includes details of their activities, progress towards benchmarks, and challenges. More than 58 courses were directly impacted by redesign efforts at the eight sites. These courses enrolled well over 50,000 undergraduate students, the large majority of which were freshmen and sophomores. Around 150 tenure-track or tenured faculty and a nearly equal number of non tenure-track faculty, as well hundreds of lecturers and graduate and undergraduate assistants, were involved in instruction for these courses. Note that some courses were offered multiple times and in multiple sections, and these figures separately count each time a student or instructor was involved in course offerings.

A substantial amount of additional support for work at the individual project sites has been provided by the institutions themselves as well as outside sources including the Bill & Melinda Gates Foundation, the Sloan Foundation and the National Science Foundation.

The AAU project team has visited all campuses. In addition to providing significant insights into the nuances of the work on individual campuses, site visits reinforced the visibility and credibility of people working on the projects, especially with their campus leadership. Below provides a partial summary of project site activities organized by the Framework:

Pedagogy

Course Redesign: Each site worked on redesigning a handful of introductory STEM courses. These courses spanned at least two departments, many sites committing to inter-departmental collaboration during the redesign. As examples, Brown University incorporated an element of student collaboration and problem-solving sessions to practice course concepts, and the University of Arizona redesigned five STEM classes modeled on a centerpiece reformed course sequence in general chemistry.

Student Learning: Through pre- and post-test methods, many of the sites, including University of California-Davis, University of Arizona, and University of

North Carolina-Chapel Hill, gathered baseline data on the learning outcomes of students in redesigned courses, in addition to the baseline data requested by AAU.

Scaffolding

Faculty Support: All sites indicated a commitment to supporting faculty in evidence-based teaching techniques with varying approaches. Examples of approaches to support improved faculty instruction include developing mentoring and apprenticeship programs at the University of North Carolina-Chapel Hill, as well as collaborating with teaching and learning centers to provide training for faculty (examples include the University of Pennsylvania and Washington University in St. Louis). In addition to providing training and learning community groups for faculty, several project sites also have developed assessment tools aligned with the AAU Framework to monitor changes in faculty instructional practices. For example, Michigan State University has developed two new assessment protocols and is using video recordings to observe effectiveness of teaching and to make improvements. Likewise, the University of Colorado Boulder (with shared technology from University of California-Davis) is developing department-level toolkits for measurement analytics.

Culture Change

Institutional Commitment: Project sites report institutional support from leaders in many forms. At University of Pennsylvania, statements about the universities long-term support for the initiative are manifest in investments made by the central administration in buildings, classroom spaces and other infrastructure on campus. At Washington University in St. Louis, the redesign of the introductory curricula has support from the Office of the Provost, and a committee was established to ensure this curriculum is offered in fall 2015.

Teaching Excellence: Each project site made some effort to incentivize faculty to engage in pedagogical reform. Several institutions studied how to make formal reward and faculty evaluation systems align with a commitment to student-centered pedagogy. The level of detail varied substantially, as did the extent to which campus teams made explicit the difference between written policy pertaining to the importance of teaching and the way in which the policy was actually implemented within departments on their respective campuses. In the annual reports, four of the eight sites addressed the promotion and tenure system explicitly. Two made a serious effort to align their promotion and tenure policies with practice when it comes to valuing and recognizing faulty investment of time in pedagogic reform.

AAU AND INSTITUTIONS AGREE ON BASELINE DATA ELEMENTS AND BEGIN DATA COLLECTION

As discussed in Chapter E4, AAU developed a set of baseline measures to study the improvement of STEM undergraduate education. These measures were tested for validity, reliability, and acceptability to each project site. After several months of back and forth, each site agreed to use the instrument to gather baseline information. All project sites have submitted these data to AAU. Though we are still in the process of assessing this data, two main observations include: (1) awareness of good instructional practices is more likely than their use (that is, faculty attitudes and practices are not fully aligned), and (2) although institutions are conveying the message about the importance of teaching, instructors do not believe that teaching is given high importance in their own reviews and promotion/tenure processes. This information comes from a survey of instructors across the eight project sites. Over 1,000 instructors responded to the survey; about two-thirds are tenured or tenure-track faculty, but instructors, lecturers, and graduate students are all well-represented. Respondents span the disciplines; about 60% referred specifically to a lower-division course they had taught.

We also collected data on campus infrastructure. There was variation between institutions, and between departments at single institutions, but departments identified lack of flexible space for teaching and learning, and lack of staff support for teaching as the weakest areas. The ability of classrooms to accommodate special needs, the presence of learning centers, and support for electronic resources were identified as strengths overall.

The process for collecting baseline data from the project sites included a request that the chairs of all impacted departments write a summary of the evaluation of teaching for salary increases and for promotion and tenure. Thirty-one department chairs from across seven of the sites responded with statements from one to three pages in length. The statements had much in common, including strong assertions that teaching is highly valued. All departments make use of student evaluations at the end of courses, and some level of peer observation. Many have some kind of annual award for excellence in teaching. Most provided conventional descriptions of review processes and the provision of feedback to faculty members.

From most of the statements it would be impossible to discern whether attention to student-centered, active or evidence-based pedagogy was recognized or required. Only six of the thirty-one had some form of explicit statement that included "introduction of innovative methods," or "introduction of active learning techniques," among the key criteria for excellence in teaching for tenure-track faculty. Interestingly, two more included such criteria for their lecturers but not their tenure-track faculty. Three of the six were explicit about their encouragement of active learning methods, via department discussion or department funding of attendance at one of the faculty trainings provided by their discipline. Another seven of the 31 had some statement that could be classified as permissive, for example, "the committee will review and consider any other elements the faculty member includes in their personal statement," or "publications or presentations on education may also be considered among the criteria for excellence in teaching," or "the time taken to introduce new methods is factored into the consideration of total workload," or "attendance at local or national meetings on education is taken as evidence of commitment to teaching." One explicitly acknowledged that student evaluations might drop in the first run of a new approach, and that this is taken into account.

AAU AND DEMONSTRATION UNIVERSITIES SHARE FINDINGS IN WAYS THAT ARE USEFUL EXAMPLES FOR OTHER INSTITUTIONS

An illustrative example occurred in May 2014 when teams from project sites were brought together at AAU. Campuses were urged to bring representatives of campus leadership, as well as faculty who are part of their teams. The goals of this meeting were to build wider awareness among the campuses of approaches being taken, to facilitate connections among groups using similar approaches, and to create a forum in which there could be candid conversation about challenges. One outcome of this meeting was the realization by project sites about why they were selected as part of the initiative. They learned that AAU selected them both for the promise of their proposals and because they represented the full array of AAU institutions. Project sites learned that they were all starting from distinct points in the innovation process. This realization improved subsequent cross-project site communication.

As evidence of further momentum, we hosted an AAU STEM Network conference in July 2014. We had an excellent turnout—representatives from 41 (out of 62) AAU universities. During the opening night reception, participating campuses presented posters showcasing their undergraduate STEM education reform efforts relevant to the Initiative. During the remainder of the conference, AAU STEM Initiative project sites facilitated interactive workshop sessions on common themes campuses confront when reforming undergraduate STEM education.

AAU publicly launched a website for the STEM Initiative: www.aau.edu /STEM. The website supports communication among member universities and profiles institutional efforts to reform undergraduate STEM teaching and learning at AAU member universities.

CONCLUSION

AAU is embarked on a systems approach to improving undergraduate STEM education. Based upon our assessment, it is clear that the initiative is having a positive impact. It has catalyzed institutional action toward reforming undergraduate STEM education, enhanced communication and collaboration on campuses, leveraged campus support (financial and other resources) from all levels of institutions, and aligned to some degree efforts to improve undergraduate STEM education within campuses. It is too early to see the longitudinal impact on items such as student learning and other barometers of cultural change, such as promotion and tenure. This is not unexpected—achieving cultural reform is difficult and long-term. And any measures of cultural change will be difficult to directly attribute to the initiative, but AAU through the initiative is attempting to achieve widespread and sustainable change to undergraduate STEM education at its member campuses.

We understand that the role of a national association such as AAU is a poorly understood actor in STEM reforms. And while we think we know what is working (as discussed above), we are studying the role of AAU in the reform process both to improve the theoretical understanding of organizational reform and to provide evidence for future endeavors of this kind.

As the senior co-author of this paper has observed, I have never been more optimistic that the various components needed for successful institutional transformation are in place.

REFERENCES

American Association for Advancement of Science. (2011). *Vision and change in biology education*. Washington DC: American Association for Advancement of Science. http://visionandchange.org/files/2013/11/aaas-VISchange-web1113 .pdf

American Chemical Society. (2011). *Chemistry education: Transforming the human elements*. www.aacu.org/pkal/documents/ACS_000.pdf

American Society for Engineering Education. (2013). *Transforming undergraduate education in engineering*. www.asee.org/TUEE_PhaseI_WorkshopReport.pdf

Anderson, W. A., Banerjee, C. L., Drennan, S. C. R., Elgin, I. R., Handelsman, J., Hatfull, G. F., Losick, R., O'Dowd, D. K., Olivera, B. M., Strobel, S. A., Walker, G. C., & Warner, I. M. (2011). Changing the culture of science education at research intensive universities. *Science, 331*(6014), 152–152. Available at http://www.sciencemag.org/content/331/6014/152.citation

Association of American Universities. (2011). *AAU 5-year initiative to improve undergraduate STEM education discussion draft.* Retrieved from: https://stemedhub.org/groups/aau/File:AAU_STEM_Initiative_Discussion_Draft.10-14-11.pdf.

Association of American Universities. (2012). *Framework for systemic change to undergraduate STEM teaching and learning.* Washington DC: Association of American Universities. https://stemedhub.org/groups/aau/File:AAU_Framework_040714.pdf.

Austin, A.E. (1996). Institutional and departmental cultures: The relationship between teaching and research. *New Directions for Institutional Research, 90,* 57–66.

Austin, A.E. (2011). *Promoting evidence-based change in undergraduate science education.* Paper commissioned by the Board on Science Education of the National Academies National Research Council. Washington, DC: The National Academies. http://sites.nationalacademies.org/DBASSE/BOSE/DBASSE_071087#.UdIxQvm1F8

Austin, A.E. (2014). *Barriers to change in higher education: Taking a systems approach to transforming undergraduate STEM education.* White paper commissioned for Coalition for Reform of Undergraduate STEM Education. Washington, DC: Association of American Colleges and Universities. www.aacu.org/CRUSE

Beach, A.L, Henderson, C. & Finkelstein, N. (2012). Facilitating change in undergraduate STEM education. *Change: The Magazine of Higher Learning, 44*:6, 52–59. http://dx.doi.org/10.1080/00091383.2012.728955

Corbo, J. C., et al. (submitted 9 December 2014). Sustainable change: A model for transforming departmental culture to support STEM education innovation. *Physics Education Research.* http://arxiv.org/abs/1412.3034

Dancy, M., & Henderson, C. (2005). Beyond the individual instructor: Systemic constraints in the implementation of research-informed practices. *2004 Physics Education Research Conference, 790,* 113.

Dancy, M., & Henderson, C. (2010). Pedagogical practices and instructional change of physics faculty. *American Journal of Physics, 78,* 1056–1062.

Dancy, M., & Henderson, C., Smith, J. (2013). Understanding educational transformation: Findings from a survey of past participants of the physics and astronomy new faculty workshop. *Physics Education Research Conference,* 113.

Eckel, P., & Kezar, A. (2003). *Taking the reins: Institutional transformation in higher education.* Phoenix, AZ: ACE-ORYX Press.

Eddy, S.L. and Hogan, K.A. (2014) Getting under the hood: How and for whom does increasing course structure work? *CBE Life Sci Educ,*13, 453.

Eiseman, J., & Fairweather, J. (1996). *Evaluation of the National Science Foundation Undergraduate Course and Curriculum Development Program: Final report.* Washington, D.C.: SRI International.

Fairweather, J. (1993). Faculty rewards reconsidered: The nature of tradeoffs. *Change, 25,* 44–47. doi: 10.2307/40165072

Fairweather, J. (2008). Linking evidence and promising practices in STEM undergraduate education. *National Academy of Sciences (NAS) White Paper.* http://www.nsf.gov/attachments/117803/public/Xc--Linking_Evidence--Fair weather.pdf

Fairweather, J. (2009). Work allocation and rewards in shaping academic work. In J. Enders & E. deWeert (Eds). *The changing face of academic life: Analytical and comparative perspectives* (pp. 171–192). Issues in Higher Education. New York: Palgrave Macmillan.

Fairweather, J., & Beach, A. (2002). Variation in faculty work within research universities: Implications for state and institutional policy. *Review of Higher Education, 26*: 97–115.

Fisher, D., Fairweather, J., & Amey. M. (2003). Systemic reform in undergraduate engineering education: The role of collective responsibility. *International Journal of Engineering Education, 19*: 768–776.

Fisher, P., Zeligman, D., & Fairweather, J. (2005). Self-assessed student learning outcomes in an engineering service course. *International Journal of Engineering Education, 21*: 446–456.

Freeman, S., Eddy, S., McDonough, M., Smith, M., Okoroafor, N., Jordt, H. & Wenderoth, M. (2014). Active learning increases student performance in science, engineering, and mathematics. PNAS. http://www.pnas.org/content/early/2014 /05/08/1319030111?tab=author-info

Freeman, S., Haak, D., & Wenderoth, M. P. (2011). Increased course structure improves performance in introductory biology. *CBE Life Sciences Education, 10*(2), 175–186. doi:10.1187/cbe.10-08-0105

Fry, C.L. (2014). *Achieving systemic change: A sourcebook for advancing and funding undergraduate STEM education.* Washington, DC: Association of American Colleges and Universities. www.aacu.org/CRUSE

Handelsman, J., Ebert-May, D., Beichner, R., Bruns, P., Chang, A., DeHaan, R., Gentile, J., Lauffer, S., Stewart, J., Tilghman, S.M., and Wood, W.B. (2004). Policy forum: scientific teaching. *Science,* 304, 521–522. Available at http://www .bioquest.org/science_vol304_pgs521_522.pdf

Henderson, C., & Dancy, M. (2007). Barriers to the use of research-based instructional strategies: The influence of both individual and situational characteristics. *Phys Rev Spec Top-Ph* Vol. 3 Issue2.

Henderson, C., Dancy, M., Niewiadomska-Bugaj, M. (2010). Variables that correlate with faculty use of research-based instructional strategies. *2010 Physics Education Research Conference* 1289, 169.

Kezar, A. (2001). *Understanding and facilitating organizational change in the 21st century: Recent research and conceptualizations.* ASHE-ERIC Higher Education Report, 28 (4). San Francisco: Jossey-Bass.

Kezar, A. (2011). What is the best way to achieve broader reach of improved practices in higher education? *Innovative Higher Education, 36,* 235.

Lattuca, L. R., Terenzini, P. T., & Volkwein, J. F. (2006). *Engineering change: A study of the impact of EC2000: Executive summary.* Baltimore, MD: ABET, Incorporated.

Lorenzo M., Crouch, C.H., Mazur, E. (2006). Reducing the gender gap in the physics classroom. *Am J Phys,* 74(2):118–122.

National Research Council. (2009). *A new biology for the 21st century.* Washington, DC: National Academies Press. www.nap.edu/catalog.php?record_id=12764

National Research Council. (2013). *The mathematical sciences in 2025.* Washington, DC: National Academies Press. www.nap.edu/catalog.php?record_id=15269 http://www.aacu.org/sites/default/files/files/publications/E-PKALSourcebook .pdf

National Research Council. (2015). *Reaching students: What research says about effective instruction in undergraduate science and engineering.* Washington, DC: The National Academies Press. http://sites.nationalacademies.org/DBASSE /BOSE/Reaching_Students_Effective_Instruction/index.htm

National Science and Technology Council, Committee on STEM Education (2013). *Federal science, technology, engineering, and mathematics (STEM) education 5-year strategic plan.* Washington, DC: National Science and Technology Council.

President's Council of Advisors on Science and Technology (PCAST). (2012). *Engage to excel: Producing one million additional college graduates with degrees in science, technology, engineering, and mathematics.* Retrieved from: http://www .whitehouse.gov/sites/default/files/microsites/ostp/pcast-engage-to-excel-fi-nal_feb.pdf

Singer, S. R., Nielsen, N. R., & Schweingruber, H. A. (Eds.). (2012). *Discipline-based education research: Understanding and improving learning in undergraduate science and engineering.* Washington, DC: National Research Council. http:// www.nap.edu/catalog.php?record_id=13362

Valente, T. (1995). *Network models of the diffusion of innovations.* Cresskill, NJ: Hampton Press.

ABOUT THE AUTHORS

Emily R. Miller is the Project Director of the AAU Undergraduate STEM Education Initiative at the Association of American Universities in Washington, D.C.

James S. Fairweather is the Professor of the Educational Administration Department at Michigan State University in East Lansing, Michigan.

4

Increasing Student Success in STEM: An Overview for a New Guide to Systemic Institutional Change

Susan Elrod and Adrianna Kezar

For the past 20 years, countless reports have been issued calling for change and reform of undergraduate education to improve student learning, persistence and graduation rates for students in STEM; however, by many measures, recommendations in these reports have not been widely implemented (Seymour 2002; Handelsman, et al. 2004; Fairweather 2008; Borrego, Froyd and Hall 2010). Aspirational student success goals in STEM have been set most recently by the President's Office of Science and Technology (PCAST) report, entitled *Engage to Excel: Producing One Million Additional College Graduates in Science, Engineering, Technology and Mathematics* (2011). This report states that STEM graduation rates will have to increase annually by 34% to meet this goal. On most campuses, the persistence and graduation rates of underrepresented minority (URM) and first-generation students still lag behind those of their majority counterparts. Thus, in order to reach the aspirational graduation rates called for in national reports, a focus on URM and first-generation student success is imperative.

Many change efforts have been started, almost always at the departmental level; however, few have reached the institutional level of entire programs, departments, or colleges in the STEM disciplines, described as necessary in these recent reports. There is growing recognition that reform in STEM is an institutional imperative, rather than only a departmental one. Student advising, faculty professional development, student research mentoring, academic support programs, clear STEM-focused institutional articulation agreements, external partnerships with business and industry related to internships and other research experiences, among other critical areas, are often overlooked within reform efforts and have been identified as central to student success. Research has emerged that demonstrates the importance of a broader vision of STEM reform for student success—moving from programs and departments to an institutional effort. For example, institution-wide implementation of high-impact practices (HIPs) has shown to dramatically improve the graduation rates of URM students (Kuh and O'Donnell, 2013).

The Meyerhoff Scholars Program at the University of Maryland, Baltimore County, epitomizes this type of institution-wide effort and combines specific academic, social and research support interventions that have resulted in dramatic improvements in graduation of minority STEM students (Lee and Harmon, 2013).

The Keck/PKAL model outlined in this article provides both a process and content scaffold for campus leaders to plan, implement, assess and evaluate change efforts in undergraduate STEM education in a way that goes beyond redesign of a single course or isolated program. We have learned from our own work, as both researchers and practitioners, that institutional change is best executed by a cross-functional team working together. Support of leaders across campus is critical, including grassroots faculty leadership, mid-level leadership among department chairs and deans, and support from senior leaders in the administration.

This article provides an overview of the Keck/PKAL project, which was generously supported by the W.M. Keck Foundation and involved 11 campuses in California, including California State Universities, private liberal arts colleges and research universities (Box 1).

The project was aimed at helping California colleges and universities promote institutional-level STEM-reform efforts. Campus case studies that describe institutional STEM-reform journeys are posted on the project website and will be published in the Spring 2015 issue of *Peer Review* (aacu.org) and referenced in a detailed guidebook to be published by AAC&U, entitled *Increasing Student Success in STEM: A Guide to Systemic Institutional Change* (Elrod and Kezar, in press). The guidebook is for campus leaders who have convened (or will convene) teams comprised of faculty members, department-level leaders, student affairs professionals, appropriate central administration officers, institutional research and/or undergraduate studies offices. Each case study highlights the elements of the Keck/PKAL model for effective institutional change for increasing student success in STEM that was developed during this three-year project.

BOX 1: Participating Institutions

- California State University, East Bay
- California State University, Fullerton
- California State University, Long Beach
- California State University, Los Angeles
- San Diego State University
- San Francisco State University
- W.M. Keck Science Department of Claremont McKenna, Pitzer and Scripps Colleges
- University of San Diego
- University of La Verne
- The California State University Chancellor's Office
- University of California, Davis

While the soon-to-be-published guidebook provides additional background information, in addition to detailed case studies and tools for guiding institutional transformation, this paper summarizes key elements of the Keck/PKAL model and highlights some of the major lessons learned from this project.

THE KECK/PKAL MODEL

The Keck/PKAL model for effective institutional change outlines both a process and content that will lead to increased student success in STEM. Although focused on STEM, it is applicable to any change process that is focused on improving student learning and success. The elements of the model are illustrated in Figure 1 and described in Table 1.

The model is shown in the context of a river because the flowing nature of a river represents both the flowing nature of change and the dynamic and powerful process of change. The flow (change process) encounters obstacles (challenges presented by certain aspects of the change process) that may result in an eddy where the flow circles around the obstacle until it can break free. The resulting eddy motion is an apt analogy for the circular swirl, or iterative process, that campus teams experience when they encounter resistance or challenges along their path toward reform. When this happens, teams must work through the issue by determining the nature of the challenge and figuring out how to get the flow going again in the desired direction. Travelers on the river may enter at various points or "put out" at certain locations to rest and regroup. New travelers may enter and join a party already on a journey down the river. In the same way, teams working on a systemic change initiative may start at different points, alter membership, or even stop out for periods of time because other campus priorities emerge, team members take on other duties, campus leadership changes, or other factors. Like paddlers on a river, change doesn't always go in one direction. Teams can paddle up or downstream, although the general flow will ultimately go downstream toward action and success.

Our approach to change in this project is based on practices of organizational learning. Within this approach, information gathering and data analysis play a central role in helping individuals identify directions and appropriate interventions for making strategic forward progress. Participants in an effective organizational learning process must engage the data, and find time for reflection, and dialogue. Forming non-hierarchical teams enhances learning and developing innovative approaches. This means having campus teams look at data related to student success in order to determine the specific challenges and problems, and to orient themselves toward a vision for change. But an organizational learning model also focuses on learning throughout the change

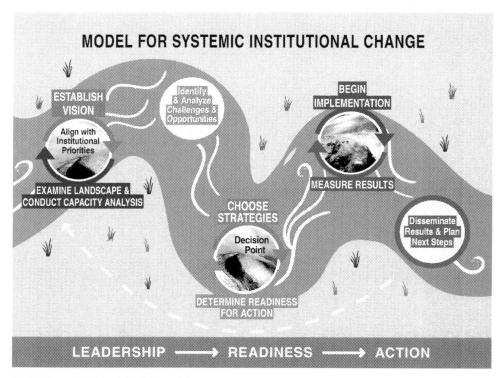

FIGURE 1. The Model—Visual Diagram

process. The model is focused on facilitating organizational learning, but it also incorporates key ideas from other research on change, such as: the need to address politics, developing buy-in and a shared vision, understanding the power of organizational culture, and helping campus leaders unearth underlying assumptions and values that might create resistance to change.

Our work with the change process at several campuses elicited a number of common challenges and barriers, including a rush to implementation and the presence of implicit biases. The most common obstacle was that campus leaders wanted to start by immediately implementing a strategy that they read about in a report or publication. While news of a successful program may motivate change, it is important to check in with campus vision and landscape analysis before jumping into implementation of the latest published student success strategy. It may or may not fit your campus situation, student population, faculty expertise or resources. In our project, campuses that jumped right into a strategy without first assessing the fit of the strategy for their campus context struggled with purpose, outcomes, implementation, and impact measurement. These quick starters ended up going back to their vision, refining it and doing

TABLE 1. Model Elements

Model Element	Description
1. Vision	The vision represents the direction that the campus is aimed in terms of altering its STEM experience to support student success. We encourage a vision that is clear and shared.
2. Landscape and capacity	A direction forward is typically best created through an analysis of the existing landscape (internal campus data, as well as external reports on STEM reform), as well as a review of current capacity to engage in change generally—such as history of reform, leadership, and buy-in and ownership among faculty. This stage focuses on the collecting of data and information to conduct analysis.
3. Identify and analyze challenges	The landscape and capacity information needs to be analyzed in order to identify both challenges and opportunities for the campus. This phase often brings in politics and culture that might be sources of both opportunities and challenges.
4. Choose strategies/ interventions/ opportunities	Campuses need to familiarize themselves with a host of strategies or interventions that they might choose from to address the challenges identified. They can examine these strategies in light of the capacity of the campus, as well as opportunities identified earlier.
5. Determine readiness for action	In addition to reviewing campus capacity and opportunities, there are key issues that emerge when implementing specific strategies. These include resources, workload, institutional commitment, facilities, timeline and other areas. Review of these issues is essential in order to ensure that the campus is ready to move forward with effective implementation of a particular strategy. Campuses will be able to identify opportunities, such as a newly established special campus projects fund, a new faculty hire with appropriate expertise, etc., that can be leveraged in support of effective implementation. This phase also involves exploring campus politics and culture.
6. Implementation	Implementation involves drafting a plan for putting the intervention or strategies in place. The plan builds on the ideas from the readiness for action, capacity of the campus, and opportunities identified. All of these, as well as a process for understanding challenges as they emerge, should be built into the plan. In addition to a well-laid-out plan, campuses may decide to pilot an initiative first and then consider how to modify and scale it after an initial trial.
7. Measure results	Campuses will also create an assessment plan to inform whether the intervention is working and ways they can be changed over time to work better.
8. Disseminate and plan next steps	In order to prevent our work from existing in silos, it is important for campuses to think about dissemination opportunities on campus, as well as off campus—regionally, statewide or nationally. Deliberate planning for next steps is also necessary to keep the momentum going.

more landscape analyses, which ultimately slowed progress, but improved long-term success.

Another common barrier we identified was that campus team members held implicit theories of how change happens that were contradictory and often contrary to the project's vision and goals. For example, a common assumption among STEM faculty is that meaningful change can only happen in departments. Faculty that hold this belief will resist examining potential levers outside the department that are important to address. These external levers can include: mathematics preparation, success in a prerequisite course in another department, level of study skills, and institutional support that is critical for sustaining change in the long term. Implicit biases can only be revealed through conversations about beliefs, values and practices. Therefore, we encourage teams to make their first meeting a discussion about how change occurs and to make their implicit theories explicit. What makes this process hard is that implicit theories are often unconsciously held. Many people may not be able to articulate a theory of change or understand why the model is hard for them to work with. It can help just to have the candid discussion among your team members: "What do you think it will take to start an undergraduate research program here?"

Other common barriers encountered were:

- Faculty beliefs about their roles as "gatekeepers" or as the "sage on the stage," as opposed to "gateways" or as "guides on the side."
- The lack of faculty expertise in evidence-based STEM education teaching and assessment methods.
- A misguided belief that all faculty and staff share the same vision.
- Failure to examine all the implicit assumptions about the problem, possible solutions and approaches; team members' implicit theories of change that may prevent them from engaging in aspects of the work.
- A lack of capacity for data collection and analysis in terms of support from centralized offices of institutional research.
- Inadequate incentives and rewards for faculty participation in STEM-reform projects.
- Inadequate planning to secure appropriate buy-in, approval or support from relevant units, committees or administrators.
- Inadequate resource identification or realization.
- Unforeseen political challenges, such as tension regarding department "turf," resource and faculty workload allocation.
- Shifts in upper-level leadership leading to stalled support or redirection of efforts to new campus initiatives (e.g., quarter to semester conversion).

- Changes in team membership because of sabbatical leaves or other assignments.
- Failure to connect STEM reform vision at the departmental level to institutional priorities to secure support and resources.
- Lack of consideration about how students will be affected by and/or made aware of the changes, including the rationale for them. In order for students to fully participate, they need to understand how they will benefit from the changes or new opportunities.

The forthcoming guidebook describing this model includes detailed information about each element of the model accompanied by an explanation, key questions to consider, highlights from campus case studies, challenge alerts (mistakes to avoid or pitfalls to be aware of), and timeline considerations. The guidebook also contains specific tools to help campus leaders and teams plan and manage change initiatives, such as:

- Tools to help campus leaders and teams determine how to get started in the process.
- A readiness survey to help teams determine whether they are prepared to move forward with implementation of their chosen strategies and interventions.
- A rubric to help campus teams gauge their progress in the model phases.
- Examples of data analyses to conduct, and implementation strategies to address common challenges facing STEM programs.
- Suggestions for how to build effective teams, develop leadership capacity, and sustain change.

These tools are also included in a practical workbook that is intended for use by teams to actively work through the elements of the model. This workbook and the full-length case studies are available on the project website:

http://www.aacu.org/pkal/educationframework

REFERENCES

Borrego, M., J. E. Froyd, and T.S. Hall. (2010). Diffusion of engineering education innovations: A survey of awareness and adoption rates in U.S. engineering departments. *Journal of Engineering Education,* 99(3):185–207.

Elrod, S. and A. Kezar. (in press). *Increasing student success in STEM: A guide to systemic institutional change.* Washington, DC: AAC&U.

Fairweather, J. (2008). *Linking evidence and promising practices in science, technology, engineering, and mathematics (STEM) undergraduate education.* http://www7.nationalacademies.org/bose/Fairweather_CommissionedPaper.pdf Accessed June 12, 2011.

Handelsman, J., Ebert-May, D., Beichner, R., Bruns, P., Chang, A., DeHaan, R., Gentile, J., Lauffer, S., Stewart, J., Tilghman, S., and Wood, W. (2004). Scientific teaching. *Science,* 304: 521–522.

Kuh, G. and K. O'Donnell. (2013). *Ensuring quality and taking high-impact practices to scale.* Washington, DC: AAC&U.

Lee, D.M. and K. Harmon. (2013). The Meyerhoff Scholars Program: Changing minds, transforming a campus. *Metropolitan Universities,* 24(2):55–70.

PCAST. (2011). *Engage to excel.* Washington, DC: President's Council of Advisors on Science and Technology.

Seymour, E. (2002). Tracking the processes of change in U.S. undergraduate education in science, mathematics, engineering, and technology. *Science Education,* 86 (1): 79–105.

ABOUT THE AUTHORS

Susan Elrod is the Interim Provost and Vice President for Academic Affairs at California State University in Chico, California.

Adrianna Kezar is the Professor of the Rossier School of Education and Co-Director of the Pullias Center for Higher Education at the University of Southern California in Los Angeles, California.

SECTION B

Case Studies—Projects at the Institution Level

The collection of chapters in this section reflect a variety of approaches to institutional transformation and represent change initiatives across a continuum from early stage (Hogan; Reinholz; Potter; Marker) to mature (Chasteen; McManus; Hastings). Some initiatives, such as those described in the chapter led by Hogan, focus on developing models to enhance faculty pedagogical expertise. The chapters led by Burd and Franklin examine mechanisms for changing the culture of research-intensive universities to better support and value teaching excellence. Transformation efforts, as described in this collection, include initiatives that began as grassroots efforts and grew, such as the CASTLE project described by Franklin and the mini grant approach described by Bunu-Ncube. Other chapters describe intentional initiatives grounded in specific models of change, such as the deliberate bottom-up, top-down model described by Potter; the fledgling efforts described by Reinholtz based on systemic change models; the successful Science Education Initiative described by Chasteen; and the CACAO change facilitation model borrowed from the business world that Marker presents. The chapter by McManus chronicles the purposeful transformation of an entire university to more appropriately meet the needs of 21st century learners. The collection is rounded out by Hastings' chapter that reflects on the key elements of long-term successful innovations, and Kirkup's chapter that examines strategies for fostering large-scale change in teaching pedagogies across a number of institutions using a government-sponsored fellowship as a tool for change.

1

Advancing Evidence-Based Teaching in Gateway Science Courses Through a Mentor-Apprentice Model

Kelly A. Hogan, Jennifer Krumper, Laurie E. McNeil,
and Michael T. Crimmins

Despite a preponderance of evidence in the educational literature supporting the efficacy of evidence-based instructional practices in STEM education (Nielson, 2011; President's Council of Advisors, 2012; Singer, 2012), widespread adoption of these approaches remains a significant challenge. Such adoption requires educational institutions to change in fundamental ways—adjusting core beliefs, behaviors, and structures—and should be expected to meet with significant resistance (Henderson, 2005, 2009).

Our approach to implementing widespread change in the culture of teaching and learning in the sciences at UNC-Chapel Hill featured an individual-centered, largely emergent strategy (Henderson, 2008; Singer, 2012). We focused on professional development via the support of mentor-apprentice relationships within the context of teaching reformed curricula in gateway courses. In our model, a mentoring teacher (often, a term faculty member, who may have a background in discipline-based educational research [DBER]) would partner with an apprentice teacher (often, a tenure-track or tenured faculty member) in an individual course (chemistry, biology) or as a part of a larger group of faculty running coherent, multiple sections of a single course (physics).

Communities of faculty members working together within a discipline are exceptionally effective in promoting large-scale educational reform (Sirum, 2010). Thus, a key element of our strategy was to leverage social networks in the form of faculty learning communities (FLCs) to enact cultural change. Evidence suggests such networks might be necessary for the dissemination of reformed practices and development of emergent practices (Dancy, 2013; Kezar, 2014), by providing social incentives for greater change work (Bouwma-Gearhart, 2012). In our model, all mentor-apprentice teams participate in interdepartmental FLCs (and some in intradepartmental FLCs). These networks provide social support, a vehicle for exchange of best practices, and a safe space for program participants to voice concerns about challenges to adoption at the department level (Bouwma-Gearhart, 2012).

Strong leadership *distributed* across ranks, organizational units and roles is required for change to propagate successfully through an institution (Kezar, 2011; Palmer, 1992). As such, faculty members from varied ranks (term faculty, and assistant, associate, full, and distinguished professors) and three academic units (physics, chemistry, and biology) were invited to participate in the AAU program. Further, campus administrators at all levels (dean, provost, chancellor) were brought into the project at the ground level to add institutional and financial support for the project agenda.

In addition to leadership, incentives must exist to foster educational reform (Henderson, 2010). Repeatedly found in the literature and with our own internal data, faculty report the most significant barrier to change is finding time (Dancy, 2005, 2010; Henderson, 2007). Thus, our project provided: (1) course release time for the development of reformed practices; (2) access to mentors with pedagogical, technical, content, and course management expertise (reducing time investment for change); (3) shared course resources (again, saving time); and (4) shared duties for course administration (saving time).

Our model is consistent with the AAU framework for change. At the center of our model is reform of pedagogy of gateway courses for students in chemistry, physics, and biology. Faculty involved in these courses are supported with mentorship, social networks, and time to engage. Lastly, strong leadership and strong social networks provide a cultural context that will support the sustainability of the reforms.

OUR INITIAL IMPLEMENTATION OF THE APPRENTICESHIP MODEL

Our approach to the apprenticeship model varied in each of the three departments due to the context of the redesigned courses within the department, the teaching culture of each department, and the ranks of the apprentices relative to their mentor. Thus, our initial implementation allowed us to examine the model in triplicate. In all three departments, the structure was similar: The mentor-apprentice teams were co-assigned to a single section. From this common structure, each department implemented the idea slightly differently. The biggest difference between our three implementations was in the amount of time each apprentice was in front of the class as the "lead teacher".

Before detailing each department's implementation, it is worthwhile to note some differences among department culture and course formats. In chemistry and biology, the course selections do not yet have common syllabi, and the objectives and pedagogical methods vary widely by section. This project was the first attempt at formally transferring common pedagogical ideas from one instructor to another. In contrast, physics and astronomy consider their courses to be "owned" by the department and sections are highly coordinated with shared objectives, activities, and assessments. Physics and astronomy had a decade of experience formally teaching as a team before this project, and already had faculty in place with a background in DBER when the project began. Physics also had a more diverse pool of faculty ready to be mentors, as opposed to lecturers in the chemistry and biology departments (Table 1), which only had fixed-term faculty members (lecturers) to take this role. Class formats differed, too: In biology and chemistry, the courses were large lectures in fixed seats, whereas physics and astronomy use a lecture/studio format designed to maximize the time spent in active engagement.

Chemistry: The apprentice provided the vast majority of in-class instruction, and was therefore viewed by students as the primary instructor. The mentor largely observed and provided suggestions and feedback to the apprentice after class. Occasionally, the mentor offered in-class suggestions to the apprentice or taught briefly to demonstrate a teaching technique. In this model, much transfer of information about course structure occurred prior to the start of the semester and continued in weekly meetings. The apprentice used materials provided by the mentor (lesson plans and most course materials) and focused on learning the technological and pedagogical tools for high structure teaching in a large class. The mentor coordinated course logistics (scheduling review sessions, training and coordinating undergraduate learning assistants, etc.) and electronic resources (learning management system administration and electronic homework).

TABLE 1. Ranks of Apprentices and Mentors in the Three Departments for Mentor-Apprentice Relationships From the First Three Semesters of AAU Funding

Department	Apprentice	Mentor
Chemistry	Distinguished Professor	Lecturer
Chemistry	Assistant Professor	Lecturer
Chemistry	Distinguished Professor	Lecturer
Biology	Assistant Professor	Lecturer
Biology	Professor	Lecturer
Biology	Assistant Professor	Senior Lecturer
Biology	Associate Professor	Lecturer
Physics	Assistant Professor	Associate Professor
Physics	Professor	Distinguished Professor
Physics	Assistant Professor	Lecturer
Physics	Assistant Professor	Professor

Biology: The apprentice and mentor co-taught the course, and thus, both instructors were viewed as primary instructors. Both the apprentice and mentor took turns leading activities in each class period. There were many similarities to the chemistry implementation. For example, the apprentice used lesson plans provided by the mentor, and weekly meetings were used to learn how to engage the students in upcoming activities and to transfer knowledge about typical student misconceptions. Additionally, the mentor coordinated course logistics so that the apprentice could focus on learning the technological and pedagogical tools.

Physics & Astronomy: Unlike the typical large lectures of biology and chemistry, the lecture/studio format in physics and astronomy typically contains two large 50-minute lectures and two separate smaller 110-minute studio sessions per week. In this format, the apprentice was not seen as the lead teacher, as the apprentice's initial assignment in the team was to lead a studio section, paired with a more-experienced studio instructor (often an experienced graduate student). The mentor delivered most of the lectures, keeping them tightly coupled with the studio activities. The faculty apprentice also attended the lectures and delivered some of them, working with the mentor to make them as interactive and effective as possible. As part of the instructional team for the course, the apprentice participated in the writing of exams and homework assignments and in the improvement of the studio activities. Expectations were clear that the apprentice would serve as lead instructor for the course in the following semester and mentor an apprentice, in turn, in a future semester. Besides the mentor and apprentice, the team for each course also included a studio supervisor who handled most of the logistical details of the course. Faculty members with DBER backgrounds also played a supporting role, assisting the team with exam preparation and teaching assistant training, and performing observations of all instructors.

EARLY RESULTS FROM THE PROJECT

Pedagogy

While this short report does not focus on the development of redesigned courses or the student outcomes, pedagogy is central to the AAU framework and our project. The apprentices and mentors carry out the transfer of pedagogical innovation within redesigned introductory courses. Table 2 demonstrates our progress in transforming the introductory courses in our three departments into high-structure, active learning formats. In addition to four high-structure courses ready before the project funding began, we have redesigned six more courses, with a seventh in transition. A current challenge is that not

TABLE 2. Courses Redesigned into High-Structure, Active-Learning Courses in the Three Departments

Course number	Course name	Sections taught in high- structure, active-method	When redesigned?
Physics 116	Mechanics	All sections	Before AAU project
Physics 117	Electricity and Magnetism	All sections	Before AAU project
Biology 101	Principles of Biology	All sections	Before AAU project
Chemistry 101	General Chemistry I	Some sections	Before AAU project
Physics 104	General Physics I	All sections	First year of AAU funding
Physics 105	General Physics II	All sections	First year of AAU funding
Chemistry 102	General Chemistry II	Some sections	First year of AAU funding
Chemistry 261	Organic Chemistry I	Some sections	First year of AAU funding
Chemistry 262	Organic Chemistry I	Some sections	First year of AAU funding
Biology 202	Genetics and Molecular Biology	Some sections	First year of AAU funding
Biology 201	Ecology and Evolution	None yet	Second year of AAU funding

Approximately 3,700 enrollments have been affected by these changes in the first year of AAU funding (many students enroll in more than one gateway course in a single year).

all sections are taught in this format in biology and chemistry; conversation is ongoing in both departments about how to spread evidence-based pedagogies more widely. Physics moved past these conversations before this project began and has taken a team approach to their introductory courses, regarding them as forming a single course with a single curriculum, pedagogy and assessment. Although not presented here, we have evidence that not only are our students showing increases in achievement and learning in redesigned courses, but that there are disproportionate benefits for first-generation college students and Black students (Eddy, 2014).

Support

The AAU framework recognizes that support is necessary for reforms in pedagogy. Our project focuses on two major areas of this scaffold: apprenticeships and faculty learning communities.

Apprenticeships

Having evidence that professional workshops alone don't always lead to permanent change (Dancy, 2013; Ebert-May, 2011) our apprenticeship model provides continual individualized support during a semester-long teaching experience. To date, course releases have allowed six tenured or tenure-track research-active faculty to apprentice one-on-one with mentors across the three departments. These include three assistant professors (one each in biology, chemistry and physics-astronomy) and one associate professor, one professor and one distinguished professor. During the current academic year, we anticipate an additional ten apprenticeships that will include participation by four lecturers, one assistant professor, one associate professor, two professors and two distinguished professors. Interviews with mentors and apprentices have been conducted, and while the picture is not yet complete, an initial report suggests issues to consider as we move into new phases of the project. There may be a need for:

- more clarity in role expectations for both mentor and apprentice;
- earlier communication with the apprentice about expected benefits from the experience;
- more conversations about differences in priority given to the work, given other demands on time and attention; and
- formal mentor training to provide a toolbox of strategies to identify/ address challenges working with different apprentices with different personalities/temperaments/time demands.

Faculty Learning Communities (FLCs)

Social networks can play a dual role in professional development and in promoting cultural change in a community. In the first year, 15 participants (mentors and apprentices from the three departments) participated in monthly FLC meetings organized by our Center for Teaching and Learning (CTL). Meetings were co-facilitated by a staff member of the CTL and a mentor. Participants found conversations around teaching challenges and strategies very helpful and also benefited from discussions conducted after peer observations of their classes. For this reason, we have adapted published rubrics to allow for formative observation without a need to include critical evaluation and will use the rubrics with each FLC going forward. Due to the increase in numbers of participants for the second year, we have created three separate FLCs for this academic year, based on their level of experience. We have one FLC primarily for mentors (nine participants), one for the first cohort of apprentices (ten participants) and one for the newest apprentices in this current academic year (seven participants). Department-level FLCs have taken different forms in each department, but include most of the AAU project participants, as well as others who have become interested in active learning.

Culture

Is reformed teaching "sticking"? One clear measure of success with our apprenticeship model will be whether apprentices are still using reformed methods semesters after they were trained. To assess this, we will follow apprentices for several semesters, performing at least two unannounced evaluative observations with the COPUS rubric (Smith, 2013). Despite being a nascent project, our early results are promising and may also demonstrate that trained apprentices transfer these skills to other non-project courses they teach.

Measuring cultural change is difficult, but our plan to assess faculty teaching attitudes and practices and institutional culture includes faculty interviews (with the same open-ended questions about teaching attitudes and practices) with both participating and non-participating faculty. Additionally, the AAU's faculty survey will capture the attitudes and practices of a wider group of UNC faculty. Lastly, in 2013, we surveyed faculty in the College of Arts and Sciences on teaching issues: Nearly half of our faculty members responded. We plan to administer a set of identical questions in Spring 2016 to take another "pulse" of the faculty and compare our science faculty responses in 2013 to 2016.

Leadership plays a role in ensuring these methods become part of the culture of teaching. In the past two years, the university leadership has provided the resources to hire nine lecturers in various STEM departments (including two in Chemistry, two in Biology, and two in Physics and Astronomy) who have expertise in high-structure, active learning course design and implementation. These new faculty members are jump-starting the project, so that there are enough mentors to sustain the initial years of the model.

As is the case with any emergent strategy, project leaders are finding other evidence that culture is changing, but they have no systematic plan for capturing these unexpected indications. For instance, departmental faculty meetings have led to difficult and impassioned discussions about the pros and cons of moving to active learning approaches, discussion that would not have occurred without an ongoing AAU project. Another example of change is that numerous science faculty members *not* formally part of the AAU project have begun to radically change their teaching (some with the help and guidance of experienced faculty, and some with no plan).

INITIAL RECOMMENDATIONS FOR IMPLEMENTING THE APPRENTICESHIP MODEL AT ANOTHER INSTITUTION

With the financial means to provide course release time, other institutions may find this is a model they are interested in executing. While our evaluation is far from complete in this young project, our mentors and apprentices involved in the first year implementation assembled a list with advice. Recommendations are presented in no particular order and pertain primarily to implementation in lecture format:

Advice for Mentors

- Be flexible and ready to mentor each apprentice differently based on their prior experiences and personalities.
- Remember that you are the mentor. Your apprentice may significantly outrank you in the university hierarchy and they may have more expertise in your course's research discipline than you do. Remember that you have greater pedagogical expertise; in this partnership it is your job to set the direction of the team.
- Do not overwhelm your apprentice with suggestions, ideas, and tasks. Remember that the time and attention of research-active faculty are limited, and change is incremental. Pick a few key goals for the partnership (e.g. "maintain student-centered approach through the semester") and focus on working in those areas.
- Set clear expectations *together* for what this collaboration will look like. For example, will you meet twice a week? Who will answer student e-mails? How would the apprentice prefer feedback to be delivered?
- Do not mentor a course unless you have already taught it at least once in a high-structure format. Your apprentice will need materials prepared, polished, and vetted well in advance.
- Treat each other as equals, both outside and inside the classroom. Students don't need to know that one of their teachers is "in training."
- Have explicit lesson plans for each class session that can be communicated clearly. Items that should be part of the lesson plan: each objective, the activities and formative assessments that align with the objective, and the expected amount of class time needed for each activity.
- Communicate to students clearly and responsively. Guide apprentice instructors to do the same. For example, a PowerPoint slide displaying an in-class multiple choice question to students might say in the instructor notes, "Choice A is the common misconception for this question because. . . ."
- Allow your apprentice to make some mistakes. The temptation will be to step in during class discussions to correct, moderate, or simply do it your way. Be careful to not undermine your apprentice's authority in the classroom; any immediate concerns or advice for moving the class in a different direction can be done quietly without drawing attention.

Resist also the temptation to "perfect" the apprentice's assessments. Minor flaws in question design will allow the apprentice to discover pitfalls themselves after seeing student data.

- Take care of course logistics, office hours, planning, and student concerns, etc., so that the apprentice can focus on preparing for each class session and learning to write assessments. Provide enough guidance and practice with these tasks, however, that the apprentice will be able to complete them in subsequent semesters.
- Give the apprentice tools to succeed in the future: Ask the apprentice to complete a small course development project that you can both use in the future, such as an in-class or a recitation activity. Mentor them through this process as well.
- Invite criticism of the design of course materials by your apprentice: (a) to open lines for future collaborations, and (b) to create an emergent product that is better for both members' contributions.
- Keep open communication in future semesters and check in with their next teaching experience (even if a different course). Ask if you can observe and give feedback via an observation rubric.

Advice for Apprentices

- Know the class/instructor you are going to be working with, observe classes taught by your mentor the semester before you begin teaching (several weeks of observation would be optimal). Make good notes on how they manage class time, discussions, activities, etc., and write down any questions about their approach that arise so you can talk to your mentor about them before you begin teaching the following semester.
- Ensure you are comfortable with the course material (the scope and the details) well in advance of classes beginning; this will ensure you can spend most of your energies focused on learning the teaching methods and class management skills of the student-centered classroom.
- Even if you are not the lead instructor, consider this "your" class as well as your mentor's; imagine how you would lead each class if your mentor was unable to make it that day and compare that to how it actually goes.
- Hold your own office hours or Q&A sessions on course content at least once a week—this is where you will best come to understand what topics are confusing for students in the class.
- Familiarize yourself with the technology available in advance of class so that you are not learning during class time.

- Be aware of the time goals in the lesson plan and endeavor to stay on track during class.
- Learn to summarize activities for students as the "expert" in the room, such as providing a "take-home" message or a summary of the approach they can use to think through problems.
- Be brutally self-reflective.
- Encourage your mentor to be frank about how you can improve.
- As much as possible, learn how your mentor approaches the administrative side of managing the class (i.e., sit in on their office hours, learn how they deal with student concerns, how to deal with grading, managing grading data, etc.).
- Keep the relationship open in future semesters. Ask for advice and invite your mentor and other instructors to observe you periodically.

Not all institutions interested in this model will have the ability to provide course releases. We recommend a few low-cost ideas that can be implemented from our model. Faculty wanting to make changes to their teaching need a model, and thus should be observing other faculty on campus who are using evidence-based teaching. It might also be possible for two or more faculty members to form a team to redesign a course that both could teach in future years, lessening the burden of redesign on individuals who have not received course release. Model materials developed for classroom use by practitioners of evidence-based teaching can also be found in digital libraries maintained by professional societies in specific disciplines; these may serve as starting points for redesign of specific lessons. Faculty should be encouraged to ask for guidance, and should invite mentors to their classroom for feedback on classroom interactions and course structure. Lastly, faculty learning communities can meet a few times a semester and provide a foundation for important supportive relationships.

LOOKING FORWARD

The funds provided by the AAU have launched new ways of thinking about teaching at UNC-Chapel Hill for many faculty. The project reinforces three major ideas: that good teaching can be taught (mentors help apprentices), it takes time to implement changes (funds provide a course release), and that teaching need not be a private endeavor (faculty meet in learning communities). These ideas have quickly taken hold in our culture within the first two years of the project, although their implementation throughout each department's course offerings is still far from universal. As we seek to broaden the use of evidence-based teaching beyond the courses that are the focus of our AAU project (all of which are foundational, lower-division courses), we will need to find ways to help faculty members embrace appropriate methods for active engagement in more advanced courses. There is less research available on the use of such pedagogy in courses beyond the introductory level, and fewer classroom materials and model instructors currently exist. This makes the task more difficult, but we are confident that participants in our project will begin to re-examine other kinds of teaching they do and seek to incorporate active engagement throughout the educational spectrum in each department. When the AAU funding is complete, the three departments will sustain the course releases formally for another two years. Beyond the five years of funding, peer visits of each other's classes, teaching mentors for all new faculty, and learning communities can be sustained. The idea that teachers at any stage can benefit from "coaching" in evidence-based pedagogies will be a belief that "sticks" in our culture.

REFERENCES

Bouwma-Gearhart, J. (2012). Research university STEM faculty members' motivation to engage in teaching professional development: Building the choir through an appeal to extrinsic motivation and ego. *Journal of Science Education and Technology*, 21, 558–570.

Dancy, M. H. & Henderson, C. (2005). Beyond the individual instructor: Systemic constraints in the implementation of research-informed practices. *2004 Physics Education Research Conference*, 790, 113–116.

Dancy, M. H. & Henderson, C. (2010). Pedagogical practices and instructional change of physics faculty. *American Journal of Physics*, 78, 1056–1063.

Dancy, M. H., Henderson, C., Smith, J. H. (2013). Understanding educational transformation: Findings from a survey of past participants of the physics and astronomy new faculty workshop. *Physics Education Research Conference*, 113–116.

Ebert-May, D. Derting, T. L., Hodder, J., Momsen, J. L., Long, T. M., Jardeleza, S. E. (2011). What we say is not what we do: Effective evaluation of faculty professional development programs. *Bioscience*, 61, 550–558.

Eddy, S. L. & Hogan, K. A. (2014). Getting under the hood: How and for whom does increasing course structure work? *CBE Life Sciences Education*, 13, 453–468.

Henderson, C., Beach, A., Finkelstein, N., Larson, R.S. (2008). Facilitating change in undergraduate STEM: Initial results from an interdisciplinary literature review. *2008 Physics Education Research Conference*, 1064, 131–134.

Henderson, C., Beach, A., Famiano, M. (2009). Promoting instructional change via co-teaching. *American Journal of Physics*, 77, 274–283.

Henderson, C. & Dancy, M. H. (2005). Teaching, learning and physics education research: Views of mainstream physics professors. *2004 Physics Education Research Conference,* 790, 109–112.

Henderson, C. & Dancy, M. H. (2007). Barriers to the use of research-based instructional strategies: The influence of both individual and situational characteristics. *Physical Review Special Topics Physics Education Research,* 3.

Henderson, C., Dancy, M. H., Niewiadomska-Bugaj, M. (2010). Variables that correlate with faculty use of research-based instructional strategies. *2010 Physics Education Research Conference,* 1289, 169–172.

Kezar, A. (2011). What is the best way to achieve broader reach of improved practices in higher education? *Innovative Higher Education,* 36, 235–247.

Kezar, A. (2014). Higher education change and social networks: A review of research. *The Journal of Higher Education,* 85, 91–125.

Nielsen, N. R. (2011). National Research Council of the National Academies. *promising practices in undergraduate science, technology, engineering, and mathematics education: A summary of two workshops.*

Palmer, P. J. (1992). Divided no more: A movement approach to educational reform. *Change,* 24, 10–17.

President's Council of Advisors on Science and Technology. (2012). *Engage to excel: Producing one million additional college graduates with degrees in science, technology, engineering, and mathematics.* Available at: https://www.whitehouse. gov/administration/eop/ostp/pcast/docsreports

Singer, S. R., Nielsen, N. R., Schweingruber, H. A. (2012). National Research Council (U.S.). Committee on the Status Contributions and Future Directions of Discipline-Based Education Research. *Discipline-based education research : Understanding and improving learning in undergraduate science and engineering.* Available at: http://www.nap.edu/catalog/13362/discipline-based-education-research-understanding-and-improving-learning-in-undergraduate

Sirum, K. L. & Madigan, D. (2010). Assessing how science faculty learning communities promote scientific teaching. *Biochemistry and Molecular Biology Education : A Bimonthly Publication of the International Union of Biochemistry and Molecular Biology,* 38, 197–206.

Smith, M. K., Jones, F. H. M., Gilbert, S. L., Wieman, C. E. (2013). The classroom observation protocol for undergraduate STEM (COPUS): A new instrument to characterize university STEM classroom practices. *CBE Life Sciences Education,* 12, 618–627.

ABOUT THE AUTHORS

Kelly A. Hogan is the Director of Instructional Innovation for the College of Arts and Sciences and Senior STEM Lecturer in Biology at the University of North Carolina, Chapel Hill in Chapel Hill, North Carolina.

Jennifer Krumper is a Lecturer of the Department of Chemistry at the University of North Carolina, Chapel Hill in Chapel Hill, North Carolina.

Laurie E. McNeil is the Bernard Gray Distinguished Professor of the Department of Physics and Astronomy at the University of North Carolina, Chapel Hill in Chapel Hill, North Carolina.

Michael T. Crimmins is the Mary Ann Smith Distinguished Professor of the Department of Chemistry at the University of North Carolina, Chapel Hill in Chapel Hill, North Carolina.

2

Developing Faculty Cultures for Evidence-Based Teaching Practices in STEM: A Progress Report

Gail D. Burd, Debra Tomanek, Paul Blowers, Molly Bolger,
Jonathan Cox, Lisa Elfring, Elmer Grubbs, Jane Hunter, Ken Johns,
Loukas Lazos, Roman Lysecky, John A. Milsom, Ingrid Novodvorsky,
John Pollard, Edward Prather, Vicente Talanquer, Kay Thamvichai,
Hal Tharp, and Colin Wallace

Faculty resistance to changing their STEM teaching practices is widely acknowledged as a barrier to use of evidence-based, active-learning instructional approaches (AAAS, 2011). What accounts for this resistance? Changes in faculty members' teaching practices are strongly influenced by *personal influences*, their own beliefs and concerns about factors, such as their level of knowledge about new teaching approaches, how dissatisfied they are with their current practices, their beliefs about whether new methods actually improve student learning, and the access they have to peer-support activities during periods of teaching transitions (Brownell and Tanner, 2012; Ebert-May et al., 2011; Gess-Newsome et al., 2003; Handelsman el al., 2004; Hatvia, 1995; Henderson and Dancy, 2007; Henderson et al., 2012; Lynd-Balta et al, 2006; Miller et al., 2000; Sirum et al., 2009; Winter et al., 2001). Changes in faculty members' practices are also influenced by their perceptions of what is valued by members of their department, college, university, and national/international professional societies. We might think of these as *cultural influences*. For example, decisions about how to teach may be heavily influenced by concerns about tenure, seeming lack of interest in teaching among research colleagues, perceptions of their status in the field, and lack of incentives for reformed teaching (Brownell and Tanner, 2012; Gibbs and Coffey, 2004; van Driel et al., 1997).

One part of the University of Arizona (UA) AAU Undergraduate STEM Education Project is centered on redesigns of foundation courses in the departments of Chemistry and Biochemistry, Physics, Molecular and Cellular Biology, Chemical and Environmental Engineering, and Electrical and Computer Engineering. Another major part of the grant is focused upon addressing the personal and cultural influences on change in the instructional practices of STEM faculty members. This part of the project is working to support the transformation of STEM faculty and departmental cultures to practice and sustain evidence-based, active-learning instruction in their classes (Figure 1). This chapter describes our work in pursuing this goal.

FIGURE 1. UA AAU Undergraduate STEM Education Project

BOTTOM-UP AND TOP-DOWN APPROACHES TO CULTURE CHANGE

About 20 years ago, faculty members in the College of Science at the University of Arizona increased their interest in working with undergraduates in their research laboratories. These activities were supported and encouraged by department heads and the college dean. Today, the culture among the faculty and administrators in the science and engineering colleges is to embrace educating undergraduates in authentic research, with the percent of students involved in independent research projects ranging from 60–100% in College of Science departments. We believe the success and sustainability of undergraduate research in faculty laboratories is largely due to the bottom-up interest and top-down support.

Similarly, a grassroots effort to improve teaching and learning in foundational science and engineering courses has been ongoing at UA for several years. However, the initiative lacked focus, had minimal cross-departmental collaboration, and had little top-down support or encouragement. This is changing in several departments at UA as a result of the AAU Undergraduate STEM Education Project. Our project leadership team is composed of the faculty involved in the redesigns of five foundational STEM courses, along with three professionals from the Office of Instruction and Assessment, an associate department head from engineering, a research postdoctoral associate, and the senior vice provost for academic affairs. This group meets weekly to discuss the goals of our project, share science education research findings and reports in the literature, and to discuss best approaches to engage additional faculty and departments in efforts to improve learning and deep understanding in STEM education. Everyone participates as an equal in these discussions without regard to rank or position at the university. This has provided significant leadership development for the faculty, and better understanding by professionals and administrators of the challenges, rewards, and successful approaches to improve student learning and student success.

BUILDING INDIVIDUAL FACULTY CAPACITY FOR EVIDENCE-BASED TEACHING

Faculty Learning Communities: Over the past few years, the UA Office of Instruction and Assessment has worked with groups of faculty from several disciplines interested in learning about best practices in areas such as outcomes assessment and online instruction. The faculty members study research results, discuss ideas, and develop projects to implement in their disciplinary areas and to bring back to the group for peer discussion and reflection. The work of these self-regulated professional development groups, called Faculty Learning Communities (FLCs), have a history of success in U.S. colleges and universities (Angelo, 2000; Cox, 2004; Beach and Cox, 2009). A key common feature of successful FLCs is the way they are structured around knowledge that is sought by faculty members, rather than knowledge and skills that others feel faculty members should have. In this way, interest in learning is a bottom-up, rather than a top-down, starting point for potential change. Our campus experiences with facilitating the FLCs led to its central role in our AAU STEM project. In this project, the FLCs were developed with two purposes in mind: (1) to engage faculty members in a long-term activity focused on learning about evidence-based, active-learning teaching approaches in STEM; and (2) to create a community in which the interested learners discuss and test new ideas about teaching with STEM faculty peers, at greater depth than discussions about teaching that generally occur in many departments.

A total of 80 STEM faculty members over years one and two have joined the AAU STEM project FLCs. In year one, only three faculty members dropped out of the FLC between fall and spring semester, due mainly to schedule conflicts. FLC groups of eight to ten members are formed, each led by a faculty member from our leadership team who currently teaches STEM courses. A FLC group meets six to seven times per semester and focuses on a curriculum that is developed each year by the AAU STEM Project Leadership Team. The curriculum includes reading and discussing research articles on best practices and using the learning-cycle framework (Lawson, 2001) to design teaching projects implemented twice during the semester. The projects usually involve strategies with which the faculty members have little prior experience. The follow-up FLC meetings are sites of rich conversations about the challenges and successes associated with individuals' project implementations. Additionally, throughout the semester, the FLC meetings are times for much questioning and reflection on broader issues related to concerns about their own teaching and their students' learning. Year two activities also included some peer observations of teaching among FLC members, an activity requested by several year one FLC members in an end-of-year survey.

Teaching Talks and Workshops: The project has also developed short-term activities designed to build STEM faculty members' knowledge of active-learning approaches to teaching. Teaching Talks were developed as a set of one-hour discussions on strategies often associated with evidence-based instruction in STEM. In the project's first semester, campus STEM faculty were invited to four talks, at a central campus location, on the following topics: collaborative learning activities, the role of student predictions in promoting learning, the value of pre-post assessment, and daily assessment strategies. The discussions were led by two OIA staff members with expertise in STEM education who are members of the AAU STEM Project Leadership Team. Participant engagement in discussion was high at the Teaching Talks, but attendance was low (ranging from six to nineteen). This year, we decided to take the Teaching Talks *to* the departments, believing that the change might increase attendance. We distributed a call to STEM department heads to request either a 20-minute presentation (e.g., at a department meeting) or a 60-minute seminar/discussion (e.g., for a department colloquium). We have visited six STEM departments.

The project also offers half-day workshops for STEM faculty who wish to gain hands-on experience with evidence-based strategies that actively and intellectually engage students. The workshops are offered by an AAU STEM Project Leadership Team member with broad experience nationally as an active-learning professor and instructional workshop facilitator. The workshops model active engagement, with participants gaining first-hand experience with methods, such as rapid feedback assessment to small group collaborative learning activities, and the Peer Teaching Assistant Model. Our workshop attendance was 51 in May 2014 and over 50 in November 2014.

Collectively, faculty from 35 departments across campus have participated in one of our workshops, an FLC, or a Teaching Talk.

SUPPORTING TRANSFORMATION WITHIN DEPARTMENTS

A Case Study in Chemistry: At two years into our STEM education project, UA science and engineering department heads have different levels of engagement with our project. In one department, Chemistry and Biochemistry, the department head is fully supportive. Furthermore, the faculty voted for a three-year pilot to teach all sections of General Chemistry 151 and 152 (~3,500 students) with a new curriculum and active-teaching approaches. The course was redesigned and taught by two faculty members in the department over the last three years with significant improvements in student learning as measured by the American Chemical Society Conceptual Exam. Students who took both semesters of the reformed course earned higher overall grades and higher final exam scores in the second semester course.

One of the research studies in our AAU project includes a qualitative investigation of the Department of Chemistry and Biochemistry as this pilot unfolds over the next two years. We want to know the challenges and successes of the faculty and TAs teaching the reformed general chemistry course for the first time and during subsequent semesters. Our team has observed the faculty teaching the redesigned course and used the COPUS observation tool (Smith et al., 2013) to record our observations. We also want to know the impressions, attitudes, and opinions of the instructors who are implementing the redesign, faculty in the department who are not involved in teaching this course, the opinions of the department head and the associate head, and the undergraduate preceptors involved in the course. We are collecting survey, interview, and focus group data from all these individuals related to attitudes toward the full implementation of the course redesign. Insight from the faculty involved in the course redesign implementation in chemistry may help us implement redesigned courses in other departments.

We already know that some faculty members in Chemistry and Biochemistry do not support the course redesign and are opposed to the teaching approach. Fortunately, these individuals are in the minority, but they have not been persuaded by very good learning outcomes, positive student satisfaction survey results, and better student performance in the following course (organic chemistry) resulting from the previous three-year development period of the general chemistry course. Other preliminary findings from our study suggest that more communication should be encouraged among the faculty teaching the new curriculum. In addition, the outcomes from our chemistry pilot may be different from some others in that, except for one tenured full professor involved in the redesign, only one of the other faculty teaching general chemistry was on the tenure track.

Noticing Change in Physics: About ten years ago, the attitude in physics depart-ments toward teaching seemed to either support progressive teaching methods or suggest that it was up to the students to learn what was presented in lec-ture, with little support from the faculty (Henderson and Dancy, 2009). Our physics department was more aligned with the second attitude, but that is now changing.

In fall 2013, a professor in astronomy and his postdoctoral associate offered to teach a section of introductory physics with calculus (mechanics) using evi-dence-based teaching methods and active-learning pedagogies. The astronomy postdoc, trained in science education, would teach one section of Physics 141 in the spring using evidence-based teaching methods and active-learning peda-gogies, and an associate professor would teach another section as he had in the past using primarily lecture as a teaching approach. In addition, the postdoc would have only two traditional class meetings, while during the other class meeting the students would work in small groups on activities provided for them. A graduate teaching assistant was available to check their work and help them when they were having difficulty. During the fall semester, the astronomy teaching team worked with the physics professor to agree on a set of the ques-tions that would be given during each of the exams and the final. Thus, students in both sections would have a set of common exam items graded by the same set of TAs.

After teaching the physics course in the spring, the student learning out-comes on the common test items were compared. The students in the course taught by the postdoc with active-learning pedagogies significantly outper-formed the lecture instruction of the physics professor on the common final exam. What happened when the physics professor saw the student learning out-comes? He was surprised and wanted to know why the students taught by the postdoc did better. When showed the pie chart from the COPUS observations, it was clear that the students were much more engaged in the active-learning classroom, and we know that student engagement is a strong predictor for in-creased learning (Hake, 1998; Prince, 2004; Ruiz-Primo et al., 2001; Smith et al., 2005; Wood, 2009). The evidence of improved learning and the discussion about what the postdoc did to improve learning led the physics professor to say he wants to change the way he teaches. Thus, as a result of this study, more faculty in physics are active in our STEM project's Faculty Learning Communi-ties, redesigning other physics courses, and paying attention to publications in physics education.

Unlike chemistry and biochemistry, the sections of introductory physics are taught by professors of various ranks and by lecturers. It will be interesting to see if improvements in teaching practices in introductory courses can be sustained more or less with a mix of faculty, tenure-track vs. lecturer, teaching the course. However, at this early stage in the project, a few observations are worth noting. In fall 2014, the faculty member serving in the role of director of the undergraduate program for the Physics Department and a member of our AAU STEM project, gave a presentation on the project at the department's annual retreat. Additionally, he has invited colleagues to use his problem-solving activities as ideas for their class planning, a step that he feels is particularly important since lack of time for developing new materials is regularly mentioned by colleagues as a barrier to their use of reformed teaching strategies. Since this faculty member is the primary decision-maker for teaching assignments in the department, he is also in a unique position to assign and support new faculty members interested in learner-centered teaching to key teaching positions in introductory courses.

SUPPORTING TRANSFORMATION ACROSS THE INSTITUTION

Teaching Awards, Institutional Media Coverage, and Increased-Emphasis on Teaching During P&T: Our project has additional activities to decrease the barriers to cultural transformation. These include new STEM teaching awards for faculty who use evidence-based teaching methods and active-learning pedagogies, and focus on learner-centered teaching approaches. In addition, we are working closely with the associate vice president for university relations to put out news stories about our project. Within the last year, 14 news stories have appeared on our university website and four videos were produced with interviews of our faculty teaching redesigned courses and about the project. In addition, the promotion and tenure guidelines have been changed to require peer observations of teaching and a teaching portfolio. These new P&T guidelines alone won't be enough to increase the emphasis of quality teaching in the promotion decisions, but we expect that the provost will support teaching considerations in these decisions. Furthermore, the chair of the faculty (elected position of the Faculty Senate) plans to use a report from a task force on teaching quality to increase emphasis among the faculty for quality teaching.

Collaborative Learning Space Project: Faculty members in our leadership team have been able to teach using active-learning pedagogies in large lecture halls that range from 100 to 700 students, but for most of them, the lecture hall is not an optimal teaching environment. Once again, the bottom-up inspiration and top-down support approach of our project is allowing us to make changes. An impromptu meeting with a faculty member in our leadership group, the senior vice provost, and the dean of library quickly led to the development of a pilot classroom project in an underutilized room in the science and engineering library. The time from inspiration to teaching in the pilot space was only three and a half months and required significant work and collaboration on the part of the library staff, University Information Technology Services (UITS), faculty on the AAU STEM Project, and the senior vice provost. The Collaborative Learning Space in the library was used for six weeks in the fall semester to teach up to 260 students at a time in a room with round tables, monitors and short-throw projectors, speakers and microphones, notebook computers, an enhanced internet network, along with videography to capture student interactions and faculty use of the space. Eight courses used the space for two to twelve class sessions. Students, TAs, preceptors, and faculty were surveyed, interviewed, and videotaped. The outcome: the pilot use of collaborative space in the library was a very successful experiment with the space in the library converted to a large collaborative classroom. Now, a total of five new collaborative learning spaces will be used for teaching in fall 2015.

Collaborative Learning Spaces Workshop: The UA AAU Undergraduate STEM Education Project sponsored a two-day workshop on classroom space redesign. In addition to the use of the reading room in the library for an active-learning classroom, we reviewed what could be done with three additional large flat spaces at UA. The workshop was coordinated by a principal of the Learning Spaces Collaboratory in Washington, D.C., and an architect from a firm in Los Angeles. Updating classrooms is an item in the UA strategic plan, and the outcomes of our workshop helped us direct the redesign of five collaborative learning classrooms.

FINAL THOUGHTS

The UA AAU Undergraduate STEM Education Project developed from a *bottom-up* set of interests in advancing evidence-based, active-learning approaches to instruction. The project's early momentum and increasing visibility across campus have been enhanced by *top-down* support from STEM department heads, the Office of the Senior Vice Provost for Academic Affairs, and the staff of the Office of Instruction and Assessment. Early accomplishments have been described in this chapter. However, some key challenges remain for the project.

Challenges Related to Personal Influences on Faculty Change

Only four faculty members who recently completed the fall semester CWSEI Teaching Practice Inventory (Wieman and Gilbert, 2014) indicated they had been FLC members both year one and year two of our project. The self-reported practices by these individuals indicate they were most likely to have increased their evidence-based practices in these areas: encouraging more student collaboration on assignments, more frequent use of graded assignments, use of pre-assessments, use of instructor-independent pre-/post-tests (e.g., concept inventories), and highlighting student learning and teaching difficulties at TA meetings. We wondered whether these changes represent areas that these FLC members wanted to learn more about in the FLCs. The FLC facilitators from our project leadership team regularly commented on the tension between our intended FLC curriculum, focused on evidence-based approaches to teaching (e.g., use of learning cycles), with the tendency for FLC members wanting to focus on practical strategies and methods easily implemented in lectures (e.g., think-pair-share). This tension is an ongoing challenge that our leadership team continues to address as it further develops a FLC curriculum with the potential to foster real change in teaching practices.

The literature on change in teaching practices indicates that short-term efforts to increase faculty members' knowledge of new teaching approaches result in little change in individual practices or departmental cultures (Gibbs and Coffey, 2004; Henderson et al., 2011). If this is true, are the efforts made by the project to offer short-term professional development activities, such as Teaching Talks and workshops, the best use of our limited time and resources? Yet, our spring workshop had high attendance and enthusiastic response. Our challenge is to identify the role of short-term workshops and presentations in promoting real change in faculty members' teaching practices.

Challenges Related to Cultural Influences on Faculty Change

In a year-end survey of FLC members at the end of the spring semester 2014, only three of 13 respondents reported having talked with colleagues in their departments or elsewhere about changes in their teaching. Additionally, these three individuals indicated that the conversations were either brief or very general. The project must address the important work of enabling and supporting our FLC members (and leadership team members) to become voices of change in their home departments.

The different approaches and pace of change occurring in the Physics and Chemistry and Biochemistry departments suggests that culture change is not a one-size-fits-all approach. This challenge requires increased attention by project leaders to the ways in which all STEM departments involved with the project show signs (or not) of culture change.

ACKNOWLEDGEMENTS

We acknowledge and thank the Association of American Universities and The Leona and Harry A. Helmsley Charitable Trust for funding this work.

REFERENCES

American Association for the Advancement of Science. (2011). *Vision and change in Undergraduate biology education: A call to action, final report.* Washington, DC: AAAS.

Angelo, T. A. (2000). Transforming departments into productive learning communities. In A.F. Lucas (Ed.), *Leading academic change: Essential roles for department chairs* (74–89). San Francisco: Jossey-Bass.

Beach, A. L. and Cox, M. D. (2009). The impact of faculty learning communities on teaching and learning. *Learning Communities Journal*, 1(1), 1–6.

Brownell, S. E, & Tanner, K. D. (2012). Barriers to faculty pedagogical change: Lack of training, time, incentives, and . . . tensions with professional identity? *CBE Life Sciences Education*, 11(4), 339–346.

Cox, M. D. (2004). Introduction to faculty learning communities. In M.D. Cox (Ed.), *Building faculty learning communities* (5–23). San Francisco: Jossey-Bass.

Ebert-May, D., Derting, T. L., Hodder, J., Momsen, J. L., Long, T. M., & Jardeleza, S. E. (2011). What we say is not what we do: Effective evaluation of faculty professional development programs. *BioScience*, 61(7), 550–558.

Gess-Newsome, J., Southerland, S. A., Johnston, A., & Woodbury, S. (2003). Educational reform, personal practical theories, and dissatisfaction: The anatomy of change in college science teaching. *American Educational Research Journal*, 40(3), 731–767.

Gibbs, G., & Coffey, M. (2004). The impact of training of university teachers on their teaching skills, their approach to teaching and the approach to learning of their students. *Active Learning in Higher Education*, 5(1), 87–100.

Hake, R.R. (1998). Interactive-engagement versus traditional methods: A six-thousand-student survey of mechanics test data for introductory physics courses. *American Journal of Physics*, 66(1), 64–74.

Handelsman, J., Ebert-May, D., Beichner, R., Bruns, P., Chang, A., DeHaan, R., Gentile, J., Lauffer, S., Stewart, J., Tilghman, S. M. & Wood, W. B. (2004). Scientific teaching. *Science*, 304(5670), 521–522.

Hativa, N. (1995). The department-wide approach to improving faculty instruction in higher education: A qualitative evaluation. *Research in Higher Education*, 36(4), 377–413.

Henderson, C., Beach, A., & Finkelstein, N. (2011). Facilitating change in undergraduate STEM instructional practices: An analytic review of the literature. *Journal of Research in Science Teaching*, 48(8), 952–984.

Henderson, C., & Dancy, M. (2007). Barriers to the use of research-based instructional strategies: The influence of both individual and situational characteristics. *Physical Review Special Topics-Physics Education Research*, 3(2), 020102-1–020102-14.

Henderson, C., Dancy, M., & Niewiadomska-Bugaj, M. (2012). Use of research-based instructional strategies in introductory physics: Where do faculty leave the innovation-decision process? *Physical Review Special Topics-Physics Education Research*, 8(2), 020104-1–020104-15.

Lawson, A. E. (2001). Using the learning cycle to teach biology concepts and reasoning patterns. *Journal of Biological Education*, 35(4), 165–169.

Lynd-Balta, E., Erklenz-Watts, M., Freeman, C., & Westbay, T. D. (2006). Professional development using an interdisciplinary learning circle: Linking pedagogical theory to practice. *Journal of College Science Teaching*, 35(4), 18.

Miller, J. W., Martineau, L. P., & Clark, R. C. (2000). Technology infusion and higher education: Changing teaching and learning. *Innovative Higher Education*, 24(3), 227–241.

Prince, M. (2004). Does active learning work? A review of the research. *Journal of Engineering Education*, 93(3), 223–231.

Ruiz-Primo, M. A., Briggs, D., Iverson, H., Talbot, R., and Shepard, L. A. (2011). Impact of undergraduate science course innovations on learning. *Science*, 331(6022), 1269–1270.

Sirum, K. L., Madigan, D., & Kilionsky, D. J. (2009). Enabling a culture of change: A life science faculty learning community promotes scientific teaching. *Journal of College Science Teaching*, 38(3), 38–44.

Smith, M. K., Jones, F. H., Gilbert, S. L., & Wieman, C. E. (2013). The classroom observation protocol for undergraduate STEM (COPUS): A new instrument to characterize university STEM classroom practices. *CBE-Life Sciences Education*, 12(4), 618–627.

Smith, K. A., Sheppard, S. D., Johnson, D. W., and Johnson, R. T. (2005). Pedagogies of engagement: Classroom-based practices. *Journal of Engineering Education*, 94(1), 87–100.

Van Driel, J. H., Verloop, N., Van Werven, H. I., & Dekkers, H. (1997). Teachers' craft knowledge and curriculum innovation in higher engineering education. *Higher Education*, 34(1), 105–122.

Wieman, C. E., & Gilbert, S. L. (2014). The teaching practices inventory: A new tool for characterizing college and university teaching in mathematics and science. *CBE Life Sciences Education*, 13(3), 552–569. doi:10.1187/cbe.14-02-0023

Wieman, C. E., Perkins, K., & Gilbert, S. L. (2010). Transforming science education at large research universities: A case study in progress. *Change: The Magazine of Higher Learning*, 42(2), 6–14.

Winter, D., Lemons, P., Bookman, J., & Hoese, W. (2012). Novice instructors and student-centered instruction: Identifying and addressing obstacles to learning in the college science. *Journal of the Scholarship of Teaching and Learning*, 2(1), 14–42.

ABOUT THE AUTHORS

Gail D. Burd is the Senior Vice Provost for Academic Affairs and a Distinguished Professor in the Department of Molecular and Cellular Biology at the University of Arizona in Tucson, Arizona.

Debra Tomanek is the Associate Vice Provost of the Office of Instruction and Assessment and a Professor of the Department of Molecular and Cellular Biology at the University of Arizona in Tucson, Arizona.

Paul Blowers is a Distinguished Professor of the Department of Chemical and Environmental Engineering and Associate Professor of the Department of Public Health at the University of Arizona in Tucson, Arizona.

Molly Bolger is an Assistant Professor of the Department of Molecular and Cellular Biology at the University of Arizona in Tucson, Arizona.

Jonathan Cox is the Research Associate of the Department of Molecular and Cellular Biology at the University of Arizona in Tucson, Arizona.

Lisa Elfring is an Associate Professor of the Department of Molecular and Cellular Biology and the Department of Chemistry and Biochemistry at the University of Arizona in Tucson, Arizona.

Elmer Grubbs is an Instructor of the Department of Electrical and Computer Engineering at Northern Arizona University in Flagstaff, Arizona.

Jane Hunter is an Associate Professor of Practice of the Office of Instruction and Assessment at the University of Arizona in Tucson, Arizona.

Ken Johns is the Associate Department Head and Professor of the Department of Physics at the University of Arizona in Tucson, Arizona.

Loukas Lazos is an Associate Professor of the Department of Electrical and Computer Engineering at the University of Arizona in Tucson, Arizona.

Roman Lysecky is an Associate Professor of the Department of Electrical and Computer Engineering at the University of Arizona in Tucson, Arizona.

John A. Milsom is a Senior Lecturer of the Department of Physics and the Director of Undergraduate Studies at the University of Arizona in Tucson, Arizona.

Ingrid Novodvorsky is the Director, Teaching, Learning and Assessment of the Office of Instruction and Assessment at the University of Arizona in Tucson, Arizona.

John Pollard is an Associate Professor of Practice of the Department of Chemistry and Biochemistry at the University of Arizona in Tucson, Arizona.

Edward Prather is an Associate Professor of the Department of Astronomy at the University of Arizona in Tucson, Arizona.

Vicente Talanquer is a Distinguished Professor at the Department of Chemistry and Biochemistry at the University of Arizona in Tucson, Arizona.

Kay Thamvichai is a Professor of Practice of the Department of Electrical and Computer Engineering at the University of Arizona in Tucson, Arizona.

Hal Tharp is an Associate Professor and the Associate Department Head of the Department of Electrical and Computer Engineering at the University of Arizona in Tucson, Arizona.

Colin Wallace is a Lecturer of the Department of Physics and Astronomy at the University of North Carolina at Chapel Hill in Chapel Hill, North Carolina.

3

From Grassroots to Institutionalization: RIT's CASTLE

Scott V. Franklin

In April of 2010, an assistant professor of biology at Rochester Institute of Technology had a pivotal annual review with the dean of the College of Science. The faculty member, a geneticist with an impressive track record of involving undergraduates in research, was disturbed by new Institute tenure policies that emphasized national prominence in scholarship. Given her department's undergraduate-only population and RIT's limited infrastructure, this was unlikely in a competitive, expensive and fast-moving a field as human genetics. The assistant professor worried about her prospects for tenure, and emphasized the deep interest in student learning that had brought her to RIT.

The dean suggested the faculty member speak with physics faculty involved in discipline-based education research about biology education as a viable scholarship alternative. Thus began a chain of events that ultimately resulted in the creation of the Center for Advancing Science & Math Teaching, Learning & Evaluation (CASTLE), a grass-roots initiative nurtured with timely administrative support. The bottom-up development allowed faculty to control the center's growth, maintaining a focus and coherence that helped define CASTLE to administrators. This paper presents CASTLE as a case study, noting the decisions—conscious or realized in hindsight—that guided its development. In the context of models for institutional change, CASTLE serves as an example that combines top-down and bottom-up elements and serves as a guide for others contemplating broad culture change.

MODELS FOR INSTITUTIONAL CHANGE: TOP-DOWN, BOTTOM-UP, AND OUTSIDE-IN

Models for institutional change (e.g., DiMaggio, 1998) have been investigated in business (Lounsbury and Crumley, 2007), law (Smets, Morris and Greenwood, 2011) and other industries (Kellogg, 2009), and attention is now turning to academic institutions (Henderson, Beach & Finkelstein, 2011). Within physics departments, Henderson and Dancy (2007) surveyed faculty for the contextual environment in which they chose whether to implement student-centered, research-based physics materials, and in 2008 Henderson evaluated national new faculty workshops for efficacy at implementing change. In the broader academic landscape, Seymour (2002) categorized change strategies, noting that neither purely "top-down" nor "bottom-up" strategies were likely to succeed, and instead suggested a combination of the two would be required for successful change, with the department as the fundamental unit of change.

Recently, Reinholz et al. (2015) articulated two intriguing approaches that combine top-down and bottom-up elements. In the *middle-out* approach, departmental culture shifts first, often as a result of an individual or small group of champions, developing strategic visions and activities around reformed teaching and only then converting other faculty and influencing upper administration. The contrasting *outside-in* approach combines change at the faculty and upper administrative level to provide impetus for departmental change. For example, individual faculty may use research-based course assessments to demonstrate the efficacy of reformed teaching that, combined with an administrative desire for increased retention, then drives departments to adopt better pedagogies.

The role of a *champion* in bringing about culture change has been well-documented (e.g., Chasteen et al., 2011). Such an individual is an early adopter of transformational methods and a persistent advocate for change; Henderson and Dancy (2011) found "STEM change agents strongly favor the individual" category of change. In the middle-out approach, the primary role of the champion is to demonstrate effectiveness and potential *within his own culture* (i.e., the home department). This can take the form, for example, of investigating new pedagogies or teaching methods, incorporating them into a class, and establishing through assessment that the change has improved student learning. The result is to convince enough colleagues within the department that the innovation is worth adopting, at which point normal departmental mechanisms for decision-making (e.g., faculty meetings, curriculum committees, etc.) engage and the department consciously decides to move forward with the change.

In the outside-in approach, the champion, while continuing to advocate locally, more importantly serves to establish and maintain communication between faculty and upper administrative levels. This bi-directional flow of information allows faculty to convey upward the nature of their activities, while administration provides support and guidance that ensures these activities are aligned with institutional goals. **Unlike in the middle-out model, there are not established mechanisms (e.g., faculty meetings) for this communication, and so the champion becomes responsible for seeking out opportunities and contexts in which to facilitate this transfer of information.**

CASTLE is a case study of the efficacy of the outside-in approach, and it is through that lens that we interpret the historical narrative.

INITIAL EFFORTS:
THE SCIENCE & MATH EDUCATION RESEARCH COLLABORATIVE

Figure 1 shows a timeline of CASTLE's development. Shortly after the biology faculty member's meeting with the dean, a cohort with four additional faculty—two tenured physicists in physics education research, another pre-tenure biologist interested in biology education research, and a chemist deeply familiar with chemistry education research—formed a discipline-based education research (DBER) journal club. Although similar journal clubs had been started previously in the College of Science, several factors enabled this attempt to succeed.

First, *the junior faculty had a strong self-interest in the group's success.* Because the group had formed at the dean's urging, it was understood that the ensuing scholarship would be considered favorably during the tenure process. This was formally endorsed in faculty plans of work and, ultimately, in a revised College Strategic Plan. The junior faculty realized that the group's success would be instrumental in their achieving tenure, and they jumped at the chance to

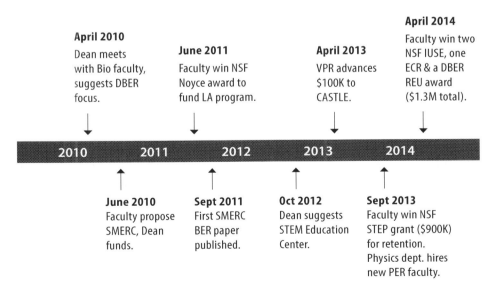

April 2010
Dean meets with Bio faculty, suggests DBER focus.

June 2011
Faculty win NSF Noyce award to fund LA program.

April 2013
VPR advances $100K to CASTLE.

April 2014
Faculty win two NSF IUSE, one ECR & a DBER REU award ($1.3M total).

2010   2011   2012   2013   2014

June 2010
Faculty propose SMERC, Dean funds.

Sept 2011
First SMERC BER paper published.

Oct 2012
Dean suggests STEM Education Center.

Sept 2013
Faculty win NSF STEP grant ($900K) for retention. Physics dept. hires new PER faculty.

FIGURE 1. Development milestones in the life of the Center for Advancing Science/Math Teaching, Learning & Evaluation (CASTLE).

develop research projects, travel to related conferences and write papers and grant proposals.

Second, *the authority of the group was distributed.* The presence of multiple members with experience and familiarity with DBER meant that no one person was the resident expert. Group meetings were true discussions, with an atmosphere of equality pervading all interactions. The hour-long journal clubs were quickly expanded to include an additional hour-long group meeting, a practice that continues to this day, and members remark that it's the only meeting that they are excited to attend and reluctant to leave.

A third feature, *a devotion to rigor,* would prove important not only in maintaining the group's integrity but also in advocating to upper administration. Speculative statements about student learning were challenged for supporting data, often leading to new research questions. Guided by information from the administration, the group realized that it would be judged on its disseminated scholarship—peer-reviewed publications, conference presentations and grant proposals—and the idea of rigor suffused every discussion. By June 2010, the biologists had identified a new research question involving student understanding of meiosis, and begun the qualitative and quantitative research that would result in their first publication (Wright and Newman, 2011).

The group's activities aligned with an expansive institutional definition of scholarship as part of an overall effort to increase faculty productivity of peer-reviewed work. DBER could be considered scholarship of discovery, fundamental research that adds to a body of theoretical or experimental knowledge, or scholarship of teaching/pedagogy, the study and investigation of student learning to develop strategies that improve learning. The presence of both as legitimate activities that would merit tenure gave needed latitude to initial investigations, which explored both foundational and practical aspects of student learning.

In June 2010, the faculty submitted a proposal to the dean to form the *Science & Math Education Research Collaborative (SMERC).* The proposal (see Appendix I) laid out an ambitious three-year plan to integrate SMERC faculty into their new research fields through conference travel and an external speaker seminar series. In exchange, the group set measurable goals for both annual (multiple external grant proposals and presentations at national meetings) and cumulative (multiple peer-reviewed publications and at least one new externally funded project) evaluation. The proposal also described a future "Phase II," and many of the ideas therein would later come to fruition in the formation of CASTLE. The college allocated space for the group and, as the activity increased, the physical footprint grew, demonstrating a significant college commitment.

The dean agreed to fund $40,000, with an additional $20,000 coming from departmental travel funds. In the context of the outside-in model, we can identify the following faculty-level and administrative-level actions that initiated the transformation:

TABLE1. Faculty/Administration Actions in Creating SMERC

Faculty-Level Actions	Administration-Level Actions
1. Created journal club	1. Endorsed DBER for tenure and promotion
2. Initiated DBER research projects	2. Contributed financial resources for travel,
3. Traveled to DBER conferences	external seminar speakers and equipment
	3. Provided laboratory/meeting space

COMMUNICATION ACROSS INSTITUTIONAL LEVELS

A challenge of the outside-in approach is maintaining communication between the faculty and administrative levels. The dean's approval and funding of the three-year plan led to quarterly updates, provided by a senior SMERC member. Already tenured, this faculty member could devote time to strategic outreach and planning, working with the dean to develop white papers and strategic plans. In response to a request from the provost, SMERC established the Upstate New York regional network of Project Kaleidoscope, a national organization dedicated to transforming STEM classrooms, organizing and hosting the first two biannual meetings. This demonstrated the alignment of SMERC activities with administrative agendas and raised the group's profile. While quarterly meetings and other activities ostensibly updated the administration on SMERC activities, in practice they also served to develop a shared language, agreement on acceptable measures of success and consensus on a long-term vision.

It was recognized that there might be institutional resistance to discipline-based education research, particularly if it was interpreted as "soft" or not rigorous. The group made a conscious effort to develop a consistent description of their activities, replacing the common "science education" or "scholarship of teaching and learning" with the less familiar "discipline-based education research." The new term allowed the group to define itself as engaging in scholarship of discovery, the classification previously reserved for traditional research, as opposed to the scholarship of pedagogy, which was seen as less rigorous. Group reports and presentations consistently used this language, which was eventually adopted by the dean in the college's strategic plan. The external seminar series also played an important role in conveying this message. Speakers were scheduled time with school heads, the dean and, occasionally, the provost during which they could speak about how SMERC and RIT aligned with the national trend in DBER (e.g. Singer, Nielsen and Schweinguber, 2012). These meetings validated the group and further elevated its profile in the eyes of the administration.

The administration sent clear messages about the necessity of peer-reviewed publications for tenure considerations. Because DBER journals were unfamiliar to the college and administration, the group undertook a study to rank the relevant journals by impact factor and draw attention to the peer-reviewed nature of many conference proceedings. As a result, the group's dissemination efforts became recognized as demonstrating necessary rigor, with supporting statements in faculty plans of work and annual reviews.

THE CREATION OF CASTLE

By the Fall of 2012, SMERC was firmly entrenched within the College of Science. Members had attracted a number of undergraduate research students, demonstrating demand among students and the suitability of such projects for undergraduate research, produced peer-reviewed publications and earned favorable reviews on grant submissions. Faculty began to receive national recognition, with one recognized as a Bioscience Education Network Fellow and another as a CREATE for STEM Fellow. The School of Physics and Astronomy initiated a search for a tenure-track appointment in physics education, the first new position ever dedicated exclusively to discipline-based education research. The dean consistently cited SMERC as a model of faculty scholarship and, because of SMERC's success, created a new college-wide award to motivate faculty in other research areas to form collaborative groups. Finally, much of the language contained in Phase II of the SMERC proposal about discipline-based education research and K–12 teacher recruitment and preparation was incorporated into the College Strategic Plan, thus institutionalizing the planned expansion.

SMERC was one of several education-related initiatives that had recently formed within the College of Science. NSF funding was awarded to create a learning assistant program, with the dual goals of both transforming courses across the college and also recruiting and training undergraduates to careers in secondary education. The dean formed the *Women in Science* (WISe), a faculty-led group dedicated to increasing representation of women at all levels in the sciences. The WISe group led "graduate school boot camps" for undergraduates interested in pursuing graduate work, faculty luncheons and roundtables to address issues faced by female faculty and began a Summer Math Applications in Science with Hands-on (SMASH) Experience for rising eighth-grade girls. SMERC members collaborated with chemistry faculty to apply for, and win,

NSF funding to reform an organic chemistry classroom. And, finally, SMERC faculty headed the institute's successful proposal to the NSF STEM Talent Expansion Program to address retention of deaf/hard-of-hearing and first-generation STEM majors. Together these, and other, initiatives suggested to the dean that an organizational structure—a center—could build a coherence and synergy and raise the visibility of all, and charged the director of SMERC with developing the proposal.

At RIT, the designation "Research Center of Excellence" recognizes and supports clusters of faculty and staff with a common scholarship vision. Designated Research Centers/Laboratories that demonstrate a well-defined interdisciplinary research focus involving multiple externally funded projects are eligible to receive 20% of all indirect cost payments from externally funded projects. A significant step toward CASTLE's realization was the vice-president of research's agreement to designate the nascent center as a provisional Center of Excellence and advance $100,000—to be reclaimed from later overhead return—for administrative purposes.

With dean and administrative encouragement, SMERC faculty began to make contact with leaders of the other initiatives. They followed closely a model used at The University of Colorado at Boulder to form the Center for Science Learning, which shared white papers, annual reports and other strategic documents and greatly assisted CASTLE's development. Recognizing the wide range of faculty activities in STEM education, an inclusive approach was taken to defining CASTLE as "a network of affiliated faculty, projects, and programs engaged in scholarship surrounding science and math education." CASTLE's vision is broad, with three objectives that aim to build community by:

- nurturing a community of faculty, administrators, and staff interested in science and math education and pedagogy by facilitating dialog about evidence-based practices, discipline-based education research, and methods of assessment and evaluation;
- establishing a robust and sustainable infrastructure that transforms science educational practices, supports discipline-based education research, and promotes PreK–20 faculty recruitment, preparation, professional development, and outreach; and
- fostering innovations in education by integrating an interdisciplinary community of scholars; promoting, sustaining, and evaluating reform efforts; advocating for diversity and access; and influencing policy, fundraising, and public outreach.

The group recognized different ways faculty could be affiliated with the center, including:

- *Members:* For faculty whose primary activities closely align with CAS-TLE, involvement is written into plans of work, and department heads solicit a letter from the CASTLE director to include annual reviews.
- *Project Affiliates:* Faculty can affiliate on a per-project basis, for example a grant or project involving other CASTLE affiliates.
- *Program Affiliates*: Directors of affiliated programs (e.g., the Women in Science) are affiliated as liaison between the center and their respective program.
- *Collaborators*: Collaborators participate in ongoing interactions (e.g., regular participation in CASTLE seminars, journal clubs, and other activities), but need not explicitly include CASTLE in a plan of work or annual report.

The center's creation therefore was also an example of faculty and upper administration working in tandem, with significant communication across the levels. The various actions pertaining to the creation are:

TABLE 2. Faculty/Administration Actions in creating CASTLE Center

Faculty-Level Actions	Administration-Level actions
1. SMERC faculty reached out to colleagues, identifying potential collaborations for CASTLE	1. Dean encouraged faculty to consider center creation
2. Faculty created bylaws, working together to define center mission, vision and affiliation types	2. Vice-president of research contributed $100,000 as an advance on future F&A

FACILITATING TRANSFORMATION ACROSS THE COLLEGE

With CASTLE's official formation, faculty began seeing connections between existing programs. CASTLE didn't necessarily create new initiatives, but it enhanced the impact of existing ones. Figure 2 shows the collaboration that arose between the different CASTLE activities. For example, learning assistants, recruited and trained in the LA program, became involved with developing materials for both a summer math program for middle-school girls and a retention program for first-generation and deaf/hard-of-hearing RIT undergraduates. Faculty originally recruited to mentor the learning assistants began attending SMERC journal clubs and seminars, thus broadening their knowledge of education research and assessment strategies. *LivePhoto*, an NSF-funded project to develop video vignettes in physics, spawned a similar initiative in biology involving curriculum development and biology education research. The

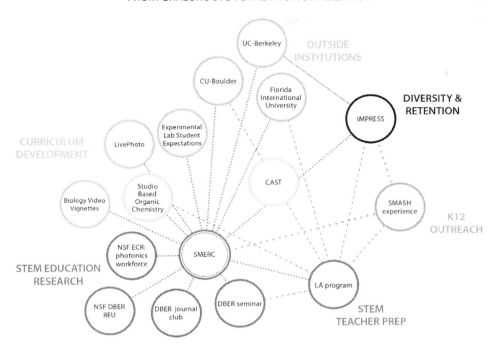

FIGURE 2. Connections and synergistic activities between CASTLE programs. We emphasize not the individual programs, which pre-dated the center's existence, but rather the multiple-connectedness of the network. In particular, the Science & Math Education Research Collaborative (SMERC) and Learning Assistant (LA) program impact virtually every other CASTLE activity.

"spin-off" project received NSF funding in 2014, and is an example of how connections brought about by CASTLE lead to new initiatives.

There has been significant growth in the number of faculty affiliated with center activities. As seen in Figure 3, the number of faculty affiliated with the center has grown seven-fold. Much of this is driven by the learning assistant program, which aggressively recruits traditional STEM faculty to mentor undergraduates in the classroom. This has had an incredibly important, positive impact. Because the LA's role is purely supportive to faculty in their efforts to transform courses, faculty opinion about the LAs is positive and these good feelings transfer to CASTLE. Course transformation brought about by LAs is also a natural incubator for curriculum development projects, and the connection with CASTLE provides faculty with support should they choose to pursue external funding. Critical to its consideration as a scholarship center, CASTLE has also seen growth in the amount of secured external funding (see Figure 4),

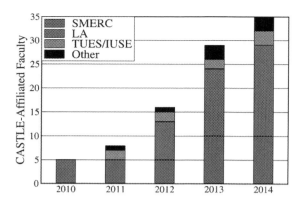

FIGURE 3. Growth in faculty affiliated with CASTLE activities. CASTLE's formal creation in 2013 led to a dramatic increase, spurred on by faculty seeing value in the connected nature of center activities.

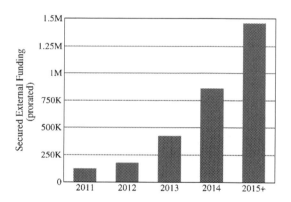

FIGURE 4. Growth in secured external funding for CASTLE activities.

providing valuable financial return that is used on administrative staff to further CASTLE activities.

CONCLUSIONS AND LESSONS LEARNED

Several lessons can be derived, in retrospect, from the story of SMERC and CASTLE. The fortuitous alignment of changing institutional expectations, an expansive definition of scholarship, interest from individual faculty, and commitment to pursuing a new area of research created an environment ripe for the formation of a new community. It took explicit administrative approval, encouragement, recognition and financial support to convince faculty that this community would benefit their path to tenure and promotion. Careful attention paid to increasing lines of communication across all levels—faculty, school heads and upper administration—ensured that the agendas of all remained aligned, even as specific wording and definitions evolved. And the focus on creating connections between many different programs created a broad network of faculty affiliates, including many from traditional disciplinary backgrounds, whose interest in course transformation connects them to the center. The network breadth is a critical element of institutional stability, and CASTLE's place in the college becomes more solid as new faculty become affiliated.

And what of the faculty whose meeting with the dean started it all? Four years later, she enters the tenure process with eight peer-reviewed publications in her new area, numerous internal and external grants to support her research, and recognition in the college and across the country as a leader in biology education research. More important, however, is the sense of satisfaction, community and joy she and her colleagues have found. It is this sense of shared vision and commitment that drove the group forward and now spreads throughout the college.

REFERENCES

DiMaggio, P. (1988). Interest and agency in institutional theory. In L. G. Zucker (Ed.), *Institutional patterns and organizations: Culture and environment*, Ballinger publishing, Cambridge, MA. 3–21.

Chasteen et al. (2011). A thoughtful approach to instruction: Course transformation for the rest of us. *Journal of College Science Teaching*, 40(4), 70–76.

Henderson, C. (2008). Promoting instructional change in new faculty: An evaluation of the physics and astronomy new faculty workshop. *American Journal of Physics*, 76(2), 179–187.

Henderson, C. and Dancy, M. (2007). Barriers to the use of research-based instructional strategies: The influence of both individual and situational characteristics. *Physical Review Special Topics: Physics Education Research*, 3(2), 020102.

Henderson, C., Beach, A., & Finkelstein, N. (2011). Facilitating change in undergraduate STEM instructional practices: An analytic review of the literature. *Journal of Research in Science Teaching*, 48(8), 952–984.

Henderson, C. and Dancy, M. (2011). Increasing the impact and diffusion of STEM education innovations. White paper commissioned for the *Characterizing the Impact and Diffusion of Engineering Education Innovations Forum*, Feb. 7–8, 2011. New Orleans, LA

Kellogg, K. C. (2009). Operating room: Relational spaces and microinstitutional change in surgery. *American Journal of Sociology*, 115(3): 657–711. 2009.

Lounsbury, M. and Crumley, E. T. (2007). New practice creation: An institutional perspective on innovation. *Organization Studies*, 28(7): 993–1012.

Singer, Susan R., Nielsen, Natalie R. and Schweinguber, Heidi (Eds.). (2012). *Discipline-based education research: Understanding and improving learning in undergraduate science and engineering*. Washington, DC: National Academies Press.

Reinholz, D. et al. (2015). Towards a model of systemic change in university STEM education. Unpublished manuscript.

Seymour E. (2002). Tracking the processes of change in US undergraduate education in science, mathematics. *Science Education*, 86(1) 79–105.

Smets, M., Morris, T. and Greenwood, R. (2011). From practice to field: A multilevel model of practice-driven institutional change. *Academy of Management Journal*, 55(4) 877–904.

Warrick, D.D. (2009). Developing organization change champions. *OD Practitioner*, *41*, 14–19.

Wright, L.K. and Newman, D. L. (2011). An interactive modeling lesson increases student understanding of ploidy during meiosis. Biochemistry and Molecular Biology Education, 39(5): 344–351.

ABOUT THE AUTHOR

Scott V. Franklin is a Professor of the School of Physics & Astronomy and Director of the Center for Advancing Science/Math Teaching, Learning & Evaluation (CASTLE) at Rochester Institute of Technology in Rochester, New York.

4

Towards a Model of Systemic Change in University STEM Education

Daniel L. Reinholz, Joel C. Corbo, Melissa H. Dancy,
Noah Finkelstein, and Stanley Deetz

Converging evidence concludes that certain types of teaching practices are most likely to improve student outcomes in undergraduate STEM courses (e.g., Freeman et al., 2014). Despite efforts to document and disseminate such practices, they are still not widely adopted (Henderson & Dancy, 2009). This lack of adoption suggests the need for new models and approaches towards institutional change (PCAST, 2012). This paper advances such a model and describes our approaches to implementing it.

INSTITUTIONAL CHANGE IN HIGHER EDUCATION

While institutional change models are well-developed in business and government settings (Real & Poole, 2005), similar models in higher education are only beginning to emerge (e.g., AAAS, 2011; Chasteen, Perkins, Beale, Pollock, & Wieman, 2011; Henderson, Beach, & Finkelstein, 2011). These emergent models focus primarily on changing *practices* associated with teaching and learning. Our model builds on these efforts by adding a focus on culture, in addition to practice, which we argue is required to effect sustained change. We situate our model with respect to prior efforts, particularly the Science Education Initiative (SEI), and highlight points of divergence.

The SEI is a course transformation effort aimed at STEM departments across two institutions (Chasteen et al., under review, 2011), one of which we are presently working with. The SEI focuses on transforming individual courses across a department using a three-component model: (1) defining learning goals for a course, (2) identifying areas of student difficulty, and (3) developing materials to help students meet the now-established learning goals. Science Teaching Fellows, disciplinary experts with educational training, were hired into each department to help promote and guide this transformation process. It typically took two to three semesters to develop learning goals collaboratively and to implement and refine new instructional approaches. While the SEI is largely considered successful, it did not explicitly focus on changing culture within departments. With respect to this, our faculty interviews have provided evidence of "slippage" in departments where reforms were made, due to the end of funding and new faculty who were not involved with the SEI teaching the transformed courses. In our work with such departments, we take the positive impact of SEI as a starting point for our own change efforts, with careful attention to sustaining and improving reforms.

Our change model is built on the following principles, elaborated below:

1. We focus on both prescriptive and emergent components of change (Henderson et al., 2011).
2. We pay explicit attention to sustaining the change process, focusing on continued improvement (Phillips, 1977).
3. We recognize the existing culture and institutional constraints, while focusing on reforming incentive structures to seed and sustain change (AAAS, 2011).
4. We take the department as a key unit of change (AAAS, 2011), while recognizing the need to target the university ecosystem at multiple levels.

These principles derive from and reflect findings from prior STEM educational transformation efforts. Henderson et al. (2011) classify such efforts across two dimensions: the primary target of change (individuals vs. environments) and whether the outcome was known in advance (prescribed vs. emergent). The dissemination of best practice curricula (prescribed-individuals) and top-down policy making (prescribed-environments) were found to be ineffective in isolation. For instance, efforts targeted at faculty can be limited by a highly traditional environment (Henderson & Dancy, 2007). These findings suggest that change efforts should target the university at multiple levels (Henderson et al., 2011).

Our experiences with the SEI highlight the need to continuously sustain change efforts. In general, efforts heavily driven by external support tend to regress after the support is removed (i.e., education problems do not stay solved; Phillips, 1977), unless there are structural changes that are difficult to reverse (e.g., replacing lecture halls with SCALE-UP style, or studio, classrooms; cf. Beichner et al., 2007). In contrast, efforts that result in cultural change, rather than just shifts in practice, may help a department sustain its efforts without continued external support. Nevertheless, fundamental changes to institutional reward structures, providing adequate incentives for improving teaching and learning, are key to sustaining reform efforts (AAAS, 2011).

We take the department as a key unit of change; this allows for efforts to be integrated across the curriculum, rather than being implemented piecemeal in isolated courses (AAAS, 2011). Given the complexity of the university (Henderson et al., 2011), change strategies focused solely on individuals, not the systems they are embedded in, are unlikely to succeed; our model addresses the university ecosystem at multiple levels.

CORE COMMITMENTS

Our change model targets the development of departmental cultures that are aligned with six core commitments, which we believe are emblematic of the culture of highly effective departments that value undergraduate education:

1. Educational experiences are designed around clear learning outcomes.
2. Educational decisions are evidence-based.
3. Active collaboration and positive communication exist within the department and with external stakeholders.
4. The department is a "learning organization" focused on continuous improvement.
5. Students are viewed as partners in the education process.
6. The department values inclusiveness, diversity, and difference.

Although academic departments are not typically viewed as networks, they share important characteristics with Networked Improvement Communities (NICs; Bryk, Gomez, & Grunow, 2011). NICs are collaborative networks organized to address complex, persistent problems in education. Departments also attempt to address such problems; this requires: (1) a clear goal, (2) gathering of evidence to evaluate proposed solutions, (3) mechanisms for positive coordination and collaboration, and (4) mechanisms for sustained learning and improvement (Bryk et al., 2011). These four statements, aligned with our first four core commitments, establish a functional problem-solving process. Moreover, addressing complex problems requires diverse skills and perspectives (Bryk et al., 2011), so the perspectives of students (commitment 5), particularly those from diverse backgrounds (commitment 6), are of significance.

A MODEL FOR INSTITUTIONAL CHANGE

Our change model addresses the university system as a whole, with careful attention to connections across three levels: faculty (as individuals and groups), departments, and administration. Activities at each level are synergistic; changes in a department influence both individual faculty and the administration (our middle-out approach), and efforts at the outside levels (faculty and administration) influence the department itself (our outside-in approach). Our two approaches (see Figure 1) are synergistic, and most effective when used together.

FIGURE 1. The outside-in approach (blue arrows) involves providing external support to the administration and faculty to impact the department. The middle-out approach (orange arrows) focuses on the department directly. Because all levels of the university are linked, both approaches are intended to affect the university at all levels.

Nevertheless, we have begun by studying and implementing these approaches separately, to better develop theory and practice before using them together in an integrated effort.

The Outside-In Approach

The outside-in approach combines efforts at the faculty level and administrative level to shift department culture. By working with groups of faculty, as in the SEI (Chasteen et al., 2011), we aim to reform educational practices and shift beliefs about education. Contrasting with prior approaches (Henderson et al., 2011), our efforts focus explicitly on cultural change.

Our approach involves creating Departmental Action Teams (DATs), which consist of faculty working collaboratively to address a shared issue of departmental interest. Like a faculty learning community (FLC; Ortquist-Ahrens & Torosyan, 2009), DAT faculty have agency to choose the educational issue they will work on, and *learning* and *community* are considered central to the DAT. DATs differ from FLCs insofar as they focus on a common, shared issue in a single department. Our DATs embody our six core commitments (e.g., use of evidence, clear outcomes) and engage faculty in a collaborative process aimed at shifting how faculty engage in scholarship of teaching and learning (SoTL; Huber & Hutchings, 2005). We see DATs as a mechanism for local cultural change. To sustain our efforts, DATs must be institutionalized through departmental support.

At the administrative level, the outside-in approach focuses on shifting university incentive structures and resources. At most research universities, there is little incentive for faculty to invest the time and effort required to teach effectively, because such investment is viewed as conflicting with research productivity. When individuals (or even collections of individuals) do engage, they often do so in isolation, developing their own tools (Glassick, Huber, & Maeroff, 1997). Thus, to sustain educational transformation, shifts in institutional incentive structures and resources are required. Our approach involves working with a variety of institutional structures to: (1) prioritize research-based teaching practice, and (2) provide resources for faculty, departments, and administration to do so. Similarly, working with administration we seek to promote a faculty-developed framework for defining (and celebrating) teaching excellence that can be adopted and contextualized by individual departments. Such a framework and approach could provide needed, locally relevant tools to shift the promotion and tenure guidelines and culture, if promoted by institutional leadership (cf. Iowa State University, 2014). In parallel with efforts to promote effective educational practices (e.g., technological tools, easily accessed data on educational outcomes, and coordinated support for pedagogical development; cf. Berret, 2014), we seek to simultaneously provide incentives and resources that promote a culture of educational excellence.

The Middle-Out Approach

The middle-out approach involves a sustained change process focused on aligning department culture with our core commitments. We adopt strategies from the organizational change literature (e.g., Conversant, 2014) that have been successful in systems similar to academic departments to guide departments through a process involving five inter-related components. These include:

1. developing a department vision,
2. revising assumptions about teaching and learning,
3. developing capacity to meet learning goals (within and across courses),
4. integrating teaching and learning goals systematically with research and other departmental functions, and
5. developing a collaborative process for continuous assessment and innovation.

We take our six core commitments above as a starting point to creating a shared vision, but allow for the department to build upon and interpret them within their local context. In essence, our core commitments lay out basic parameters for what a collaborative process might achieve. Through individual faculty interviews, we create "mental maps" of their beliefs around teaching and learning. The maps help identify areas for productive change efforts and also barriers to faculty embracing the shared vision (Borrego & Henderson, 2014). The mental maps are also tools for intervention: By sharing them with the faculty, we can shift faculty beliefs towards alignment with the vision by revealing assumptions and incongruities that the faculty were not previously aware. Steps 3–5 of our process involve the department building capacity, integrating meaningful teaching and learning across all department activities, and developing collaborative processes for ongoing assessment and innovation.

The activities associated with these change processes include: increase awareness of competing values, inculcate more productive maps of faculty work and student learning, and increase department capacity to meet multiple goals. As a result, appropriate student outcomes emerge without being prescribed by the change process as faculty shift their practices to align with the new culture they are co-creating. This shift in culture also has the potential to impact the administration (e.g., demonstrated success of the process could convince the administration to encourage other departments to engage in a similar process). These changes at the faculty and administration level can then further influence department culture, creating a sustained feedback cycle.

We anticipate this change effort would last one to two years. The process would involve, at minimum, a one- to two-day departmental retreat to develop a vision, mental maps, assessment criteria, and a process going forward that includes 30-day, 90-day, and one-year goals. The retreat would establish working groups to complete different tasks (e.g., establishing learning goals for the program, creating a more supportive environment for innovations and positive relationships, and revising reward systems). At regular intervals, the department would meet to assess progress, reflect on successes and lessons learned, and adjust its process to move forward. The members of our project team will be involved as facilitators of this change process.

SAMPLE APPLICATIONS

Our preliminary efforts to implement these two approaches offer several insights. We are currently engaged in our second year of activities associated with the STEM Institutional Transformation Action Research (SITAR) Project. SITAR is a three-year grant-funded project to implement and study institutional change at a large research university in the USA. Our project is presently involved with four STEM departments, with plans to scale up over time. For this brief paper, we report on our efforts with two departments: the Runes Department and the Charms Department (actual names redacted for confidentiality). These case studies highlight our approaches. Both of these departments were prior recipients of funding through the SEI.

In all departments involved in the project, we administered a survey focused on teaching practices, beliefs, professional development opportunities, and promotion and tenure guidelines (all measured on Likert scales). In Runes, we conducted nine, one-hour interviews with individual faculty about their teaching and their perceptions of the department; we also engaged in one-on-one consultations with two faculty around education projects. In Charms we conducted a faculty survey and held several meetings with departmental leadership to determine readiness for engagement in our change processes.

The Runes Department: The Outside-In Approach

In Runes, 31 of 34 faculty responded to our survey, for a response rate of 91%. Faculty indicated that social interaction (4.25 out of 5), active participation (4.75 out of 5), interactive learning (4.66 out of 5) and engagement (4.1 out of 5) were very important aspects of learning. Despite these beliefs, use of small group work (2.6 out of 4), regular opportunities for students to talk (2.8 out of 4), and opportunities for students to explore content before formal instruction (2.3 out of 4) were mixed. Most telling, instructors indicated that the statement "I guide students through major course topics as they listen and take notes" was mostly descriptive of their teaching (3.2 out of 4). These survey responses seem to indicate a discontinuity between professed beliefs and actual teaching practices, indicative of the institutional culture and incentive structures. This may be the result of SEI's focus on the practices of faculty in this department with no corresponding changes at the administrative level.

Runes is spread across multiple buildings on two campuses. As a result, many of the faculty we interviewed reported feeling isolated; through the creation of a DAT, we aimed to create more community around education. Our DAT consists of six faculty members (a mix of tenure-track and non-tenure track faculty), and two facilitators from our project team; many of the DAT members were identified through our individual faculty interviews. Our DAT began this fall and will continue meeting regularly through the spring. The DAT has received department support, in the form of an instructor course buy-out for the department lead, service credit for all members, and the sanction of the department chair and teaching committee.

In the outside-in approach, our work at the administrative level is intentionally lagging the work at the departmental level. At the administrative level, the senate has shifted its calls in its campus-wide teaching awards to focus on evidence-based practices (University of Colorado, 2008). Additionally, working with the faculty senate, we have revised the tools for evaluating "excellence" in teaching awards (e.g., to include evidence of scholarship in teaching and learning, measures of student learning and engagement; cf. University of Colorado, 2013). Once this framework has been established and interpreted by departments and once the DATs have begun the process of normalizing evidence-based conversations about education, we will then work with the senior administration to require evidence of student learning in tenure and promotion decisions. In this sense, "top-down" efforts are phase-shifted relative to our "bottom-up" work; only once there is sufficient faculty buy-in would such mandates be implemented. This would serve to institutionalize the changes already happening at the departmental level.

The Charms Department: The Middle-Out Approach

In Charms, thirteen faculty members responded to a survey gauging interest in participating in our cultural change process. Twelve of these faculty indicated that they were interested or very interested, with only one faculty member reporting no interest. Based on this survey, the decision to participate in our process was brought to a faculty vote; faculty unanimously agreed to participate. We have begun our preliminary data collection efforts this fall, and will begin the change process in the spring.

To measure the impact of our process, we are currently revising the PULSE vision and change rubrics (Bianco, Jack, Marley, & Pape-Lindstrom, 2013) to better capture shifts in culture, not just practice. We also intend to use surveys at the individual faculty level (inspired by Henderson, under development). The PULSE will serve both as a measurement tool (for pre/post testing department culture) and as a formative intervention tool for facilitating discussions around cultural change.

CONCLUSION

There is an urgent need for cultural change in STEM departments. Our change model aims to effect cultural change in alignment with six core commitments for productive departmental culture. We provide two synergistic approaches to using the model, which address the university ecosystem as a whole. Ultimately, these two approaches should be used in conjunction to effect systemic, sustained reform in STEM departments. These approaches may be used simultaneously or sequentially; for instance, initial efforts and success with the outside-in approach might prompt a department to seek a more holistic change process through the middle-out approach. As we continue to study and implement our approaches, we hope to validate and refine them, providing productive starting points for change efforts at other institutions.

ACKNOWLEDGEMENTS

We thank the Association of American Universities and the Helmsley Charitable Trust for funding this work.

REFERENCES

AAAS. (2011). *Vision and change in undergraduate biology education: A call to action*. Washington, D.C.: American Association for the Advancement of Science.

Beichner, R. J., Saul, J. M., Abbott, D. S., Morse, J., Deardorff, D., Allain, R. J., . . . Risley, J. S. (2007). The student-centered activities for large enrollment undergraduate programs (SCALE-UP) project. *Research-Based Reform of University Physics*, 1(1), 2–39.

Berret, D. (2014). Dissecting the classroom. *The Chronicle of Higher Education*. Retrieved from http://chronicle.com/article/Dissecting-the-Classroom/144647/

Bianco, K., Jack, T., Marley, K., & Pape-Lindstrom, P. (2013). *PULSE vision and change rubrics*. Retrieved from http://www.pulsecommunity.org/page/pulse-and-vision-change-v-c

Borrego, M., & Henderson, C. (2014). Increasing the use of evidence-based teaching in STEM higher education: A comparison of eight change strategies: Increasing evidence-based teaching in STEM education. *Journal of Engineering Education*, 103(2), 220–252.

Bryk, A. S., Gomez, L. M., & Grunow, A. (2011). Getting ideas into action: Building networked improvement communities in education. In M. T. Hallinan (Ed.), *Frontiers in sociology of education* (pp. 127–162). New York, NY: Springer.

Chasteen, S. V., Perkins, K. K., Beale, P. D., Pollock, S. J., & Wieman, C. E. (2011). A thoughtful approach to instruction: Course transformation for the rest of us. *Journal of College Science Teaching*, 40, 70–76.

Chasteen, S. V., Wilcox, B., Caballero, M. D., Perkins, K. K., Pollock, S. J., & Wieman, C. E. (under review). Educational transformation in upper-division physics: The Science Education Initiative model, outcomes, and lessons learned.

Conversant. (2014). High-performance conversation. Retrieved from http://conversant.com/about/approach

Freeman, S., Eddy, S. L., McDonough, M., Smith, M. K., Okoroafor, N., Jordt, H., & Wenderoth, M. P. (2014). Active learning increases student performance in science, engineering, and mathematics. *Proceedings of the National Academy of Sciences*, 201319030. http://doi.org/10.1073/pnas.1319030111

Glassick, C. E., Huber, M. T., & Maeroff, G. I. (1997). *Scholarship assessed: Evaluation of the professoriate. Special report*. ERIC. Retrieved from http://eric.ed.gov/?id=ED461318

Henderson, C., Beach, A., & Finkelstein, N. (2011). Facilitating change in undergraduate STEM instructional practices: An analytic review of the literature. *Journal of Research in Science Teaching*, 48(8), 952–984.

Henderson, C., & Dancy, M. H. (2007). Barriers to the use of research-based instructional strategies: The influence of both individual and situational characteristics. *Physical Review Special Topics—Physics Education Research*, 3(2), 020102.

Henderson, C., & Dancy, M. H. (2009). Impact of physics education research on the teaching of introductory quantitative physics in the United States. *Physical Review Special Topics—Physics Education Research*, 5(2), 020107.

Huber, M. T., & Hutchings, P. (2005). The advancement of learning. *Building the Teaching Commons*. San Francisco, CA: Jossey-Bass.

Iowa State University. (2014). *Faculty advancement and review*. Retrieved from http://www.provost.iastate.edu/help/promotion-and-tenure

Ortquist-Ahrens, L., & Torosyan, R. (2009). The role of the facilitator in faculty learning communities: Paving the way for growth, productivity, and collegiality. *Learning Communities Journal*, 1(1), 29–62.

Phillips, M. (1977). Interview with Dr. Melba Phillips. Retrieved from http://www.aip.org/history/ohilist/4821.html

President's Council of Advisors on Science and Technology. (2012). *Engage to excel: Producing one million additional college graduates with degrees in science, technology, engineering, and mathematics. Report to the President.* Washington, DC: Executive Office of the President.

Real, K., & Poole, M. S. (2005). Innovation implementation: Conceptualization and measurement in organizational research. *Research in Organizational Change and Development*, 15, 63–134.

University of Colorado. (2008). *University of Colorado at Colorado Springs promotion and tenure proposal (Iowa State Model)*. Retrieved from http://www.colorado.edu/ptsp/initiatives/tenure.html

University of Colorado. (2013). *The Boulder Faculty Assembly Awards for Excellence in Teaching*. Retrieved from http://www.colorado.edu/facultyassembly/awards/teaching.html

ABOUT THE AUTHORS

Daniel Reinholz is a Research Associate of the Center for STEM Learning at the University of Colorado at Boulder in Boulder, Colorado.

Joel C. Corbo is a Research Associate of the Center for STEM Learning at the University of Colorado at Boulder in Boulder, Colorado.

Melissa H. Dancy is a Research Faculty of the Department of Physics at the University of Colorado at Boulder in Boulder, Colorado.

Noah Finkelstein is a Professor of the Department of Physics and the Center for STEM Learning at the University of Colorado at Boulder in Boulder, Colorado.

Stanley Deetz is is a Professor Emeritus of the Graduate School at the University of Colorado at Boulder in Boulder, Colorado.

5

The Science Education Initiative:
An Experiment in Scaling Up Educational
Improvements in a Research University

Stephanie V. Chasteen, Katherine K. Perkins,
Warren J. Code, and Carl E. Wieman

ABOUT THE SEI

Motivation

The SEI was created as an experiment in generating large-scale sustainable change in STEM education at an institutional scale. This model of change focuses resources at the departmental level, includes an explicit focus on course transformation, and provides human resources in the form of discipline-based postdoctoral education specialists.

Research has demonstrated that faculty often discontinue the use of instructional innovations, or use them in ways that may not be effective (Dancy & Henderson, 2010; Henderson, Dancy, & Niewiadomska-Bugaj, 2012), primarily due to lack of support and structure for innovation. Many existing change strategies are overly reliant on a "development and dissemination" model of change (Dancy & Henderson, 2008). We sought a new model that provided some coherence to the collective efforts of faculty and administrators, rather than individual faculty re-inventing the wheel as they develop materials and learn new instructional approaches.

The SEI was generally effective in impacting courses and faculty across the institutions, but many—primarily local—factors affected success in individual departments. In this article, we review the process and progress of the SEI, and reflect on lessons-learned in the SEI as a whole. These results provide useful guidance in the creation of scalable, institutionally-supported models of educational change.

An Overview of the SEI Program and Model of Change

Initially led by author Wieman, the SEI operated at the department level. Departments submitted proposals to a small "SEI Central" unit for funding; this funding was used primarily to hire postdoctoral fellows (Science Teaching Fellows, STFs[1]) to support course transformation. SEI Central acted as a highly-involved funding agency, providing advice, training, and administration. Funding was committed by the university higher-administration.

The main assumptions of the model are (a) that courses transformed by faculty collaborating with the STF will be "departmentally owned," with shared and sustained expectations for how these courses are taught; and (b) that changing faculty practices will lead to a shift in departmental norms favoring the use of active-learning techniques. If these assumptions hold, then we should see improvements in student learning, faculty capacity, and departmental and institutional norms and practice. This model is shown graphically in Figure 1 below.

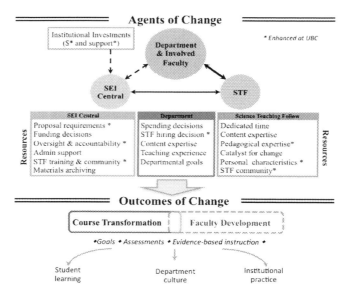

FIGURE 1. The SEI Model and its intended impacts. The strength of interactions between players is indicated by the weight of the connecting arrow. Course transformation was the explicit focus of the program; faculty development was equally important as a program goal, but implicit in the model. Areas which were enhanced at UBC are indicated with *.

The SEI was implemented at two institutions: the University of Colorado Boulder (CU)[2] and the University of British Columbia (UBC)[3]: See Table 1.

---

1. At UBC, STFs were termed Science Teaching and Learning Fellows, or STLFs. For simplicity, they are termed STFs in this paper.

2. http://colorado.edu/sei

3. http://cwsei.ubc.ca

TABLE 1. Key Aspects of the SEI Programs

	CU	UBC
Period of operation	2005–2014	2006–2016 (projected)
Total funding amount	$5.3 M (USD)	$12 M (CAD)
Funding per department	$150–$860 K (Ave $650 K[1])	$300 K–$1.75 M (CAD) (Ave 1.45 M[4])
Funding source	University	University and private donors
Number of departments	6 + 1 small pilot	6 + 1 small pilot
Total number of STFs	24	50+

[1]Averaged among the six fully-funded departments at each institution respectively.

IMPLEMENTATION

Funding and Proposal Structure

Internal funding was found to be the only viable funding option for initiating such a program. However, at UBC, early success led to additional funding through private donors and commitments of ongoing institutional support from the dean. At CU, some departments supplemented the effort with external grants.

Funding was distributed through a competitive proposal process, initiated through two separate calls for proposals. Funding was allocated based on the strength of the proposal—i.e., a large fraction of engaged faculty, with clear plans for sustaining changes. Departments were given substantial leeway in how to allocate their funding, though all chose to use the majority of funding to hire STFs.

Administration

SEI Central consists primarily of a director and/or associate director, and administrative staff; At UBC, this consisted of two to three FTEs in the earlier years, and one FTE in the final years. SEI Central responsibilities included:

- Soliciting and reviewing proposals
- Administration and oversight of funding and budgets
- Advising on hiring of STFs
- Training of STFs
- Support of STF community
- Monthly meetings with STFs and departmental directors
- Soliciting and providing feedback on annual reports from departments
- Coordinating annual SEI sharing event with presentations and posters

- Website maintenance, including instructional resources and course archive
- Collaboration with other institutions.

Role of Departments

Departmental faculty partnered with STFs in a variety of ways. STFs often developed learning goals in collaboration with faculty working individually or in groups. Some departments collectively identified courses in need of change, and faculty worked with the STF to transform that course. In other departments, STFs and departmental directors identified individual faculty who were most interested in the SEI, and partnered on transformation of that faculty member's courses. Each department identified a faculty member as a "departmental director" to serve as a liaison with SEI Central and as the immediate supervisor of the STFs. The director oversaw the hiring of STFs, guided STF work while helping them navigate faculty working relations and departmental structures and politics, and served as an advocate for the SEI work within the department.

Science Teaching Fellows (STFs)

Here we provide more detail and lessons learned about the roles of the central agents of change: the postdoctoral STFs.

Selection and hiring

STFs were hired as members of the department, and so the department was responsible for the search and hiring. STFs had typically earned a recent PhD in the discipline, and were interested in a discipline-based, education-focused postdoctoral experience. With their PhD level content-expertise, the STFs could meaningfully engage in content-specific work with the faculty—such as discussing learning goals, creating homework, and designing assessment items.

Most STFs were classified as postdocs[4], which naturally provided a somewhat restricted role for them within the departments. However, in a few cases, the status of the STF was redefined by the chair, providing a clearer, more elevated position for the STFs within the department hierarchy: For example, by framing STFs as members of the faculty (with the concomitant responsibilities), or actually hiring STFs as temporary instructors on a one-year contract.

---

4. For a short document detailing STF roles and skills, as used in STF training, see http://www.cwsei.ubc.ca/resources/STLF-develop.htm

Roles

STFs engage in a wide variety of activities (Table 2) to organize, facilitate, and enable the department faculty to achieve the changes set forth in their department's proposal to the SEI.

TABLE 2. Roles of Science Teaching Fellows

	Facilitating faculty communication and consensus building
	Collecting, distilling, and communicating data to support and guide faculty efforts
	Developing curricular materials and teaching approaches in collaboration with faculty
	Serving as a departmental resource for education research and evidence-based teaching methods
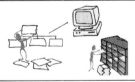	Facilitating sustainability by archiving and disseminating materials

Training and community

New STFs attended regular training sessions on education and cognitive psychology, research-based instructional practices, education research, and working with faculty effectively (e.g. communication and negotiation skills). Additionally, all STFs met regularly in semi-structured meetings. Over time, this training course was more formalized and carefully designed[5], and senior STFs were recruited to help teach the training course and run regular meetings.

---

5. See http://www.cwsei.ubc.ca/resources/STLF-develop.htm for an outline of the weekly training sessions.

The SEI Approach to Course Transformation

The SEI approach to course transformation uses a backwards design model (Wiggins & McTighe, 2001), in which specific, assessable learning objectives drive the assessment and instruction in a course. Figure 2 shows the essential questions used as guidance for SEI course transformation projects.

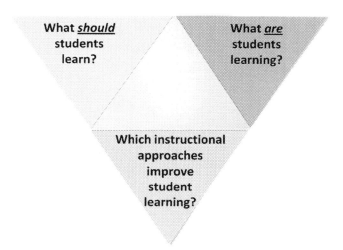

FIGURE 2. The SEI Course Transformation Philosophy

In particular, this resulted in the development and documentation of:

- **learning goals,** preferably through faculty consensus;
- **student difficulties,** based on observation and/or literature;
- **instructional materials,** based on learning goals and student difficulties, such as clicker questions, in-class activities, recitation materials, labs, and homework;
- **conceptual assessments,** typically validated or unvalidated pre-post tests; and
- **dissemination** of course materials through online archives[6].

We have written at length in other publications about the SEI course transformation approach (Chasteen et al., in press; Chasteen, Perkins, Beale, Pollock, & Wieman, 2011; Chasteen & Perkins, 2014), and refer the interested reader to those publications for more detail.

---

6. See http://www.colorado.edu/sei/fac-resources/course-archives.htm and http://sei.ubc.ca/materials/Welcome.do

RESULTS

In this section, we briefly review some of the documented outcomes from the project—a full analysis is beyond the scope of this paper. Evaluation is ongoing, and further publications are expected in the near future. Details of the evaluation methods, and more discussion of the outcomes, are available elsewhere (Chasteen et al., in press).

As shown in Table 3, a sizeable fraction of courses, faculty, and students were impacted by the SEI work. Our data indicate that most approaches did follow a backwards design model (though as we will discuss later, this wasn't necessarily the best way to get faculty involved in course transformation.) The available publications (and unpublished data) show a positive effect on student learning, but data are limited. The challenge of collecting baseline data for comparison, and of publishing data on student outcomes, is one of the lessons learned in the program. It was particularly challenging to administer assessments to traditionally-taught courses. There was often little incentive for faculty to devote time and energy to this activity, and an eagerness to begin work on the course. Often, assessments were not available until after the course approach had already changed. In some cases, collection of baseline data created tension in the department because faculty felt that they were being set up for failure.

TABLE 3. Courses, Students, and Faculty Impacted by the SEI

	CU	UBC
Courses impacted		
Number of courses with SEI involvement	103	167
Percent of undergraduate courses (in SEI departments) with SEI involvement	35%	~33%
Students impacted		
Total annual enrollment in courses with SEI involvement	18,000	43,000
Percent of annual enrollment (in SEI departments) in courses with SEI involvement	50%	67%, (78% without math)
Faculty impacted		
Number of faculty making some use of the STF	190	Not available
Percent of faculty (in SEI departments) making SEI supported changes	49%	48% (57% without math)

Note. Student enrollment represents the number of student seats, not unique students. Within any one department, the percent of courses ranged from 15–55%, the percent of student enrollments ranged from 30–85%, and the percent of faculty ranged from 10–93%. Mathematics at UBC is removed from some data as noted, as it is anomalous among SEI programs, due to cultural approaches towards teaching in the department.

The impacts of the program on departmental norms—and the sustainability of changes made through the SEI—are still under evaluation. At CU, 77% of N=97 faculty surveyed indicated that the number of conversations they have with their colleagues about teaching has increased due to the SEI. Additionally, in interviews, many faculty have lamented the loss of the STF, and have indicated that they would like the SEI to continue—in many cases indicating that this would be a worthwhile use of precious departmental funds.

The success in each department depended strongly on the timing of the proposal, the departmental culture and organizational structure, how teaching assignments are handled, the department chair, and the departmental director and STF. In one particular department case (Huber & Hutchings, 2014), however, these various factors conspired to generate a highly favorable environment for change, resulting in high rates of adoption of active-learning strategies and supportive infrastructure changes in the department (Huber & Hutchings, 2014). In other departments, change has not been quite so sweeping, and the SEI made varying levels of progress towards changing the culture of teaching. Across departments generally, when faculty participated in course transformations they showed faithful and sustained use of new teaching methods (Wieman, Deslauriers, & Gilley, 2013).

LESSONS LEARNED

While the SEI model was generally successful, some program structures were modified to address early difficulties. Due to its later start date, many lessons learned from the CU SEI were incorporated into the UBC SEI (see Figure 1).

Level of Funding

Program funding levels need to be sufficient to convince departmental leadership to invest the time and political capital required to get faculty consensus and spark action. SEI leadership had hoped to offer $1.5–2.0M USD per department, but was constrained by budget. In some CU cases, it appears that the level of funding was too small to create an appropriate level of urgency and action within a department. In contrast, we saw no evidence of this reaction in the departments at UBC, which had overall higher funding levels. The greater funding at UBC also allowed the hiring of a larger cadre of STFs, and more funding for administration to provide oversight and support.

Proposal Structure

We found that departments required more guidance than was originally provided in order to generate a successful proposal (and later execute the work.)

The biggest problem was a lack of specificity—what would be done, who would be responsible for doing it, and how changes would be sustained. Proposals, in many cases, had the intended effect of catalyzing departmental faculty to commit to a shared vision of change, but in some cases this failed.

The proposal process was significantly modified over time[7] to require clearer commitments to specific deliverables and timetables. At UBC, SEI Central helped departments iteratively improve their proposals, typically by providing small amounts of seed money to initiate programs, building towards a successful proposal for full funding.

Programmatic Oversight

Over time, it was found that additional oversight of departmental progress was needed, along with potential consequences for faculty failing to follow-through on commitments. One solution (at UBC) was to make continued funding contingent on progress. In reality, since withdrawal of funding would unfairly penalize the STF, limitations were placed on the replacement or hiring of new STFs in departments that did not make adequate progress. However, flexibility was needed when responding to shortcomings with respect to meeting goals and deadlines. Due to the absence of prior experience, it was difficult to know how to set realistic expectations.

The work of the STFs also required additional oversight, especially since most departmental directors did not have backgrounds in the scholarship of teaching and learning. At UBC, more explicit requirements were instituted for the STFs to attend meetings and submit regular reports. Additionally, the STF's role within the department was clarified[8]. Many of these changes were implemented through regular meetings with directors, SEI Central, and a representative from the dean's office.

Oversight from higher administration was also beneficial. The dean at UBC placed high importance on selecting department heads who supported the SEI, and contacted heads who were not following through on their commitments. This involvement had a clear impact that was not present at CU.

Faculty Incentive

Given the absence of institutional incentives to improve teaching, departments at UBC provided desirable "perks" to faculty, such as teaching buyouts and/or extra teaching/research assistants as part of the SEI. These perks provided

7. The most recent call for proposals (from 2007 and 2010) are at http://www.cwsei.ubc.ca /about/funding.htm and http://www.colorado.edu/sei/about/funding.htm.

8. For a two-page description of the STF role, along with examples, see http://www.cwsei.ubc .ca/resources/STLF-develop.htm

additional motivation to follow-through on the project goals, and compensated for the extra time needed to transform a course. When handled properly, with explicit expectations and deliverables, these perks were quite successful.

Departments, Departmental Chairs and Directors

The success of the SEI in various departments was also related to the culture and structures of the department, the interest and commitment of its faculty, chair and director. In general, departments that were more successful were those in which the mission of the SEI coincided with other departmental priorities, where the SEI was championed by the chair, and in which there was a collegial atmosphere toward teaching and learning.

Additionally, the effectiveness of a director varied by their standing and respect in the department, level of availability and interest in the SEI mission, and management ability. The choice of a director was localized within the department, and SEI Central had limited opportunity to affect this selection.

STFs: Hiring and Roles

We found that, usually, departments benefitted from additional guidance from SEI Central on how to select applicants likely to be successful in this unusual role. Factors such as teaching experience, content knowledge, interpersonal skills, familiarity with the individual department, respect for faculty, attention to detail, and work ethic all contribute to success in a complex and unique job.[9]

As stated previously, additional clarification on the potential roles[5] of the STFs was beneficial in helping departments envision what the STFs could, and could not, do. For example, STFs were originally discouraged from being a primary course instructor as their funding was neither intended to replace typical faculty duties, nor to employ them as sessional instructors. Teaching experience for STFs was later seen as valuable enough that it became common for them to be instructors of record, typically paid separately (i.e., not out of SEI funds), such as a one-semester course per year or less.

STFs: Community and Training

At UBC, the training model was more successful due to the larger number of STFs (providing a critical mass of new and existing STFs to run and participate in training each year).

With the larger number of STFs at UBC, it was also possible to build a stronger STF community—both among and within departments. STFs met

---

9. For a full description of the features of an effective STF, see the "STLF Reflections on their Job" document at http://www.cwsei.ubc.ca/resources/STLF-develop.htm

regularly as a group at both institutions, but at UBC, the expectations for attending these meetings were more explicit, as were regular written reports on activities, and included a weekly or biweekly reading group led by experienced STFs. This vibrant STF community has contributed greatly to STF capacity at UBC. We note that the skills acquired in this position have resulted in high demand for STFs, who have gone on to diverse careers (e.g., instructor, academic advisor, staff in teaching and learning centers, education research faculty, and traditional academic positions).

Course transformation and backwards design

The somewhat linear model of backwards design was not always the most effective. Development of effective learning goals was found to be surprisingly difficult, and not always highly motivating for faculty. Starting with smaller, concrete changes to classroom practice was often more motivating to faculty, especially if these addressed an existing concern about their course.

Introductory courses were also not necessarily the best starting point; these courses often involve multiple faculty and entrenched, overloaded curricula. In some departments, STFs found that they made more headway in meaningful change when focusing on a portion of a course, on smaller upper-level courses, or on courses where there was specific faculty interest.

DISCUSSION

In essence, this work provides an "existence proof" that it is possible to generate change in STEM departments, *without* changes in the institutional incentive system. That said, such institutional changes would greatly help—a major, ongoing challenge was the underlying conflict between the goals of the SEI and how teaching is assessed and rewarded at the university level. Faculty continually raised the concern that paying more attention to teaching would negatively impact their research productivity. Given the amount of effort required in course transformation, intrinsic rewards are often not sufficient to motivate faculty to engage and sustain the changes.

While such institutional changes are yet to appear, however, the SEI has shown that postdoctoral Science Teaching Fellows can fill an important need in departments. This "embedded expertise" provides valuable resources of time, information, and non-pejorative coaching to support faculty within a department. But, even with such resources, change is not easy or automatic; local factors played an important role, including levels of leadership, organization, and commitment within a department. The main lessons learned in the program are perhaps not surprising:

1. **STFs need support,** in the form of training, STF community, and supervision—which we feel is beyond the current scope of most teaching and learning centers.
2. **Explicit commitments, oversight, and accountability are needed for success**. Such a novel approach to education and teaching is too large a cultural change for most departments to automatically carry it out effectively, and so oversight and support is necessary, and must be written into the structure of the program along with adequate funding.

To place this work within a theoretical context, we consider the four change strategies identified in the literature (Borrego & Henderson, 2014; Henderson, Beach, & Finkelstein, 2011; see Figure 3):

- Disseminating Curriculum & Pedagogy
- Developing Reflective Teachers
- Developing Shared Vision
- Enacting Policy

These strategies operate at a range of levels (group vs. individual), and assume different directionality of change (prescribed vs. emergent). The original vision of the SEI model was that change would operate at all levels: institutional (commitment to invest in the SEI), departmental (faculty collaboration on the SEI proposal, leading to shared vision), and faculty (developing reflected teaching practice and knowledge about curriculum and pedagogy through the ensuing course work). Over time, the changes in faculty practice would lead to cultural shifts in what matters about teaching and learning in the department and the university as a whole—leading to large-scale changes in shared vision, reflected in policy. Such a multi-level approach is recommended by Borrego & Henderson (2014).

In practice, the success of this model depended on (a) local departmental factors described previously, plus (b) the level of support and oversight at the proposal stage and (c) institutional oversight. At CU, where (b) and (c) were not strongly present, changes in departmental culture depended strongly on the department in question. At UBC, where there was a greater critical mass of STFs for a longer time, stronger institutional support, and a larger SEI Central community and support, the SEI had a greater, though still not totally successful, capacity to influence departmental-level decisions by developing new advocates for change among the faculty, and by providing departments a more complete analysis of student performance through data.

This theoretical deconstruction is useful as it provides relevant information on where change models might add to the SEI approach—in the areas of

INSTITUTIONAL STRUCTURE   **STRATEGY**   **SEI COMPONENT**

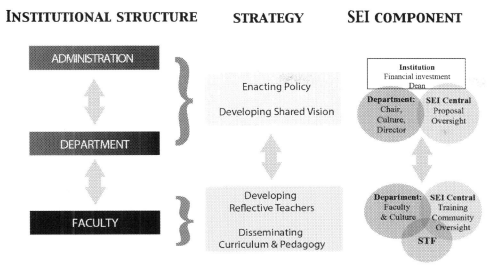

FIGURE 3. Four change strategies (Borrego & Henderson, 2014) operating at the group (department and institution) and individual (faculty) level. Arrows indicate the direction of potential influence. The SEI component, which most successfully supported these aspects of change, is shown to the right.

developing shared vision and enacting policy. Existing literature on leadership and effective team structure (e.g. Kotter, 2012; Pentland, 2014) provides many best-practices in this area. Additionally, it may be fruitful to expand the embedded expertise model to include initial groundwork within a department *prior* to an accepted proposal. For example, an intensive process of needs analysis, such as that used in Project Kaleidoscope (PKAL; http://pkal.org) could generate more successful, collective vision within departments. Including a variety of structures, support, and oversight during the project may help future programs achieve even more substantial changes in STEM education.

REFERENCES

Borrego, M., & Henderson, C. (2014). Increasing the use of evidence-based teaching in STEM higher education: A comparison of eight change strategies. *Journal of Engineering Education, 103*(2), 220–252. doi:10.1002/jee.20040

Chasteen, S. V., Perkins, K. K., Beale, P. D., Pollock, S. J., & Wieman, C. E. (2011). A thoughtful approach to instruction: Course transformation for the rest of us. *Journal of College Science Teaching, 40*(4), 70–76.

Chasteen, S. V., & Perkins, K. P. (2014). Change from within: The Science Education Initiative. In M.A. McDaniel, R. F. Frey, S. M. Fitzpatrick and H.L. Roediger, (Eds), *Integrating cognitive science with innovative teaching in STEM disciplines* (e-reader version), 298–370. St. Louis: Washington University in St. Louis Libraries.

Chasteen, S. V., Wilcox, B., Caballero, M. D., Perkins, K. K., Pollock, S. J., & Wieman, C. E. (in press). Educational transformation in upper-level physics: The Science Education Initiative model, outcomes, and lessons learned. *Phys. Rev. ST Phys. Educ. Res.*

Dancy, M., & Henderson, C. (2008). *Barriers and promises in STEM reform.* Washington, D.C.: National Academies of Science.

Dancy, M., & Henderson, C. (2010). Pedagogical practices and instructional change of physics faculty. *American Journal of Physics, 78*(10), 1056–1063.

Henderson, C., Beach, A., & Finkelstein, N. (2011). Facilitating change in undergraduate STEM instructional practices: An analytic review of the literature. *Journal of Research in Science Teaching, 48*(8), 952–984. doi:10.1002/tea.20439

Henderson, C., Dancy, M. H., & Niewiadomska-Bugaj, M. (2012). Use of research-based instructional strategies in introductory physics: Where do faculty leave the innovation-decision process? *Physical Review Special Topics—Physics Education Research, 8*(2). doi:10.1103/PhysRevSTPER.8.020104

Huber, M. T., & Hutchings, P. (2014). *The Carl Wieman Science Education Initiative in the Department of Earth, Ocean, and Atmospheric Science, University of British Columbia.* (Unpublished benchmark case study, Bay View Alliance).

Kotter, J. P. (2012). *Leading change* (1 edition.). Boston, Mass: Harvard Business Review Press.

Pentland, A. (2014). *Social physics: How good ideas spread—The lessons from a new science.* New York: Penguin Press HC.

Wieman, C., Deslauriers, L., & Gilley, B. (2013). Use of research-based instructional strategies: How to avoid faculty quitting. *Physical Review Special Topics—Physics Education Research, 9*(2), 023102. doi:10.1103/PhysRevSTPER.9.023102

Wiggins, G., & McTighe, J. (2005). *Understanding by design* (expanded 2nd ed.) Alexandria, Virginia: Association for Supervision and Curriculum Development.

ABOUT THE AUTHORS

Stephanie V. Chasteen is the Associate Director of the Science Education Initiative and a Research Associate of the Department of Physics at the University of Colorado at Boulder in Boulder, Colorado.

Katherine K. Perkins is the Director of the Science Education Initiative and PhET Interactive Simulations and an Associate Professor Attendant Rank at the University of Colorado at Boulder in Boulder, Colorado.

Warren J. Code is the Associate Director of the Science Centre for Learning and Teaching and the Assistant Director of the Carl Wieman Science Education Initiative at the University of British Columbia in Vancouver, British Columbia, Canada.

Carl E. Wieman is a Professor of Department of Physics and the Graduate School of Education at Stanford University in Stanford, California.

6

Planning Transformation of STEM Education in a Research University

Robert Potter, Gerry Meisels, Peter Stiling, Jennifer Lewis, Catherine A. Bénéteau, Kevin Yee and Richard Pollenz

It is widely accepted that today's innovation economy requires a work force that is well-grounded in STEM (NAP, 2005 and 2010). Major public research universities play an important role in meeting this need because they educate a large fraction of all students, especially those interested in STEM. Yet we know that nearly two thirds of students who begin their studies in one of the STEM disciplines change majors to continue in non-STEM fields (PCAST, 2012). The University of South Florida (USF) is no exception: Only 33% of entering first-time-in-college students (FTIC) who have declared a STEM major graduate with a STEM degree within six years (Pollenz et al., 2014).

Decreasing this loss of initial STEM majors provides one of the best opportunities to increase the number of professionals in the STEM disciplines. A retention rate of 50% would provide nearly all the STEM-competent graduates needed by industry at this time (Carnevale et al., 2010). To achieve this end, universities must adopt evidence-based classroom strategies. However, improving classroom practice that leads to increased retention historically has not been as valued or as rewarded as have research accomplishments in research universities (Anderson et al., 2011). Changing that culture is one of our goals; achieving this requires more collaboration among faculty in teaching, professional development, support for change, alignment of policies with desired outcomes, and interaction of university faculty with secondary schools and community colleges (Labov et al., 2009; Anderson et al., 2011; Kober, 2015).

Data alone will not lead to change, but are the foundation on which effective arguments can be developed. This is our objective at USF, an urban research university that enrolls 44,000 students annually and is among the top 25 public universities in external funding for research. USF began pursuing cultural change more than a decade ago with a variety of programs, partially supported by NSF, HHMI, FL DOE, and US DOE. Programs to improve student outcomes have included:

- **Creating the USF Academy for Teaching and Learning Excellence** to support faculty professional development.
- **Establishing the STEM Education Center**, a summer program for high school students.
- **Founding the Coalition for Science Literacy** to improve teaching in grades five through 16.
- **Developing a data base** that provides detailed information on retention of STEM majors.
- **Opening the SMART Lab** (Science, Math, and Research Technology) which provides 330 computers for computer-assisted learning.
- **Participating in the STEM Talent Expansion Program (NSF-STEP)** to improve key gateway mathematics courses.
- **Joining the Leadership Alliance,** a consortium to develop underrepresented students into outstanding leaders.
- **Launching the USF-HHMI STEM Academy** to enhance student success and persistence in biology and biomedical sciences.
- **Vigorously pursuing Discipline-Based Education Research (DBER)** by hiring four additional faculty in 2013, bringing the total to eight.

Of special note is the **Student Success Initiative,** which began in 2009 with a provost-appointed task force. The group's task was to radically rethink student success with particular focus on retention, progression, and professional school and career readiness. The Office of Student Success (http://usfweb3.usf.edu/studentsuccess/) was created the next year and the faculty leader of the task force was appointed vice provost for student success. This Student Success Office has worked with deans to implement the new policies, procedures, practices and programs recommended by the taskforce. Our efforts toward STEM student success are fully in alignment with the larger university-wide goals providing a political and potential resource advantage.

USF also led a number of programs to provide content-focused professional development of teachers.

PLANNING TEAM

In early 2013, we appointed a team to create a widely accepted and supported, coherent plan for cultural change. Our goal being a culture that values teaching, as well as research, and one where faculty and GTAs employ student-centered approaches using evidence-based teaching methodologies.

The 12-member team has been key to the success of the planning process. It is broadly constituted of faculty from three colleges, college and provost-level administrators, and the director of the teaching and learning center:

- Catherine Beneteau, Mathematics & Statistics, co-PI
- Scott Campbell, Chemical Engineering
- Allan Feldman, Science Education
- Gladis Kersaint, Mathematics Education and Associate Dean, College of Education
- Randy Larsen, Chair, Chemistry
- Jennifer Lewis, Associate Chair, Chemistry, co-PI
- Gerry Meisels, Chemistry and Director, Coalition for Science Literacy, PI
- Richard Pollenz, Cell, Molecular, and Microbiology and Associate Dean, Undergraduate Studies
- Robert Potter, Chemistry/Biochemistry and Associate Dean, College of Arts and Sciences, co-PI
- Les Skrzypek, Chair, Mathematics and Statistics
- Peter Stiling, Past Chair, Integrative Biology, Special Faculty Assistant to the Provost for STEM Initiatives, co-PI
- Kevin Yee, Director, Academy for Teaching and Learning Excellence

Supported by a planning grant from NSF's WIDER program, the planning team (PT) began meeting at least bi-weekly in September 2013. The team quickly agreed upon a bottom-up/top-down approach to creating cultural change (Kezar 2012). The PT developed a common understanding of the issues, analyzed data and institutional context and began to design a change plan that continues to evolve. Efforts fall into six groups: student progression data, the seminar series, faculty learning community, the advisory board, policy and context analysis, and the current action plan.

Student Progression Data

USF admits annually a freshman (FTIC) cohort of 3,500–4,000; 26% are Hispanic or African American, 58% are female and 16% are first-generation college students. FTIC includes about 1,700 STEM majors (microbiology, marine biology, chemistry, geology, math, physics, statistics, information systems and engineering majors). Nearly 1,000 of the 1,700 STEM majors (57%) declare majors in biology (cell biology, microbiology, integrative biology, marine biology) and the biomedical sciences (BMS).

Of the STEM-FTIC who entered in 2006 and 2007, within a six-year period 33% received a degree in the STEM major they had declared initially, 3%

in another STEM major, and 27% in a non-STEM major. We evaluated the curricular progress of the 553 biology and biomedical science majors who left these degree programs and earned a non-STEM degree from USF. The analysis focused on: i) credentials (Q SAT score and AP credit), ii) the number of semesters they remained in the major, iii) course progression and academic performance through the gateway and majors STEM courses, iv) number of credits, v) trackable undergraduate research experience and vi) GPA at major change and at graduation.

Three student groups of increasing academic performance were identified. Only 21% of the students who left the BIO/BMS major were struggling to complete the curriculum. Thirty percent left the STEM majors in high academic standing and having passed biology and calculus. These results suggest several areas for reform:

- New courses/experiences in the summer or first year to build community within the student population and connect them with their "major" (Pinto, 1993; Rovai, 2002; Estrada et al., 2011; Graham et al., 2013).
- Modified introductory gateway courses that engage students in the content and help make direct connections to biological sciences.
- Early exposure to undergraduate research either through revision of laboratories or through other innovative methods (e.g., UT Summer Research Institute).

The Transforming STEM Education Seminar Series: Its Critical Features and Important Role in the Transformation Plan

The seminar series has been the centerpiece of planning and the most visible outcome of the process. It built awareness of the need to improve instruction for undergraduates generally and drew attention to the special needs of students from population groups underrepresented in the STEM disciplines. The series also highlighted evidence-based instructional practices as a way to improve student outcomes, and strategies to create systemic change. The series developed a larger community of interested and motivated individuals who helped lead the change process on campus.

The planning team identified and invited high profile speakers and scheduled them in a monthly series on the same day of the week in the same location. The advisory board was very helpful in creating the following list:

- Dr. Adrianna Kezar; USC (December 10, 2013). *STEM Education, Shared Leadership, and You*

- Dr. Vasti Torres; USF (January 14, 2014). *Do Students Under-Represented in STEM Experience the Learning Environment Differently?*
- Dr. Richard Pollenz; USF (February 18, 2014). *Understanding Institutional Data Can Inspire University-Wide Adoption of Evidence-Based Practices in STEM Education*
- Dr. George Kuh; IU (March 5, 2014). *Fostering STEM Student Engagement: What Matters*
- Dr. Melanie Cooper; MSU (April 18, 2014). *Evidence-Based Approaches to STEM Education*
- Dr. E. William Wischusen; LSU (Sept. 9, 2014). *Impact of a Pre-Freshman Boot Camp on Student Performance*
- Dr. Jay Labov; NAS-NRC (Oct. 8, 2014). *The Changing National Landscape of Undergraduate STEM Education*
- Dr. Linda Slakey; Sr. Adv. AAU STEM Ed. former Director Undergraduate Div. NSF (Nov. 4, 2014). *Making Student-Centered Teaching the New Normal: Are We at a Tipping Point?*
- Dr. Shirley Malcom; Head of Education and Human Resources Programs at AAAS (Dec. 2, 2014). *Undergraduate STEM Education: Moving Diverse Populations from the Margins to the Center*
- Dr. Michael Klymkowsky; UC Boulder (Feb. 17, 2015). *The Challenges of Active Learning and Coherent Curricula in the Sciences*
- Dr. Gabriela Weaver; U. Mass. (April 21, 2015). *Shifting the Teaching Culture in a Research University to Student-Centered Approaches*

E-mail invitations to seminars were sent to over 400 individuals: all faculty, advisors and graduate students in Natural Sciences and Mathematics, faculty in the College of Engineering, and key upper level administrators. Planning team members also sent individual follow up requests using personal connections to stimulate interest. The initial invitations contained several items designed to inform and build momentum for change. E-mails began with *"You are invited to our 'fourth' event on exploring strategies to transform STEM education at USF in order to increase the retention and proficiency of STEM majors, especially of those who are members of underrepresented groups."* In addition, messages included invitations to a post-seminar discussion and a pre-seminar subject exploration and discussion of readings. The readings (suggested by the speaker) were included as attachments with the invitation. Faculty and graduate students could obtain a certificate of completion from our Academy for Teaching and Learning Excellence if they attended at least four seminars.

Speakers served three important functions. First, they provided a valuable high profile window onto the national scene. Second, they shared state-of-the-art insight on some aspect of best practices in undergraduate STEM education and on the change process. Third, speakers served as ambassadors for USF's change process. During their visit, speakers were familiarized with USF by meeting with several of the constituent groups prior to the seminar. Selected planning team members attended breakfast then speakers met our eight discipline-based education research (DBER) faculty in science and mathematics. They had lunch with members of the science instructor group and finally met with interested graduate students prior to the seminar.

Seminar attendees signed in at the door of the seminar and received name tags, a one-page bio of the speaker, and a summary of past events with take-home lessons. The list of past seminars and take-home lessons helped us emphasize continuity and reinforced key ideas adding to the momentum for change. Two examples are shown below:

Dr. Adrianna Kezar (December 10, 2013)
Professor of Higher Education and Co-Director of Pullias Center for
  Higher Education, University of Southern California
STEM Education, Shared Leadership, and You
Take-Home Lessons
- Start with vision first—change without direction won't work
- Use institutional data to determine how to proceed
- Faculty must be a central part of determining direction and impetus for change (it cannot be top-down)

Dr. Vasti Torres (January 14, 2014)
Dean of the College of Education and Professor of Education,
  University of South Florida
Do Students Under-Represented in STEM Experience the Learning Environment Differently?
Take-Home Lessons
- Students are very far from homogenous, and their unique circumstances or backgrounds may result in additional difficulties in completing a STEM major.
- An ideal way to treat students is as unique individuals, with no preconceived notions on the instructor's side. This is partly accomplished by not assuming common cultural stereotypes or references when teaching.

Response to the seminar series has been excellent, with attendance ranging from 40–80 people with 158 different people having participated in one or more events.

We have been purposeful in distributing the face of leadership for change by consistently referring to the planning team and providing all names on the seminar invitation. We have three different members of the team involved directly in each seminar. One welcomes the attendees describing the NSF grant-funded project supporting the series and its intended outcomes. Another faculty introduces the seminar speaker and one team member facilitates the interactive discussion session with the speaker after the seminar. These assignments are rotated among all team members. Finally, the PI leads the speaker's meeting with the planning team at the end of the day.

Faculty Learning Community Development

The PT formed a faculty learning community (FLC). FLCs are a widely recognized method of increasing faculty buy-in and beginning the process of culture change with those working on the front lines (Cox 2004). FLCs may take many forms, ranging from informal to incentivized, or from intentionally topic-driven to community-focused without a pre-determined course of investigation. We chose this open-topic approach which emphasizes building community and creating a cohort of faculty dedicated to change and to exploration. The absence of financial incentives for participation resulted in only those dedicated to the endeavor attending on a regular basis.

We cast a wide net; all full-time faculty in STEM departments were invited to attend the first meeting with free lunch, hosted using local university funds. More than 50 faculty attended. The group discussion addressed effective teaching practices in STEM disciplines, which led to a free-ranging discussion. At the conclusion of the event, more than two dozen indicated interest in continuing to meet every couple of weeks, and so the FLC was formed organically.

The FLCs met every month over the academic year, at first experimenting with online polls to arrange a meeting date and time, and then settling into a routine when the most participants could attend. Scheduling is a persistent issue with FLCs (Cox, 2004), particularly when participants do not originally sign up with a set schedule in mind. Attendance waned in predictable fashion over the year, yet with 15–20 participants commonly present at each meeting.

A typical FLC meeting started with a brief presentation of a technology, a technique, or a discrete piece of content (text or video), and then followed with a discussion of the pros, cons, obstacles, and workarounds to the idea being reviewed. The overall tenor of the discussions was one of collegiality, as it was recognized from the beginning that the FLC needed to function in an informal setting, with emphasis on sharing over formal presentations—a model commonly referred to as a community of practice (Wenger, 1998). It typically ended with a group decision on what to explore in the next session.

Early in the series of meetings, the director of the Academy for Teaching and Learning Excellence (ATLE) provided participants with a custom-built "menu" of practice-based (and research-tested) course interventions. These interventions were separated into categories such as "appetizers" (easy to implement), "entrees" (more encompassing to implement, but a proven high-impact strategy) (Kuh, 2008), and "desserts" (interesting, useful, and cutting-edge ideas, often technology-based, that might be fruitfully explored by experienced instructors to provide additional benefit in their teaching practices). The ATLE menu provided a range of possible topics for the FLC to explore in future meetings, such as the flipped classroom, peer learning/use of student-response systems ("clickers"), and interactive techniques to use in lecture halls.

The FLC achieved its desired goal of creating a vibrant community of engaged, committed participants dedicated to improving their teaching. Such an FLC transforms participants into teaching-interested ambassadors within their home departments. One unexpected benefit has been interdepartmental collaboration. The inherent networking of the multidisciplinary meetings led to recognition of where course content overlapped and to ideas on how to streamline, enhance, and align STEM content not just across courses but across departments. The collaboration also led to interdisciplinary research investigations, as the networking enabled like-minded researchers to find allies in other departments for planned explorations, both in disciplinary science and in the teaching of STEM content.

Advisory Board: Composition and Role in Transformation Planning

The composition of the advisory board was determined by the planning team during the development of the grant and purposely identified external thought leaders in organizational change and undergraduate STEM education. The deans of the relevant colleges were added as internal members based on their influence on faculty and their role in resource allocation.

Project advisory board

- ERIC R. BANILOWER, Senior Researcher and Partner, Horizon Research, Inc.
- MICHAEL N. HOWARD, Consultant in evaluation, research, program development, and technical assistance; mathematics, science, and technology education.
- ADRIANNA KEZAR, Professor for Higher Education, University of Southern California and Co-director of the Pullias Center for Higher Education.

- GEORGE D. KUH, Chancellor's Professor of Higher Education Emeritus at Indiana University Bloomington.
- JAY B. LABOV, Senior Advisor for Education and Communication for the National Academy of Sciences (NAS) and the National Research Council (NRC).
- ERIC M. EISENBERG, Professor of Communication and Dean of the College of Arts and Sciences, USF.
- RAFAEL PEREZ, Professor of Computer Science and Engineering and Associate Dean of Academic Affairs, the College of Engineering, USF.
- VASTI TORRES, Professor of Educational Leadership and Dean of the College of Education, USF.

The advisory board has been critical to the evolution of the plan for transformation. Board members have served as seminar speakers, provided advice on future speakers, provided input on the design of the logic model for the project, and helped identify missing elements in our preliminary plan. They have also provided options for solving some of our planning conundrums.

Policy and Context Analysis

The planning team realized early on the importance of identifying policies and systems that could either hinder or facilitate the transformation to evidence-based instructional practices. The major policy barriers were identified to be the evaluation of teaching and its role in personnel decisions, disciplinary faculty teaching assignments, space allocation and room organization. Currently, the university has too few of the flexible learning spaces necessary to accommodate the active-learning environments needed for many evidence-based practices. The team has been working with the vice provost for student success to remedy this problem and ensure that faculty and students have greater access to these types of classrooms. Teaching evaluations currently rely heavily on student judgment, while other relevant information is largely lacking. The chairs have now begun discussions in departments as to how to more effectively and fairly judge quality teaching. Models from other research institutions will be provided to seed the discussions. The plan will call for teaching assignments to encourage and reward redesign and team teaching. The resulting resource needs will have to be negotiated with the USF administration, with initial costs hopefully secured from external sources. The now four-year-old Student Success Initiative (described earlier) is a primary component of our university's strategic plan and is fully aligned with the transformation plans. This coincidence facilitates upper administrative support and increases the potential for resources. The University's new tenure and promotion guidelines require excellence in both teaching and research, adding new emphasis to the value and expectations for teaching effectiveness. Finally, the State University System of Florida has instituted a new Performance Based Funding Model almost totally based on undergraduate outcomes (University System of Florida Board of Governors 2014). The transformation to more effective evidence-based practices in STEM should improve student outcomes and hence performance metrics, making performance funding a potentially significant source of support for the change process.

Current Action Plan

As we have continued to learn from our collective experiences over the past year, the following elements of our evolving plan have emerged:

- Continue the seminar series:
  - Build awareness, understanding and a common language around reform
  - Make teaching more public and more community involved.
- Work with faculty and departments to develop a coherent, engaging and rigorous STEM experience for students:
  - Expand evidence-based practice into all gateway science and mathematics courses (involves curriculum/course content redesign).
  - Facilitate departmental and interdepartmental discussions of evidence-based practice and interconnections in curriculum.
  - Build meaningful connections between and among foundational courses across disciplines.
  - Over time, introduce evidence-based practice in upper-division courses.
  - Establish small-group student support systems, building on pre-college programs and first-year university experience courses.
  - Chronicle processes and evaluate progress (faculty and student outcomes).
- Address remaining policy issues (space, teaching evaluations).

SUMMARY

The current planning process is in reality many years in the making and has been developing through a more informal, but growing, network at USF. The WIDER planning grant brought greater coherence, systemic thinking, a greater will to act, and increased administrative and faculty attention to evidence-based practice and the national concern about undergraduate STEM education. The apparent progress and success at USF suggest that institutions wishing to develop adoption of evidence-based strategies in their STEM courses should take a three-part approach: (1) build a community of faculty who are committed to changing classroom practice; (2) adopt a bottom-up/top-down approach to change after the foundation is laid; and (3) provide adequate institutional resources for planning, leadership, and implementation.

This work was supported in part by NSF grant **DUE-1347753.**

REFERENCES

Anderson W. A., Banerjee, U., Drennan, C. L., Elgin, S. C. R., Epstein, I. R., Handelsman, J., Hatfull, G. F., Losick, R., O'Dowd, D. K., Olivera, B. M., Strobel, S. A., Walker, G. C., Walker, I. M. (2011). Changing the culture of science education at research universities. *Science*, vol. 321, Jan.14, 2011.

Carnevale, A., Smith, N., & Strohl, J. (2010). *Help wanted: Projection of jobs and education requirements through 2018.* Washington, DC: Georgetown University, Center on Education and the Workforce.

Cox, M. D. (2004). Introduction to faculty learning communities. In M. D. Cox & L. Richlin (Eds.), Building faculty learning communities (pp. 5–23). *New Directions for Teaching and Learning: No. 97*, San Francisco: Jossey-Bass.

Estrada-Hollenbeck, M., Woodcock, A., Hernandez, P., and Schultz, P.W. (2011). Toward a model of social influence that explains minority student integration into the scientific community. *Journal of Educational Psychology*, 103 (1), 206–222.

Florida State University System Board of Governors. (2014). *Performance based funding model.* http://www.flbog.edu/about/budget/performance_funding.php accessed 4/13/15.

Graham, M. J., Frederick, J., Byars-Winston, A., Hunter, A-B., Handelsman, J. (2013). Increasing persistence of college students in STEM. *Science*, 341: 1455–1456.

Kezar, A. (2012). Bottom-up/top-down leadership: Contradiction or hidden phenomenon. *Journal of Higher Education*, 83(5), 725–760.

Kober, N. (2015). *Reaching students: What research says about effective instruction in undergraduate science and engineering.* Board on Science Education, Division of Behavioral and Social Sciences and Education. Washington, DC: The National Academies Press.

Kuh, G. D. (2008). *High-impact educational practices: What they are, who has access to them, and why they matter.* Washington, DC: Association of American Colleges and Universities.

Labov, J., Singer, S., George, M., Schweingruber, H, & Hilton, M. (2009). Effective practices in undergraduate STEM education, part 1: Examining the evidence.

CBE Life Sciences Education, 8(3), 157–161, accessed 4/13/15 at http://www
.ncbi.nlm.nih.gov/pmc/articles/PMC2736016

NAP. (2005). *Rising above the gathering storm: Energizing and employing America
for a brighter economic future*. Committee on Prospering in the Global Economy of the 21st Century: An agenda for American science and technology;
Committee on Science, Engineering, and Public Policy, National Academies
Press, Washington, DC; National Academies follow up, 2007.

National Academies. (2010). *Rising above the gathering storm, revisited: Rapidly approaching category 5*. Washington, DC: National Academies Press.

PCAST. (2012). *Engage to excel: Producing one million additional college graduates
with degrees in science, technology, engineering, and mathematics*. http://www
.whitehouse.gov/sites/default/files/microsites/ostp/pcast-engage-to-excel
-final_feb.pdf
accessed 4/12/15

Pollenz, R. S., Fuentes, H. L., and Meisels, G. G. (2014). *Utilizing institutional student data to vest university stakeholders in STEM education reform*. Presented
at the AAC&U Academic Renewal Conference, Transforming STEM Higher
Education, Atlanta, GA.

Rovai, A. P. (2002). Sense of community, perceived cognitive learning, and persistence in asynchronous learning networks. *Internet and Higher Education 5th
ed*, 319–332.

Tinto, V. (1993). *Leaving college: Rethinking the curses and cures of student attrition
(2nd ed)*. Chicago: University of Chicago Press.

Wenger, Etienne. (1998) *Communities of practice: Learning, meaning, and identity*.
Cambridge: Cambridge UP, Print.

ABOUT THE AUTHORS

Robert Potter is the Associate Dean of Graduate and Undergraduate Studies of
the College of Arts and Sciences Graduate and Undergraduate Studies at the
University of South Florida in Tampa, Florida.

Gerry Meisels is a Professor of the Department of Chemistry and the Director of the Coalition for Science Literacy at the University of South Florida in
Tampa, Florida.

Peter Stiling is a Professor of the Department of Integrative Biology and Special
Faculty Assistant in the Office of the Provost at the University of South Florida
in Tampa, Florida.

Jennifer Lewis is a Professor of the Department of Chemistry and the Director
of the Center for the Improvement of Teaching & Research in Undergraduate

STEM Education (CITRUS) at the University of South Florida in Tampa, Florida.

Catherine A. Bénéteau is an Associate Professor of the Department of Mathematics and Statistics at the University of South Florida in Tampa, Florida.

Kevin Yee is the Director of the Academy for Teaching and Learning Excellence (ATLE) at the University of South Florida in Tampa, Florida.

Richard Pollenz is a Professor of the Department of Cell Biology, Microbiology, and Molecular Biology and the Associate Dean and Director of the Office for Undergraduate Research at the University of South Florida in Tampa, Florida.

7

Supporting STEM Education: Reflections of the Central Indiana Talent Expansion Project

Lisa Bunu-Ncube, Jeffery X. Watt, Howard Mzumara,

Charles R. Feldhaus, Andrew D. Gavrin, Stephen P. Hundley,

and Kathleen A. Marrs

A National Center for Educational Statistics study of six-year graduation data indicates that nationally, 59% of STEM majors fail to complete their degree, with 21% of those changing majors to a non-STEM field (NCES, 2000, 2009). The National Science Board's Science and Engineering Indicators 2000 and 2014 reports reinforce these conclusions: STEM degrees granted today remain below levels reached in the early 1990s (NSB, 2000). Yet, the U.S. Department of Labor projects that jobs requiring technical degrees will grow to an estimated 6 million job openings by the end of the decade—the majority being in computer sciences, mathematics, medical and health technology, and engineering (US Bureau of Labor Statistics, 2010). Given this talent gap, it is crucial for universities to develop strategies that encourage more students to successfully complete degree programs in STEM degrees.

The Central Indiana STEM Talent Expansion Program is a five-year (2010–2015), $2M project that enhances a central Indiana pipeline to increase the number of students from the greater Indianapolis region (central Indiana) obtaining STEM degrees that will be sustainable after the expiration of this grant. The goals of this project are to increase the numbers of students of all demographic groups who:

1. pursue STEM academic and career pathways;
2. participate in STEM research, industry internships, and honors activities;
3. graduate with an undergraduate degree in STEM fields; and
4. transition into industry, graduate and professional programs.

The program has set a target of increasing the number of STEM graduates at IUPUI by 10% per year (an increase of an additional 782 STEM graduates by 2015).

THE INSTITUTION AND DEPARTMENTS

Indiana University Purdue University Indianapolis (IUPUI) is located in downtown Indianapolis, and is the state's only urban research university, with 22 schools offering over 200 degree programs. IUPUI has a national reputation for its involvement with the City of Indianapolis and the Indianapolis public school systems through the IUPUI Urban Center for the Advancement of STEM Education (UCASE), and the STEM Education Research Institute (SERI). Created in 1969 by the legislature of Indiana, IUPUI embodies the unorthodox partnership between Indiana and Purdue Universities to serve the educational needs in the largest metropolitan region of the state, representing one-fifth of the state's population. IUPUI has grown substantially in its 40-year history, becoming the third largest campus in the state, and is the only four-year public institution of higher education in this region. More than 60% of IUPUI's 31,000 students are first-generation college attendees and 16% of its student body belongs to minority groups.

The School of Science and the School of Engineering and Technology are two of the three largest undergraduate schools by headcount at IUPUI: Both schools award Purdue University degrees. Together the two schools are known as leaders in undergraduate STEM education. Both schools have leadership roles in implementation of Project Lead the Way (Engineering and Biomedical Sciences: http://science.iupui.edu/community/projectleadway) in school districts state-wide, and faculty from both schools are regional facilitators of the Indiana STEM Resource Network: https://www.istemnetwork.org/.

The School of Science is recognized for its innovation in teaching science through the Just-in-Time Teaching (JiTT) and Peer-Led Team Learning (PLTL) projects (Marrs, 2004; Gafney, 2007). The Math Assistance Center provides a technology-rich environment for collaborative learning, peer-mentoring, and supplemental instruction for students in all levels of mathematics (Watt, 2013). The School's Project SEAM (http://science.iupui.edu/community), a collaborative effort involving fifteen central Indiana school districts and five postsecondary institutions teamed to create a "seamless" transition between high school and college for all students, has a strong record of providing science and mathematics professional development to hundreds of local high school teachers since the partnership was established in 1999.

The Central Indiana STEM Talent Expansion Program (CI-STEP) involves undergraduate majors in the School of Science's six degree-granting STEM departments (Biology, Chemistry, Computer Science, Geology, Physics, and Mathematics), and in the School of Engineering's and Technology's six engineering degrees (Biomedical, Computer, Electrical, Energy, Mechanical, and Motorsports), plus the six technology degrees (Biomedical Engineering Technology, Computer Engineering Technology, Computer Information Technology, Construction Engineering Management Technology, Electrical Engineering Technology, and Mechanical Engineering Technology). In addition, the Central Indiana STEM Talent Expansion Program is collaborating with Ivy Tech Community College Central Indiana, as part of the pipeline to increase the number of students graduating with STEM degrees. This collaboration between Ivy Tech and IUPUI is building on the articulation agreements and programs already established, including the creation of new seamless pathways for students pursuing STEM programs between the two- and four-year institutions. These two campuses (IUPUI and IVYTech Central Indiana CC) have the largest number of African American and Hispanic students of any postsecondary institutions in Indiana. Moreover, the largest number of students transferring from an Indiana community college to a four-year institution was from Ivy Tech Central Indiana to IUPUI.

In addition, the School of Engineering and Technology has developed articulation agreements with other four-year institutions, allowing students at these institutions the opportunity to transition seamlessly to the Accreditation Board for Engineering and Technology (ABET) accredited engineering and technology programs at IUPUI, or earn a dual degree at both institutions. For example, the School of Engineering and Technology has partnered with Butler University, a private institution in Indianapolis, to establish the Engineering Dual Degree Program (EDDP). The EDDP allows students to study at Butler and also have access to the engineering programs at IUPUI. Completion of this program results in two degrees, one in Engineering from Purdue University, and another from Butler University in another major. This is not a 2+2 program where students are expected to transfer after three years to IUPUI to complete their studies. Instead, the EDDP has a curriculum that integrates the engineering courses into the students' plan of study, which allows students to be full-time residents at Butler for the duration of the program.

METHOD: USE OF MINI GRANTS

The Central Indiana STEM Talent Expansion Program (CI-STEP) started its work in 2010 as a comprehensive award aimed at increasing graduation in STEM through a number of various initiatives involving structural changes within the organization using a bottom up approach. According to Burke (2011), most organizational change, especially change that impacts large, complex organizations, is unplanned and gradual. Large scale, planned change that affects the entire system is unusual; and revolutionary change in strategy, mission and culture, is extremely rare. Based on the mission of the National Science Foundation STEP program, it was clear that planned, revolutionary change was expected of awardees. One of the most impactful and successful initiatives in the CI-STEP project has been the mini grant program developed by the CI-STEP team. This move toward planned, revolutionary change involved recruiting STEM faculty from schools across the IUPUI campus to participate in meaningful change that resulted in measurable progress and success for STEM students.

A request for proposals and a proposal application template was developed by the CI-STEP team (see Appendix 2 and 3, respectively) and was distributed to all faculty members in STEM departments in May 2011. The request explicitly stated that each proposal include work that was above and beyond the normal requirements of the position, that successful achievement of the objectives or outcomes would promote retention and persistence in STEM and that clear, concise methods of assessment and evaluation be included in all proposals. Awards ranged from $5,000–$25,000. Involvement of collaborators, immediate impact on a broad range of students and demonstrated innovativeness, effectiveness and inclusiveness were also prerequisites for successful awards. An essential component of a successful proposal was the letter of support from administration and a statement of sustainability beyond the mini grant project funding. Awardees were required to complete a progress report (see Appendix 4) at the conclusion of the mini grant activity. CI-STEP attributes much of its impact to date on the increase of STEM graduates to the successfully funded mini grants and their dedication to the mission of STEP, ultimately taking a step toward institutional and cultural change on our campus.

CI-STEP MINI GRANTS OUTCOMES

A total of 11 mini grants, involving 18 faculty members, were funded in 2011 and 2012 and nearly $300K was awarded directly. Although these mini grants have had varying specific objectives, all had the main goal of increasing the number of STEM graduates at IUPUI. Appendix 1, provides a summary of the mini grants (those awarded for activities not originally included in the NSF proposal), the project objectives and outcomes.

The mini grant project titled "Tutoring Services in the Physics Learning Space (PhyLS)" had a main objective of reducing of drop, fail, and withdraw (DFW) rates from introductory physics courses. The PhyLS was designed to advance student success in introductory physics by providing tutoring/mentoring services to all students taking these courses. These gateway physics courses, typically having the highest DFW rates on campus (averaging 25.1% in 2010) and serve almost 1,500 students each year. In order to reduce the DFW rates,

PhyLS has adopted the "assistance center" model that has proven successful in math, chemistry and biology.

Student perceptions of the PhyLS were extremely positive. For instance, during the fall term of 2013, a sample of 51 students completed an online evaluation survey about the PhyLS. Students responded to Likert-type survey questions regarding the learning space, five-point response scale, with 5 being completely agree and 1 being completely disagree. Mean responses to the items are shown in Table 1 below.

Overall, students' responses demonstrate a very positive reaction to implementation of tutoring services in the PhyLS. The hours of operation of PhyLS were increased based on higher than expected usage. The success of this initiative has persuaded the school to fully sustain PhyLS after the grant expires.

The mini grant titled "Transfer Student Recruitment and Support" had a set of goals which included: recruiting transfer students to study within engineering and technology (E&T) fields of study; retaining transfer students currently studying within E&T fields of study; and building community by connecting transfer students to faculty, staff, and current students, as well as pertinent resources at IUPUI and within the Purdue School of Engineering and Technology, IUPUI. The main outcomes of this mini grant project were developing transfer student ambassadors to connect with new students and student orientations for E&T transfer students. Since there was no transfer student orientation program before the grant, and this initiative dramatically increased one-year retention by more than 16%, the school will be institutionalizing the use of student ambassadors and a formal orientation program for transfer STEM students to IUPUI.

The main goal of the mini grant project "Enhancing Student Comprehension and Success in Genetics through Problem Based Learning Experience (Recitation)" was to reduce the drop, fail, and withdrawal (DFW) rates in a genetics course and increase student comprehension of genetics. A peer mentor

TABLE 1. Mean Responses to Likert-Type Survey

Item	Mean
Was attentive and focused during the session	4.2
Provided me with appropriate, relevant information	4.2
Supportive and encouraged me to continue working to be successful	4.0
Was able to explain the tutoring/mentoring services to my satisfaction	4.0
The mentor provided me with appropriate, relevant information	4.5
The mentor encouraged me to continue working to be successful	3.9
My overall experience with the Tutor-on Duty was positive	4.3

was engaged to help students in the course. During the fall semester, 48 students (36%) attended one or more mentoring sessions, similar to the attendance in the spring. After implementing the peer mentor, a slight decrease in DFW was observed. The number of students dropping, failing, or withdrawing decreased from 16% to 14%. However, it is difficult to determine whether the mentor was the direct cause of the decrease. However, the department has agreed to continue to support the project for the near future.

"Engineering and Technology—Alliance for Retention for Multicultural Students (ETARMS)" mini grant project assisted students to be successful in engineering and technology by providing them with mentoring and resources. The mini grant project had positive outcomes that included increased use of educational resources such, as freshman engineer mentoring, math tutoring, and physics tutoring. Students also reported a greater understanding of class material and that the financial resources given to them reduced the amount of financial stress they had. This project will also be sustained by the school.

From "Studio to Student: e-Mentoring in Computer Graphics Technology" produced and disseminated fourteen videos that outlined various insights into how to break into the computer graphics (CG) field in Hollywood and beyond. This course attracted 78 students, which is unprecedented in this program with 92% completing the course successfully. This initiative was a one-time expense to produce the videos for future use, so sustaining the project is not an issue with this mini grant.

The "CHEM-C 341 Organic Chemistry Workshop Series Peer Mentoring Using the Peer Led Team Learning (PLTL) Model" mini grant project aimed to reduce the number of students dropping, failing, and withdrawing from first semester organic chemistry. As a result, the DFW rates have decreased about 10% after workshops were implemented with a 6–10% increase in positive student perception of problem-solving ability. Twenty-five percent of the peer mentors expressed an interest in teaching after this experience. Study findings to date suggested that faculty have been successful in using the PLTL approach to lower the failure rates in the workshops. Reduction of DFW rates for the chemistry course and training of additional discussion leaders (using the PLTL model) to decrease the number of students in each workshop are positive interventions for increasing the success and number of STEM graduates.

After the workshops students reported that the workshops increased their understanding of organic chemistry material, and that their problem solving skills have also increased. For example one student reported "[Participation] helped me better understand the problems and how to apply knowledge during exams." Additionally peer leaders had a positive benefit from participating in the workshop series. For example, one leader stated: "I am certainly more comfortable . . . communicating with people." The PLTL model in chemistry is growing nationally as a best practice, and CI-STEP believes that the department will be in a position to sustain this program in the future.

The mini grant titled "Peer Mentoring Using the Peer Led Team Learning (PLTL) Model in ENGR 19700: Introduction to Programming Concepts" used a peer mentor to assist other students with in-class assignments and provided help sessions each week. The 2013DFW rate is the lowest of the seven semesters shown in Table 2.

Table 3 gives final exam averages for the previous two years, as well as the spring 2013 semester for the sections taught by the investigator.

There were modest improvements in student performance and persistence in mentored sections of ENGR 19700 over previous semesters. The final exam scores for the year after the implementation of peer mentoring was statistically higher than before implementation ($p < 0.05$). Having a mentor available in class made it possible for students to receive significantly more help with their programs and thus improved student confidence. In the future, it would be desirable to extend the mentoring model initiated this semester to all sections of ENGR 19700 and to investigate its impact on performance and persistence with the larger group.

TABLE 2. DFW Rates for Sections of ENGR 19700

Semester	Number of Students	Number of Sections	DFW Rate (Percent)
Fall 2007	31	1	29
Spring 2008	61	2	36.1
Spring 2009	65	2	40
Spring 2010	65	2	29.2
Spring 2011	69	2	27.5
Spring 2012	61	2	36.1
Spring 2013	63	2	27.0

TABLE 3. Final Exam Averages for Sections of ENGR 19700

Semester	Number of Students Taking Final Exam	Number of Sections	Final Exam Average (Percent)
Spring 2011	60	2	75.04
Spring 2012	45	2	74.36
Spring 2013	56	2	76.19

The "Calculus Course Redesign—Introduction of Recitations to Increase Student Learning" project aimed at improving the DFW rate. Calculus was another course that has an unacceptably high DFW rate, indicating that a large number of students are not successful in meeting the course outcomes or attaining proficiency in quantitative skills. Calculus recitations were developed and implemented for the large lecture section of MATH 16500 (fall semesters) and 16600 (spring semesters). Recitations became a required component of the course. Results of implementing were positive. Despite the larger class size, sections of calculus with recitation sections have a significantly lower DFW rate, ~20%, than other sections of the course. Students in sections of calculus with recitations performed10 percentage points better on the departmental final exam (Watt, 2013). Although the department sees the retention value that recitation sections can provide, the department is only willing to incorporate recitations into sections of courses with enrollments over 100 students.

The Summer Residential STEM Bridge Program was designed for students who would be residents on campus. Students living in the same buildings had an opportunity to get to know one another before the semester began and there was more interaction as the semester continued. There were some issues that needed addressing: rapport with upper classman as RAs; promoting the program during orientation (since new students see a variety of different advisors at orientation); and a decrease in outreach to participants after the semester started (plans to increase outreach with the next cohort are being considered). A spinoff of the residential STEM bridge program was an overnight orientation for the next cohort of students. The number of students participating in the STEM Bridge program increased by 32% and 22% over the past two years (65 students in 2010, 86 in 2011, and 105 in 2012). Recent data indicated that STEM bridge participants have higher GPAs compared to non-participants; students participating in Summer Residential STEM Bridge have lower DFW rates compared to non-participants; and minority students (especially African Americans) participating in Summer STEM Bridge obtained higher GPAs, lower DFW rates and higher fall-to-fall retention rates compared to nonparticipating AA students.

A new "Post Enrollment Requirement Checking (PERC)" in math courses was implemented as a mini grant during the Fall 2012 semester. Many STEM students were dropping out of their intended major as a result of not being successful in the first math course, and then moving onto the next math course, and failing it. These students believed they could pass the next math course without being successful in the prerequisite. Advisors found it difficult to catch this situation before it became too late, contributing to lowering the first year retention rate. The math department worked with the Registrar's Office to develop an automatic withdrawal program that would remove enrolled students in math courses one week before the semester starts, if they did not have the proper prerequisites grade. When the Post Enrollment Check (PREC) was run one week before classes started, the identified students were withdrawn from the math course, and the student and the advisor were automatically notified by e-mail of the action. In the Fall 2012 semester, 47 students were identified as enrolling in math courses in which the prerequisite course was not passed. For the Spring 2013 semester, 84 students were identified and advised before the first day of class.

"Promoting the Math Minor to Students and Advisors" mini grant project involved actively promoting the math minor to students and advisors across campus as a way of setting a short-term goal in the pipeline to completing a BS degree. The department completed the paperwork and the registrar posted the minor on the transcript at the time of completion (usually in the sophomore year). This transcript documentation provided motivation to students that they had completed a component of their degree. Many STEM majors usually have a minor in their plan of study, or will earn the minor by selecting one more math course as an elective. The number of minors awarded each year provided an indicator of the number of STEM majors in the pipeline. The number of minors have increased each year: 32%, 14%, and 94% (44 awarded in 2008, 58 in 2009, 66 in 2010, and 128 in 2011). This rapid growth is partly due to students becoming more aware of their eligibility to obtain the minor, but it is also due to 53 students (of the 128 awarded last year), who took an additional course above their requirement (a free elective) to obtain the minor.

The "School of Science Career Development Services (CDS) Center" mini grant project was implemented by Career Services. One of the initial goals of the new director was to increase the awareness of the center, its location, and services provided. The center was promoted through various programs and methods. Although only two employees staff the center, outreach to hundreds of undergraduate and pre-professional students, has been successful. The number of students utilizing career services increased from 95 students in the first year to 327 students in 2011–12; and one-on-one advising went from 95 to 327. Educational programs include: resume development, class presentations, workshop series, social media networking, and etiquette lunch. Strategic and intentional efforts were undertaken to acquaint faculty with CDS staff and services. This initiative is one of the biggest successes of CI-STEP, and the school is already fully funding the center and has added two more full-time positions.

CONCLUSIONS

Mini grant projects are a viable way of incorporating educational interventions into the colleges and universities that have immediate and sustainable impacts not only on student learning but outcomes as well. A definite and positive outcome of the approaches used in this project was effective student engagement in their own learning and success. Students as both mentors and learners were actively engaged in learning, which translated into positive student learning outcomes. There were both intended and unintended outcomes that positively impacted the learning outcomes.

Since CI-STEP was an NSF-funded project, it was critical that the project address two issues: (1) how will the best practices implemented by the project be sustained after the grant, and (2) how has faculty culture toward STEM education been changed by the project (i.e., how likely will other faculty at the institution adopt best practices in STEM teaching in the future)? CI-STEP has been very successful on both of these issues. First, the grant allowed faculty to leverage administration with a resource match. For example, one mini grant funded the start-up costs and student stipends for the first two years of the Physics Learning Space, if administration would allocate space. With the data collected on student use and student learning outcomes from the first two years, the administration saw the value in institutionalizing the Physics Learning Space by finding ways to sustain the costs of the center with student course fees. In addition to the Physics Learning Space, six other CI-STEP initiatives will also be sustained by various departments after the expiration of the grant. Secondly, the mini grants allowed many more faculty members (18 individuals in addition to the Co-PIs of CI-STEP) at IUPUI take ownership of small pieces of the overall project, where the individual faculty member could experiment on their own STEM course to find ways to increase student success at the course level. By getting more faculty members involved in CI-STEP activities, the grant has helped change faculty culture on improving teaching. The best evidence of this change is when faculty get excited over the increase in the number of graduates from their department—and realize that small changes in teaching STEM add up quickly to more graduates.

CI-STEP has been hugely successful, meeting its graduation goal for STEM students during its first four years (10% increase per year). The increase in the graduation numbers have occurred from retention and persistence interventions, and not from increased recruitment of students. The vast majority of these interventions are now sustained by the departments and schools on campus. This sustainability becomes built into a department's budget based on data-driven evidence that retaining more students in the STEM pipeline to graduation more than pays for the intervention. Although, without increased recruitment of STEM students, a point of diminishing returns will eventually be reached as the leaks in the pipeline are addressed. Therefore, future activities will include STEM outreach activities and experiences for high school students to enter the pipeline.

REFERENCES

Burke, W. W. (2011). A perspective on the field of organization development and change: The Zeigarnik effect. *Journal of Applied Behavioral Science*, 47, 143–167.

Gafney, L. and Varma-Nelson, P. (2007). Evaluating peer-led team learning: A study of long-term effects on former workshop leaders, *Journal of Chemical Education*, 84, 535–539.

Marrs, K. A. and Novak, G. M. (2004). Just-in-time teaching in biology: Creating an active learner classroom using the internet. *Cell Biology Education*, in press.

National Center for Education Statistics (NCES). (2000). http://nces.ed.gov/

National Center for Educational Statistics (NCES). (2009). *Students who study science, technology, engineering, and mathematics (STEM) in postsecondary education*, July 2009.

National Research Council. (2000). *How people learn: Brain, mind, experience and school.* Bransford, J. D., Brown, A. L., and Cocking, R. R., Eds. Washington DC: National Academy Press, <http://books.nap.edu/books/0309070368/html/index.html> Accessed November 15, 2003.

National Science Board. (2000 and 2014). Science and engineering indicators 2000. Arlington, VA: National Science Foundation.

Watt, J. X. (2013). CI-STEP: Transforming the calculus course to increase STEM graduates. *Proceedings of the 2013 International Conference on Mathematics Education,* (2013. 11. 1-2) 635–645.

ABOUT THE AUTHORS

Lisa G. Bunu-Ncube is the Director of the Office of Assessment & Accreditation at the A.T. Still University in Mesa, Arizona.

Jeffery X. Watt is the Associate Dean and Professor of Mathematical Sciences of the Purdue School of Science at Indiana University Purdue University Indianapolis in Indianapolis, Indiana.

Howard R. Mzumara is the Director of Testing Center of the Department of Planning and Institutional Improvement at Indiana University Purdue University Indianapolis in Indianapolis, Indiana.

Charles R. Feldhaus is an Associate Professor of the Purdue School of Engineering and Technology at Indiana University Purdue University Indianapolis in Indianapolis, Indiana.

Andrew D. Gavrin is a Chair and Associate Professor of Physics of the Purdue School of Science at Indiana University Purdue University Indianapolis in Indianapolis, Indiana.

Stephen P. Hundley is a Professor of the Purdue School of Engineering and Technology at Indiana University Purdue University Indianapolis in Indianapolis, Indiana.

Kathleen A. Marrs is the Associate Dean and Associate Professor of Biology of the Purdue School of Science at Indiana University Purdue University Indianapolis in Indianapolis, Indiana.

APPENDIX 1
CI-STEP MINI GRANT PROJECTS

CI-STEP Mini Grant	Objective	Outcomes
Tutoring Services in the Physics Learning Space (PhyLS)	Reduction of drop, fail, and withdraw (DFW) rates from introductory physics courses	• Courses Served: Physics 218/219; Physics P201/P202; Physics 152/251; Physics 100; Physics 200 • Course enrollment annual count: 1,600 students • Students have positive satisfaction ratings to the learning space. However, many students noted that the area was too small and crowded. • During the fall of 2012, 1,063 distinct students used the learning space. During 2012, 80% of students returned to the center more than once. Around half of the students returned to the center five or more times. • In the spring of 2013, 778 distinct students visited the learning space. • Over the course of both the fall and spring semester the median length of a visit was one hour.

CI-STEP Mini Grant	Objective	Outcomes
Transfer Student Recruitment and Support	• Recruit transfer students to study within engineering and technology (E&T) fields of study. • Retain transfer students currently studying within E&T fields of study. • Build community by connecting transfer students to faculty, staff, and current students as well as pertinent resources at IUPUI and within the Purdue School of Engineering and Technology, IUPUI.	• Two transfer student ambassadors were added to connect with new transfer students. • Overnight orientation for E&T transfer students (eight students attended). • • 70 transfer students attended one of the five full-day orientations geared toward E&T transfer students. • Two visits to Ivy Tech Indianapolis and a visit to Vincennes University were made to recruit E&T transfer students. • Fourteen Ivy Tech students and two Ivy Tech staff visited the Purdue School of Engineering and Technology, IUPUI. Presentations were made by E&T staff, and current IUPUI students spoke. • Three summer 2013 student transfer ambassadors were added to the Office of Student Services Summer team to work with transfer orientations and connect with prospective transfer students. • Two summer 2013 orientation events were held with 28 students attending.
Enhancing Student Comprehension and Success in Genetics through Problem Based Learning Experience (Recitation)	• Reduce drop, fail, and withdrawal in genetics course BIOL-K 322 • Increase students understanding of genetic material taught in BIOL-K 322 • Strengthen content knowledge to increase the likelihood of success in future genetics courses.	• A peer mentor has been placed in the course. The students can visit the PM during office hours to receive assistance with the course. The peer mentor attends the lectures so he has a fresh and current understanding of what is going on in the course. • Overall reactions to the peer mentor were positive.
Engineering and Technology - Alliance for Retention for Multicultural Students (ETARMS)	• Assist students to be successful in engineering and technology. • Provide students with mentoring • and resources to facilitate • academic success of students.	• Students reported use of educational facilities, such as faculty resources, freshman engineer mentoring, math tutoring, and physics tutoring. • Students reported a greater understanding of class material and that the financial resources given to them reduced the amount of financial stress they had. • Students reported being able to accomplish many of the goals they had set at the beginning of the semester. For example one student reported "I was able to pay better attention in class."

CI-STEP Mini Grant	Objective	Outcomes
From Studio to Student: e-Mentoring in Computer Graphics Technology	Production and dissemination of 14 videos that outline various nsights into how to break into the computer graphics (CG) field in Hollywood and beyond.	• This course attracted 78 students which is unprecedented in this program (and 72 of them completed the course successfully). • Developed a modern, adaptable model of STEM education delivery that was a leader in its approach within the Purdue School of Engineering and Technology. • Attracted, will hopefully retain (over time), and hopefully recruited new and existing technology students to the CGT program at IUPUI. • Attracted a highly talented industry professional, and his colleagues, to the CGT program at IUPUI. • Harnessed technological lessons and a pedagogical model that will benefit CGT in the smooth adaptation of future online course offerings.
CHEM-C 341 Organic Chemistry Workshop Series	• Reduction in the number of students dropping, failing, and withdrawing from first semester organic chemistry. • Better student performance on the ACS organic chemistry final exam. • Increase students' satisfaction with the organic chemistry course.	• Thirty undergraduate peer leaders were trained to help undergraduate students with organic chemistry problem-solving skills. The leaders were trained weekly in classroom management, and organic chemistry material. Over 300 workshop sessions were held. The workshop questions and quiz questions were uploaded to a database for future students to use. • Results indicated a significant reduction in the DFW rate, and a significant increase in students' performance on the ACS final chemistry exam. • Students reported that the workshops increased their understanding of organic chemistry material, and that their problem-solving skills have also increased. For example, one student reported, "[Participation] helped me better understand the problems and how to apply knowledge during exams." • Additionally peer leaders had a positive benefit from participating in the workshop series. For example, one leader stated: "I am certainly more comfortable . . . communicating with people."

CI-STEP Mini Grant	Objective	Outcomes
Peer Mentoring using the Peer Led Team Learning (PLTL) Model	• Reduction in the number of students dropping, failing, and withdrawing from the course. • Better student performance on the ACS chemistry final exam. • Increase students' satisfaction with the chemistry course.	• Results indicated a significant reduction in the DFW rate, and a significant increase in students' performance on the ACS final chemistry exam. • Students reported that their problem-solving skills have also increased.

APPENDIX 2
REQUEST FOR PROPOSALS TEMPLATE

Announcement of Awards for
Central Indiana STEM Talent Expansion Program

The *Central Indiana STEM Talent Expansion Program (CI-STEP)* at Indiana University-Purdue seeks faculty proposals for projects that promote graduation in STEM fields. The program has set a target of increasing the number of STEM graduates at IUPUI by 10% per year (an increase of an additional 782 STEM graduates by 2015). Proposals are due by October 15, 2011 to the Principal Investigator of the CI-STEP grant, Dr. Jeff Watt. His email is: jwatt@math.iupui.edu. Projects may begin upon award and are to be completed by May 15, 2013. The awards for The *Central Indiana STEM Talent Expansion Program are* supported by the National Science Foundation.

Purpose

The purpose of the *Central Indiana STEM Talent Expansion Program (CI-STEP)* is to increase retention, persistence, and graduation in STEM disciplines through projects that employ and assess the impact of several program-wide intervention strategies on student success, leading to higher numbers of students graduating with STEM degrees. These intervention strategies include: Summer STEM Bridge Academies, peer-mentoring and academic advising support for transfer students, Peer-Led Teaching and Learning, and Just-in-Time Teaching, honor seminars, and career development services and internships. IUPUI, with its nationally recognized commitment to improving educational success for all students, has numerous support services already in place to assist with this initiative, making it possible for us to integrate research and education on effective strategies for student learning in STEM disciplines.

CI-STEP proposals are expected to help increase the numbers of students of all demographic groups who:

- pursue STEM academic and career pathways;
- participate in STEM research, industry internships, and honors activities;
- graduate with an undergraduate degree in STEM fields; and
- transition into industry, graduate and professional programs.

Proposals are welcome from all STEM disciplines. Proposals are encouraged from individuals or from faculty teams representing programs, departments; they may also involve collaborative projects with faculty members from IVY Tech and Butler University. Projects should propose significant interventions to promote retention, persistence and graduation in STEM disciplines.

Proposals must include the following three characteristics:

1. The proposal must involve work that is above and beyond the normal requirements of the individual's position(s); and
2. Successful achievement of the objectives or outcomes will promote retention and persistence in STEM; and
3. The proposal must clearly identify the methods to be used for assessing outcomes.

Proposals that have one or more of the following characteristics are especially encouraged and will receive priority consideration:

- Seeking awards ranging between $5,000–$25,000
- Involve collaborators
- Have immediate impact on a broad range of students
- Demonstrate innovativeness, effectiveness and inclusiveness

Proposals that have the following characteristics are discouraged and will receive lower priority:

- Seeking funding for capital goods
- Research
- Conferences

Eligibility

All full- and part-time faculty members at IUPUI, Ivy Tech and Butler are eligible. Individual faculty members pursuing teaching innovation projects are encouraged to include at least one other faculty or staff member in some capacity; for example, as an outside peer reviewer for the project, a consultant on instructional design, or an administrative partner to overcome barriers.

Funding

Awards will be limited to $25,000 (may range from $5,000 to $25,000).

Application/Implementation Timeline

Proposals might be implemented over one or two semesters, and may or may not include work to be done in summer.

Proposal Submission and Review Process

Step 1. Your college dean reviews and approves the proposal.

Your SUPERVISING ADMINISTRATOR, i.e., *Head of Department*, must review your proposal and *APPROVE IT OR SUBMIT A LETTER OF SUPPORT: A letter of support might show how the proposal fits college plans and priorities, as well as how it will be supported by the administrator. If the proposal is dependent on resources in addition to the award, the letter of support should explicitly indicate the dean's commitment to providing any proposed or assumed college resources. For part-time, temporary or fixed-term faculty members, the letter of support must give some assurance that the faculty member will be able to complete the proposed project within the stated timeframe.*

Step 2. The CI-STEP Award Committee reviews and approves the proposal.

The CI-STEP Project Award Review Committee will review and approve the proposal after the supervising administrator's written approval is obtained. The committee may accept, reject, or make suggestions on how to improve the proposal to make it acceptable. Rejected proposals and proposals with improvement suggestions shall be returned to the proposer(s).

The committee will use the guidelines below under "Review Criteria" in evaluating proposals. **The first three criteria are required of all awards for CI-Step proposals.**

Step 3. If approved, you begin the proposed work or project after receipt of notice from the Project PI or designee.

You will receive a letter of notification from the Project PI that your project (as proposed, or as modified, if the committee requested modifications) is approved. The letter will identify the monetary award to be paid upon successful completion.

Review Criteria

1. *Work is above and beyond the normal requirements of the individual's position(s):* It is recognized that a faculty member's normal responsibilities include classroom teaching, student advising, course evaluation, classroom preparation, the evaluation of student performance, committee assignments, classroom research, professional development, service to the college, and community service. The award for CI-STEP is to given for work:
   - performed at a time outside the normal work hours or duty days, or
   - involving an activity not normally required, or
   - encompassing a scope of activities not normally required.
2. *Promotion of Retention, Persistence and Graduation:* Goals and objectives prioritize student learning and successful achievement of students' personal and career goals in STEM disciplines. There is a clear relationship between the performance objectives or project goals, the plan of work and methods, and the intended student learning outcomes.
3. *Evaluation and Assessment Methods:* An appropriate selection of assessment strategies and tools are used throughout the proposed performance period/project. The evaluation plan is clearly connected to program or course learning goals and outcomes.

APPENDIX 3
PROJECT PROPOSAL APPLICATION

Potential applicants are encouraged to discuss their ideas with CI-STEP project administrators before applying. Please allow enough time (at least one to two weeks) before the deadline for your application to be reviewed and approved by the CI-STEP Project review committee. All applications must follow the format below.

Applications must be received by October 15, 2011 (for full consideration) and must include:

1. Cover sheet
2. Project narrative
3. Budget summary and budget narrative
4. Letter of support from administrator
5. Certification page

Submit your proposal via e-mail to: CI-STEP Awards Review Committee, c/o jwatt@math.iupui.edu

Cover Sheet
Title of the Project
Principal investigators
Departments
Date

Proposal Content

Proposal Section	Proposal should address:	Suggested Length
Amount Requested	How much funding is being sought?	
Project Description	What issue or problem is being addressed? What are your goals and methods? What activities will address your goals? How is the project innovative for your own development, your program or the college? Who will be involved: how many faculty, students, etc.?	1–2 pages
Significance/ Rationale/ Evidence	Why is the project important? How do you know it is important, what is your evidence (i.e., how has the issue or problem been documented)? What are the conditions or contexts in which the project will be taking place? What is the need, both locally and in a system or national context? How is the project linked to college and/or CI-STEP priorities and initiatives? How will this project increase the number of students who pursue STEM academic and career pathways? How will this project increase the numbers of students who participate in STEM research, industry internships and/or honors activities? How will this project increase the number of students who graduate with an undergraduate degree in STEM fields? How will this project increase the number of students who transition to industry, graduate and professional STEM programs? How will the innovation or change be sustained after the project funding has ended?	1 page

Proposal Section	Proposal should address:	Suggested Length
Anticipated Difficulties	What kinds of hurdles or limitations do you expect to encounter? How would you address them?	1 paragraph
Timeline of Activities	When are activities planned? How can you assure the project will be completed within the proposed timeframe?	1 page
Outcomes	What specific outcomes* do you want to achieve? How will your planned activities achieve these outcomes? How will your plan promote retention, persistence and graduation in STEM? *Each outcome must be matched to an assessment in your evaluation plan.	1 page
Evaluation Plan	How will success be measured? How will you know that you have achieved your outcomes? What kind of evidence will you gather? What kinds of assessments will you use? Is there an assessment matched to each outcome? What is the impact on campus or the surrounding community?	½ - 1 page
Dissemination	With whom will you share this information? How will the project be shared with others? Consider campus professional development days, conference presentations, articles, electronic portfolio. How will [list members of campus, discipline/program, or system community] be included or informed of the dissemination?	1 paragraph
Budget	How much money will the entire project require? What resources, equipment, or other funding are you requesting from other sources? How did you arrive at this budget? Does the budget include up to 12% for administrative overhead (if required by your college or university)?	½-1 page
Total		5–8 pages

Budget Narrative and Budget Summary

Please use the table below to organize your proposal's budget information, or create your own grid using the budget categories below. Write a brief description of each budget item, then attach a brief (less than one-page) budget narrative describing and justifying each item in detail. See Budget Preparation Guidelines in Appendix 1.

Budget Summary

Budget Category	Brief Description	Request	Matching Funds*	Total Budget
PI (s) Name(s):				
Other Faculty Stipends Name(s):				
Travel				
Student Stipends				
Equipment				
Materials/Supplies				
Other				
TOTAL				

* You must identify the source of any matching funds.

Certification Signatures

Based on the criteria for eligibility in the *CI-STEP* Award Guidelines, I am eligible to apply. I understand and agree that a written final report, including how the objectives and/or goals have been achieved, is due as stated in the Guidelines. I will provide a copy of my report to the CI-STEP Project Review Committee, and the Head of the Department. I understand that unless there exists a law characterizing some portion of the information submitted as private, proposals will be treated as public information on submission in accordance with the Data Practices Act.

Applicant Signature _____ Date _____

Approving
Administrator Signature _____ Date _____

Budget Preparation Guidelines

Budget Category	Guidelines & Policies
Project Manager(s)	**Funding Requested:**
Other Faculty Stipends	Stipends up to a maximum of $500 for any one incident may be requested for faculty partners on this proposal. Substantial collaboration between faculty members should be addressed by separate proposals for each individual. The separate proposals should clearly identify the collaborating partners.
Travel	Costs directly related to the project or associated with collaborating with other employees (meals, lodging, and mileage) will be considered. Requests for travel to attend conferences or training institutes are rarely funded. Professional development funds may be used as match. IDENTIFY SOURCES OF ALL FUNDS.
Student Stipends	Estimate number of student hours that will be paid (at campus rates) for assistance directly related to the project. Students may also receive nominal stipends for non-classroom activities related to the grant project. IDENTIFY SOURCES OF ALL FUNDS.
Equipment	Equipment, hardware and software that is directly relevant to the proposed project will be considered. Requests whose budgets are primarily for equipment will ordinarily not be funded without extraordinary justification. Equipment is generally worth $5,000 or more and has a usable life of at least two years. IDENTIFY SOURCES OF ALL FUNDS.
Materials/Supplies	Special project supplies, which may include printing, copying, postage, long distance telephone. IDENTIFY SOURCES OF ALL FUNDS.
Other	Other costs directly related to your project. IDENTIFY SOURCES OF ALL FUNDS.

APPENDIX 4
CI-STEP MINI GRANT PROGRESS REPORT

Today's date:

This progress report has been developed to organize and aid each mini grant in their mission to improve STEM graduation at IUPUI and IvyTech. The information provided in this document will be used for internal purposes as they relate to the CI-STEP grant's progress in the last fiscal year and may appear in summary form in the 2011–2012 annual report to the NSF. Please take as much space as necessary to answer each question as fully as possible.

Title of mini grant:

Name of all investigators listed on your mini grant:

Start date of mini grant:

Please list the top 3 objectives of your mini grant:

1.

2.

3.

Please describe in detail, the progress that has been made to date:

Please list any preliminary results/conclusions/outcomes to date:

Please detail future plans regarding the mini grant:

Please provide any supplemental materials that have been generated as they relate to the mini grant's efforts (i.e. pictures, handouts, fliers, website etc.). These can be attached to this document and listed here for accounting purposes.

The CI-STEP grant personnel thank you for your service to our mission and for your time and effort in accomplishing it. Please do not hesitate to contact us with any questions or concerns.

8

Applying the CACAO Change Model to Promote Systemic Transformation in STEM

Anthony Marker, Patricia Pyke, Sarah Ritter, Karen Viskupic,
Amy Moll, R. Eric Landrum, Tony Roark, and Susan Shadle

CONTEXT

Since its inception in the Middle Ages, the university classroom can be characterized by students gathered around a sage who imparts his or her knowledge. However, the effective classroom of today looks vastly different: First-year engineering students not only learn basic engineering principles, but are also guided to consider their own inner values and motivations as they design and build adaptive devices for people with disabilities; students in a large chemistry lecture work animatedly together in small groups on inquiry-based activities while an instructor and teaching assistants circulate and guide their learning; students learning differential equations practice explicit metacognitive skills while problem-solving in class. Even though educational research, especially research that is targeted at STEM disciplines, demonstrates what most effectively engages students and supports their learning, many of today's classrooms look much like they did a century ago, with a professor delivering a primarily one-way lecture and students passively sitting in seats bolted to the floor. At this juncture in history, colleges and universities face a public call to engage a more diverse representation of students in effective learning, persistence, and degree attainment, and to do so economically and efficiently. It is essential that institutions draw upon methods demonstrated to effectively increase student learning and success. Educational researchers have thoroughly explored the "basic" science in this area, and a body of literature documents effective evidence-based instructional practices, hereafter referred to as EBIPs.

Although EBIPs are well documented, we know far less about how to shift faculty practice and institutional culture to catalyze widespread adoption of these practices. "Applied research" is the current frontier, as propagating EBIPs has proven remarkably challenging, whether across institutions, across campus, or even down the hall in the same department. The National Science Foundation (NSF), a driving force and primary sponsor of STEM education research, has called for wider propagation of EBIPs. NSF's solicitation for the Widening Implementation and Demonstration of Evidence-Based Reforms (WIDER) program notes that "Despite the myriad advances in STEM teaching and learning know-how, it is the sense of policy makers and practitioners (and evident in accounts published in articles in academic journals) that highly effective teaching and learning practices are still not in widespread use in most institutions of higher education" (NSF, 2013, para. 67).

For this reason, identifying and assessing effective change strategies has moved to the forefront in STEM education, as evidenced by increasing scholarship activity in this area. Higher education researchers are exploring networks and other organization-level dynamics, such as "mutual adaptation and social movements [that] create ownership, sustainability, depth of adoption and spread" (Kezar, 2011, 241). Discipline-based education research has been a focus (National Research Council, 2012), and disciplinary societies are investing in propagating EBIPs. For example, since 2002 the American Physical Society and The American Astronomical Society have joined with the American Association of Physics Teachers to support, with NSF assistance, training on effective teaching for new physics faculty (AAPT, 2014). Similarly, the American Chemical Society and Research Corporation for Science Achievement provide Cottrell Scholars Collaborative (CSC) workshops for new faculty to "promote transformative change through the exploration of new pedagogies and the dissemination of proven methods. . . ." (CSC, 2014, para. 1). In geosciences, the On the Cutting Edge professional development program managed by the National Association of Geoscience Teachers (NAGT) has provided training and resources for early-career and experienced instructors through virtual and in-person workshops since 2002 (NAGT, 2015). In a special issue on transforming STEM education, guest editors for the *Journal of Engineering Education* noted that the prevailing focus of STEM educators has been on course- or curricula-level changes, and suggested new discussion "has laid some foundation for others to take the next steps and fully launch into systemic inquiries, studies and analyses of the complexities of educational transformation" (McKenna, Froyd, & Litzinger, 2014, 189).

CACAO CHANGE FACILITATION MODEL

At Boise State University, we are engaged in a project that seeks a complex, systemic solution to widespread adoption of EBIPs. This ambitious three-year project aims to identify and reduce institutional barriers to EBIP adoption across more than a dozen departments. The project was initiated in response to the NSF WIDER invitation to propose and test models to effectively support broader propagation of EPIBs and to achieve an ultimate outcome of increasing student success. Our project, WIDER PERSIST—Promoting Education Reform through Strategic Investment in Systemic Transformation, asks: Can we apply a change facilitation model from the business world to implement EBIPs more widely throughout a higher education institution? The facilitation model we chose recognizes instructional practice as only one element of the instructional climate. Other elements include institutional policies on workload and tenure and promotion, department traditions, social networks, institutional structures such as centers for teaching and learning, and faculty associations, institutional leadership, facilities, resources and other variables. Another key element of this model is that it is consistent with and allows us to leverage pedagogical transformations already underway.

The model, Dormant's **C**hange, **A**dopters, **C**hange **A**gent, **O**rganization (CACAO) model (Dormant, 2011), is a synthesis of Rogers' work (2003) on the diffusion of innovations (passive) and the work of Kotter (1990) on the purposeful implementation of designed changes (active). Dormant's model does the important work of helping us integrate and apply these concepts. She combines the approaches suggested in Rogers' work, which tends to look at change from the bottom or middle and up, and Kotter's work, which looks at change from the top down, into a single model. The model enables people using it to develop customized and purposeful change plans that take into consideration the:

- Benefits and drawbacks of the change itself
- Audience (adopter) characteristics
- Stages people go through in accepting or rejecting a change, and appropriate strategies for each stage to smooth adoption
- Leadership support and social networks that allow the group to find the right change champions
- The change agent's relationship to the change
- The creation of a well-rounded change team that is both proactive and responsive

The CACAO model provides a series of steps and strategies to guide a team toward achieving a particular change. We describe how we have applied several specific aspects of the model and our year one results. The four dimensions around which the model is organized are Change, Adopters, Change Agents, and Organization:

Change: First, the model dictates the value of collecting information about how adopters view the change. Dormant specifies the need to examine five characteristics. This examination, when complete, provides a profile that illustrates how likely adopter groups are to resist the change, and the areas in which resistance is likely to occur. As a result, the change profile provides a way of anticipating and mitigating resistance by developing strategies that make the most of the change's strengths and counteract the change's weaknesses. Those characteristics are:

Relative Advantage: the extent to which the change offers adopters advantages over the old way of doing things

Simplicity: the extent to which adopters can understand the change

Compatibility: the extent to which the change is consistent with adopter past practices

Adaptability: the extent to which adopters can adapt the change to fit local conditions

Social Impact: the extent to which the change will have little or no impact on existing social relations of the adopters.

Adopters: Second, the model looks at the stages of adoption that intended adopters typically go through when considering whether or not to implement a change. It specifies the importance of matching strategies to stages, and then provides specific strategies to most efficiently address each adoption stage (Table 1). The model further suggests that different adopter sub-groups, in this case different academic departments and groups within departments, are likely to be in different stages of adoption, mandating tailored strategies for each group.

TABLE 1. Strategies to Support Adopters

For adopters entering this age	Strategies to support adopters in this stage
Awareness	Advertise (brief)
Curiosity	Inform (detailed)
Mental Tryout	Use demonstrations
Hands-on Tryout	Provide training
Adoption	Provide support

Change Agents: Third, the model offers prescriptions for putting together an effective leadership team that includes members with expertise as organizational sponsors, content experts, change experts, grant experts, data collection and analysis experts, communication experts, training experts, and others as various needs arise.

Organization: Fourth, the model helps elucidate how to identify and manage layers of organizational hierarchy and then leverage networks of people for different roles during change implementation. The model identifies as particularly valuable people who can fulfill the roles of leadership sponsors, early acceptable innovators, opinion leaders, and traditionalists as groups that can potentially provide separate perspectives and valuable contributions. Identifying people who fit these roles and then using their contributions when and where they can most benefit the project is a crucial aspect of the change model.

CHANGE ANALYSIS: DEFINING AND UNDERSTANDING THE CHANGE

In the first year of our project, we have worked to define the change we seek and have worked with adopters to lay the groundwork for successful institution-wide change in the subsequent years of the project. One of the first tasks of the leadership team on our project was to define and communicate the intended change by developing a vision statement. This was important for two reasons: First, it provided a target against which to judge progress; and second, it served to guide task and strategy prioritization. The goal, in the case of the WIDER PERSIST project, was to increase the rate of implementation of EBIPs among university STEM faculty by directly supporting faculty and changing the culture surrounding teaching practices. By focusing on changes in the instructional culture, the project is able to encourage systemic changes, rather than strategies that simply change individual faculty behavior. Although cultural change requires a slower adoption process, it ultimately encourages sustainable practices in the long-term. Our WIDER PERSIST leadership team expressed the vision as an "end state," a new norm toward which the campus could collectively progress. The vision is that:

> The culture of teaching and learning at Boise State University will be characterized by
>
> - on-going exploration and adoption of evidence-based instructional practices,
> - faculty engaged in continuous improvement of teaching and learning,
> - dialogue around teaching supported through a community of practice, and
> - teaching evidenced and informed by meaningful assessment.

We believe the fulfillment of this vision will enhance our learning-centered culture and result in increased student achievement of learning outcomes, retention, and degree attainment; especially among underrepresented populations.

In order to both introduce the proposed change to faculty and to collect information from them about their view of the goals, we undertook extensive data collection early in the first year of the project. Doing so has informed the development of departmental change profiles; these profiles assisted us in evaluating progress and prioritizing decisions. As described earlier, when we introduced the CACAO model's four dimensions of change, change profiles provide information about perceived strengths and weaknesses of the change that might lead adopters to resist or embrace adoption. To this end, we held 17 one-hour focus groups with the staff and faculty of academic STEM departments, as well as with groups of department chairs and deans, ultimately involving a total of 194 participants. During the focus groups, participants were introduced to the vision and completed a questionnaire in which they identified and listed factors that either supported or opposed the change for each of five characteristics of change adoption (Table 2). Participants were given 5–7 min per characteristic to independently record their thoughts, which were then discussed as a group.

TABLE 2. Change Protocol: Faculty Discussion Group of the Strengths and Weaknesses of the End Etate (Vision)

Factor	Discussion Prompt
Relative Advantage	1a. Ways in which this end state is advantageous to me/my department
	1b. Ways in which this end state is disadvantageous to me/my department
Simplicity	2a. Features of our current environment & practice that make this end state easy/simple to attain and/or maintain
	2b. Features of our current environment & practice that make this end state hard/complex to attain and/or maintain
Compatibility	3a. Ways in which the end state is compatible with what I already do
	3b. Ways in which the end state is incompatible with what I already do
Adaptability	4a. In what ways might the end state allow for flexibility and individual choice (while still achieving the vision)?
	4b. In what ways might the end state limit flexibility and individual choice in order to achieve the vision?
Social Impact	5a. How will the new end state positively impact my relationships (with colleagues, with students, with administrators, etc.)?
	5b. How will the new end state negatively impact my relationships (with colleagues, with students, with administrators, etc.)?

We collected data from the faculty focus groups, which resulted in the compilation of a qualitative dataset with 1,755 drivers (positive factors) and 1,605 restrainers (negative factors) for change. The faculty results provided us with a universal set of characteristics as well as data to develop profiles and priorities for individual departments.

After our team collected the data, four researchers independently coded the barrier data according to an organizational change analysis model intended to identify the root causes of performance gaps between current practices and our envisioned goal, Gilbert's (1978) Behavior Engineering Model (BEM). The BEM (Table 3) is a 2 x 3 matrix which divides the causes for performance gaps into two main sources (rows), those originating in the environment, and those originating with the user.

For each of those sources, the model provides three types of causal areas (columns): information, instrumentation (tools), and motivation. Causes appearing in the environment are more directly under control of university leadership and can be easier to address compared to those that reside in individual adopters. Our team further categorized the causes that surfaced during our analysis as 18 commonly perceived themes (Figure 1). The majority of these themes have to do with issues of time, alignment to current assessment and metrics, classroom autonomy, resources, research-teaching balance, and institutional reward. These barriers align well with those that other research studies have previously identified and documented (Brownwell & Tanner, 2012; Henderson & Dancy, 2007; Walczyk, Ramsey, & Zha, 2007). Importantly, having the local data for our institutional context has provided the WIDER leadership with information we have used to begin devising appropriate support strategies for adopters by removing obstacles. These themes also provide fodder for discussion within departments about the barriers that impact local EBIP adoption.

TABLE 3. Behavior Engineering Model

	Lack of Information	Lack of Tools	Lack of Motivation
Causes originating in the **Environment**	✓ Data ✓ Expectations ✓ Feedback ✓ Clarity	✓ Resources ✓ Technology ✓ Space (classrooms) ✓ Tools ✓ Support ✓ TAs/Instructional support	✓ Consequences ✓ Rewards ✓ Incentives
Causes originating in the **Person**	✓ Knowledge ✓ Skills	✓ Physical capacity (incl. time) ✓ Mental capacity ✓ Flexibility ✓ Resilience	✓ Motives ✓ Affect ✓ Work habits ✓ Drive

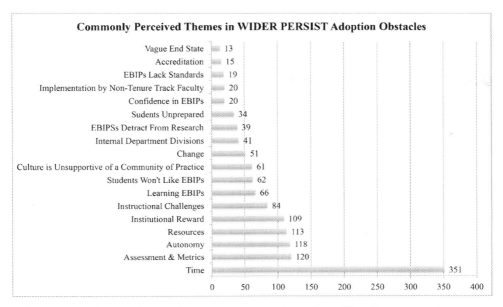

FIGURE 1: Barriers to change. Bars represent total number of faculty comments categorized in that theme.

Fortunately, in addition to change obstacles, there are very often positive drivers that encourage change. Our team is currently in the preliminary stages of analyzing these drivers. As there seems to be less research on drivers in the literature that there is on barriers, this analysis has the potential to contribute methods for accelerating change by supporting such drivers. At this early stage of analysis, the commonly recurring themes are:

- Increased opportunities for research
- Recognition of resources in place, e.g., Center for Teaching and Learning
- Enthusiasm about sharing ideas within and across departments and establishing or continuing development of communities of practice (Murray, Higgins, Minderhout & Loertscher, 2011)
- Improved student outcomes—learning, retention, graduation
- Potential for better prepared students in the classroom (engaged, participatory, and background knowledge) and workplace
- Professional recognition—becoming model departments at the university and national level

Additionally, there are a few themes that occur as both drivers and barriers. For example, potential adopters see "research" as a barrier since implementing the change will pull them away from discipline-specific research. However, at the same time, they demonstrate enthusiasm about the potential for new EBIP-driven research and grant opportunities. "Resources" is another theme that occurs in both the driver and barrier data. In the barrier data, adopters perceive a lack of resources ranging from monetary to lab equipment. In the driver data, adopters mention currently available resources such as the Center for Teaching and Learning. These perceptions of resources demonstrate both institutional and personal needs for support. Another example of a theme occurring in both the driver and barrier data is "communities of practice." In the barrier data, adopters' perceptions of communities of practice are either that the institutional culture does not support communities of practice, or that adopters are not interested in participating in communities of practice. In the driver data, adopters showed enthusiasm for participation in communities of practice as well as suggesting that the WIDER PERSIST project demonstrated an institutional interest in creating communities of practice around EBIPs.

As a next step, we have also designed an instrument based on the CACAO framework to explore department-level distributions of faculty across the adoption process. A discussion of this adopter analysis is beyond the scope of this paper and will be reported in future publications. Together, these analyses position the project team to respond to results by addressing barriers and supporting drivers. See Figure 2.

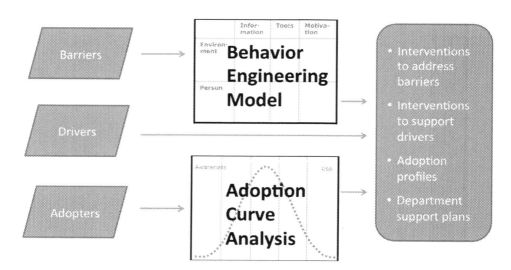

FIGURE 2. Addressing Barriers and Supporting Drivers

RESPONDING TO RESULTS

In an ideal scenario, one might prefer to have collected and thoroughly analyzed results before using them as the basis for action. However, within the CACAO model, data collection is actually part of making change happen. Focus group participants (who are prospective adopters) have demonstrated interest in the next steps. Therefore, it made sense for our team to respond to results as they came in and to refine the analysis as we progressed. We have been and will continue to respond to results in several ways.

In response to early results, we designed the Current Instructional Climate (CIC) survey. We used this instrument to collect information about faculty perceptions regarding the support for various aspects of teaching (valuing and promoting teaching, institutional conditions, unfettered teaching culture, and teaching-research balance) and in the future it will allow us to measure change in faculty attitudes. We constructed all of the items in the survey based on responses (both positive and negative) that emerged from collecting data in change conversations. For example, respondents in our change conversations indicated that the lack of appreciation for teaching in hiring decisions is a barrier to achieving the widespread adoption of EBIPs. In response, we crafted the following item for the CIC, answered using a seven-point semantic differential format: "I believe that the campus culture does not value teaching ability in hiring decisions" to "I believe that the campus culture does value teaching ability in hiring decisions." The instrument is designed to be directly sensitive to the particular barriers and drivers cited by faculty in the change conversations because CIC items were derived directly from faculty member comments. Further, we've administered our CIC instrument along with Western Michigan University's Postsecondary Instructional Practices Survey (Beach, Henderson, Walter & Williams, n.d.), which provides complementary information about how faculty perceive their current teaching practices.

Another way we've used the results is to look carefully at barriers and drivers to achieving our sought-after change. Doing so has allowed our project team to identify strategies that we think will help address particular barriers or leverage specific drivers. Several examples of this approach are summarized below (Table 4).

TABLE 4. Responding to Barriers and Drivers

Barriers	Drivers	Planned Strategy
Time		Provide teaching reductions for course redesigns; provide direct faculty development support
Uncertainty about EBIPs		Offer workshops tailored to EBIPs in particular disciplines; provide discipline-specific references and resources
Resources of classroom or materials		Influence university classroom planning/renovation process
Lack of incentive and recognition		Provide "toolkit" to tenure and promotion committees
	Interest in communities of practice	Support specific opportunities for inter and intradepartmental conversations around teaching
	Support for the outcome of increased student success	Create a "data team" to work with institutional research in order to help departments better understand how their students are doing within courses and in follow-on courses; support faculty assessment efforts
	Recognition	Create "faculty spotlight" videos to highlight faculty who are effectively implementing EBIPs

LONG-TERM VISION AND NEXT STEPS

As expressed in the vision statement, the ultimate reason for seeking change in instructional climate is to increase student achievement of learning outcomes, retention and degree attainment. To that end, a main focus of the WIDER PER-SIST leadership team effort and energy in the first year has been on working with faculty teams to implement EBIPs broadly across departments, working with university leaders to remove barriers and provide support, and putting in place systems for measuring progress. Future work and subsequent publications will describe our data collection and analysis in more detail and address the ways in which our data has been used to drive change. Involving institutions beyond Boise State University is also a major goal of the project, and the team welcomes contact from other institutions interested in applying the CACAO model on their campuses.

ACKNOWLEDGMENT

We gratefully acknowledge that this material is based upon work supported by the National Science Foundation under grant No. DUE-1347830. Any opinions, findings and conclusions or recommendations expressed in this paper are those of the authors and do not necessarily reflect the views of the National Science Foundation.

REFERENCES

American Association of Physics Teachers (AAPT). (2014). *Workshop for new physics and astronomy faculty*. College Park, MD: American Association of Physics Teachers. Retrieved from http://www.aapt.org/Conferences/new faculty/nfw.cfm

Beach, A. L., Henderson, C., Walter, E. M. & Williams, C. (n.d.). *Postsecondary instructional practices survey (NSF #1256505)*. Kalamazoo, MI: Western Michigan University.

Brownell, S. E. & Tanner, K. D. (2012). Barriers to faculty pedagogical change: Lack of training, time, incentives, and . . . tensions with professional identity? *CBE-Life Sciences Education, 11*, 339–346.

Cottrell Scholars Collaborative (CSC). (2014). *CSC New faculty workshop*. Retrieved from http://chem.wayne.edu/feiggroup/CSCNFW/cottrell-scholars -collabora.html

Dormant, D. (2011). *The chocolate model of change*. San Bernadino, CA: Author.

Gilbert, T. F. (1978). *Human competence: Engineering worthy performance*. New York, NY: McGraw-Hill.

Henderson, C., & Dancy, M. (2007). Barriers to the use of research-based instructional strategies: The influence of both individual and situational characteristics. *Physical Review Special Topics: Physics Education Research, 3*, 020102-1–020102-14.

Kezar, A. (2011). What is the best way to achieve broader reach of improved practices in higher education? *Innovative Higher Education, 36*, 235–247.

Kotter, J. (1990). *A force for change: How leadership differs from management*. New York, NY: Free Press.

McKenna, A.F., Froyd, J., & Litzinger, T. (2014). The complexities of transforming engineering higher education. *Journal of Engineering Education, 103*, 188–192.

Murray, T.A., Higgins, P., Minderhout, V., & Loertscher, J. (2011). Sustaining the development and implementation of student-centered teaching nationally: The importance of a community of practice. *Biochemistry and Molecular Biology Education, 39*, 405–411.

National Association of Geoscience Teachers (NAGT). (2015). *On the cutting edge: Strong undergraduate geoscience teaching.* Retrieved from http://serc.carleton.edu/NAGTWorkshops/index.html

National Research Council (NRC). (2012). *Discipline-based education research: Understanding and improving undergraduate learning in science and engineering.* Washington, DC: Author.

National Science Foundation. (2013). Widening implementation and demonstration of evidence-based reforms (WIDER). *Program Solicitation #NSF-13-552.* Washington, DC: Author. Retrieved from http://www.nsf.gov/pubs/2013/nsf13552/nsf13552.htm

Rogers, E. (2003). *Diffusion of innovations* (5th ed.). New York, NY: Free Press.

Walczyk, J. J., Ramsey, L. L. & Zha, P. (2007). Obstacles to instructional innovation according to college science and mathematics faculty. *Journal of Research in Science Teaching, 44,* 85–106. doi:10.1002/tea.20119

ABOUT THE AUTHORS

Anthony Marker, Ph.D., is an Associate Professor of Organizational Performance and Workplace Learning at Boise State University in Boise, Idaho.

Patricia Pyke is the Director of Research Development at Boise State University in Boise, Idaho.

Sarah Ritter is a Doctoral Student Research Assistant and McNair Scholars Interim Program Coordinator at Boise State University in Boise, Idaho.

Karen Viskupic, Ph.D., is an Assistant Research Professor and Education Programs Manager at Boise State University in Boise, Idaho.

Amy Moll, Ph.D., is the Dean of the College of Engineering and a Professor of Materials Science and Engineering at Boise State University in Boise, Idaho.

R. Eric Landrum, Ph.D., is a Professor of Psychology at Boise State University in Boise, Idaho.

Tony Roark, Ph.D., is the Dean of the College of Arts and Sciences and Professor of Philosophy at Boise State University in Boise, Idaho.

Susan Shadle, Ph.D., is the Director of the Center for Teaching and Learning and a Professor of Chemistry and Biochemistry at Boise State University in Boise, Idaho.

9

Review of the Undergraduate Science Curriculum at the University of Queensland

Michael E. McManus and Kelly E. Matthews

The University of Queensland (UQ) is a large public university located in Brisbane, Australia, that has a student body of 48,804: 36,219 undergraduates, 8,224 postgraduate coursework, and 4,361 research higher-degree students. It graduates nearly 7,000 undergraduate, 3,300 postgraduate coursework, and 700 research higher-degree students per year and has a research budget of ~$380 million per annum. In the last two decades, UQ has undergone dramatic change that has seen it go from a rather provincial institution into a research-intensive university with a truly international focus. This level of performance is borne out by UQ being consistently ranked in the top 100 of all three major global university rankings (http://www.uq.edu.au/about/docs/strategicplan/Strategic-Plan2014). This inflection point in our 100-year history is due to the recognition of four important drivers of a modern research university: (1) to educate students to understand the rapid advances in science and technology, as well as the social and historical perspectives required to give them the creativity and wisdom to apply their knowledge wisely; (2) to provide students with a rich understanding of the cultures, histories, languages and religions of the world so they are capable of appreciating the values and practices of a global society; (3) the crucial role higher education plays in the development of new knowledge; and (4) the significant role UQ can play in the growth and economic development of the State of Queensland, the nation and region as a whole. This brief summary captures how curriculum renewal of the science degree (BSc) meshes with the ambitious agenda of a renaissance university.

The key drivers were also informed by the rapid pace of globalization and the need to engage with the international community to sustain a leading position in the four areas listed above. Such a focus has seen international students from 142 different countries become ~25% of our student body, which complements the rich tapestry of >200 nationalities in modern Australia, as well as the rich heritage of our Indigenous peoples, which encompasses >250 different language groups with ~50,000 years of continuous habitation of Australia. Our international engagement further emphasizes the importance we place on diversity in creating a knowledge-centered learning community. In concert with these drivers, there were a range of other pressures influencing what we teach and how we deliver our educational programs. Like most institutions, we were actively asking ourselves whether we were truly improving student learning and were we transitioning fast enough from an instruction-driven to a learning-based paradigm. It was apparent that the didactic lecture still survived as the predominant mode of education, with summative examinations remaining as the key assessment tool. The challenge for us was to move to a learning paradigm that is: (1) learning outcome driven, (2) more mobile, (3) more personalized to match student and employer goals, and (4) universally accessible on a global stage. Another pressure was the acceleration toward universal access to tertiary education, which was changing the abilities and expectations of the student body. Unfortunately, this was happening at a time of reduced government support for higher education, resulting in students having to pay a greater proportion of the overall cost of their education.

The rapid transformation of UQ was powered by the planning and ultimate development of eight research institutes and a substantive reorganization of our academic structure from 16 faculties to seven. It was catalyzed by the synergistic interaction between our former president, John Hay, Atlantic Philanthropies Irish-American founder Chuck Feeney, and the former premier of the State of Queensland, Peter Beattie. In all of the upheaval and change, we were acutely aware of the fact that the most significant contribution a university can make to the nation and its global community is through the quality of its graduates. This backdrop of change sent a powerful message to the University's Academic Board and Executive Deans of Faculties that the review of academic programs needed to be strengthened and made bolder than previously was the case. In this context, and as part of its quality assurance approach to learning, UQ instituted a process to review all its major degree programs every seven years.

We held an international conference from November 18–19, 2004 on our St. Lucia campus entitled "Science Teaching & Research: Which Way Forward for Australian Universities?" (http://espace.library.uq.edu.au/view/UQ:152849). The purpose of this conference was to start to prepare the faculty and the University as a whole for the impending review of the BSc degree in 2006. The recommendations stemming from this conference echoed the innovative approaches to science education being championed by the U.S.A National Science Foundation (NSF) and the Boyer Commission Report entitled "Reinventing Undergraduate Education: A Blueprint for America's Research Universities" (http://naples.cc.sunysb.edu/Pres/boyer.nsf/). The former director of the NSF, Rita Colwell, and the lead author of the Boyer Report and then-president of the State University of New York at Stony Brook, Shirley Kenny, gave plenary presentations at this meeting. The conference reinforced the importance for all students at a research-intensive university to experience something special in their education. Like major NSF educational programs and the Boyer Report, the summary of the conference concluded "that students at a research university have a right to expect that they will be continually challenged intellectually and will be part of a community of learners." It went on to identify that the best way to do this is within a research-based learning system where several methods of interaction take place between lecturer and student, rather than simply the traditional "large lecture theatre with a non-accessible academic at the front." Inquiry/discovery based learning inherently implies exchange of elements in both directions between lecturer and student, to the mutual benefit of both. Students also have the right to expect that the outcome of their educational experiences at a research university will equip them, not only to be worthwhile citizens, but also with the knowledge, skills and habits of the mind to make a significant contribution to their chosen field of endeavor, and to be internationally competitive. Finally, students have the right to expect a significant amount of contact with academic researchers and scholars, who will take the role of advisors and mentors. Importantly, academic staff should not be constrained in this role because of pressures due to other aspects of their university commitments (http://espace.library.uq.edu.au/view/UQ:152849).

In late 2005, we began in earnest to prepare for the review of the BSc degree. In addition to taking on board the above recommendations, we were also guided by a 2004 report from the Faculty of Arts and Sciences at Harvard University entitled "A Report on the Harvard College Curriculum Review." A book by the former president of Harvard University, Derek Bok, entitled "Our Underachieving Colleges: A Candid Look at How Much Students Learn and Why They Should be Learning More" plus "The Future of Higher Education: Rhetoric, Reality, and the Market" by Newman et al (2004) were also very influential in shaping our higher level thinking. However, in order to connect more fully with faculty members at the scientific level, we purchased 120 copies of a 2003 National Research Council of the National Academies report entitled "Bio 2010: Transforming Undergraduate Education for Future Research Biologists." A copy of this report was distributed to all members of the committees listed below.

At first this report was a lightning rod to the mathematicians, chemists and physicists. Further, the fact it was coming from a dean who was a pharmacologist with a strong biomedical background heightened the level of anxiety. To their credit, they actually read the report, which sent a very powerful message about the enabling sciences and championed the call for all future biological scientists to have strong backgrounds in mathematics, chemistry and physics.

It was as if they had been down the road to Damascus, and we were relieved that the review process was back on track. It was necessary to go to this level of education because most faculty members had not received training in how to teach or plan a curriculum. Bio 2010 also sent a powerful message about the value of biologists, mathematicians, chemists and physicists working together and reinforced our university-wide initiative to encourage faculty to complete a Graduate Certificate in Higher Education, which includes a major project based on their own teaching. The Bio 2010 report encapsulates nicely what Handelsman, Miller, and Pfund (2007) say about the importance of scientific teaching: "Embedded in this undertaking is the challenge to all scientists to bring to teaching the critical thinking, rigor, creativity, and spirit of experimentation that defines research."

The review of the BSc degree involved significant outside input; the panel was composed of three external experts to the University, and two senior members from within UQ. The chair was an eminent Australian scientist from the University of Adelaide, with the other two prominent external members coming from the University of Michigan and the University of Newcastle, Australia. Their charge was to critique and make recommendations on the student lifecycle of the BSc degree from recruitment, transition to university, curriculum structure and content, assessment, undergraduate research experience, extra-curriculum, attainment rates, graduate employability, industry engagement, and alumni formation. The review panel met on the St. Lucia campus from November 20–24, 2006, and conducted an extensive program of interviews and discussions with stakeholders who were identified as having an interest in the BSc. Prior to this, each member of the committee received a joint submission from the three faculties teaching in the BSc: Biological & Chemical Sciences (teaches ~70% BSc degree), Engineering, Physical Sciences & Architecture (teaches ~20%), and Social & Behavioural Sciences (teaches ~10%). The panel also considered 22 submissions from academics around the university and interviewed 56 internal faculty members, 16 students and eight external government and industry representatives. From one review cycle to the next, different areas are given prominence, but the overall aim remains the same: to subject our efforts to rigorous external scrutiny and to ask whether we are truly improving student learning.

The BSc review, as we moved into 2006, quickly focused on the content and structure of the curriculum, primarily because it had ballooned in to a smorgasbord of offerings containing a large number of courses (350) that underpinned 40 majors. It was a curriculum where flexibility and choice reigned supreme. The review presented an opportunity for faculty to reassert their ownership of the curriculum, and to mold a program that would prepare students for the significant challenges of the 21st century, where most of the advances in science will come at the intersection of disciplines. First and foremost, the BSc needed to sit within the overarching educational goals of the university. To help guide this general discussion on the BSc degree from a whole-of-university perspective and to get leaders to ask the right questions, we used the following from Newman et al (2004):

- What knowledge do we expect students to acquire to be productive and effective in the workforce and as citizens?
- What does it mean when we say that students are prepared for successful participation in the economy and society?
- What knowledge and skills do our students currently have when they leave high school or enter from the workforce?
- What skills and knowledge are necessary for all students regardless of a major?
- Which teaching methods do we use, and are they producing successful outcomes for all students?
- What roles can technology play in improving teaching and learning? What roles can it play in assessing learning?
- What assessment tools should be used to demonstrate mastery of agreed-upon academic goals and knowledge levels?
- What are the priorities and appropriate balance of the faculty role at UQ among teaching, research, advising and service?
- Which bachelor of science degrees at other institutions are succeeding in achieving high levels of learning?

An emphasis on the structure and content of the curriculum, and on what and how it is taught focused the minds of the different faculties and schools involved in the delivery of the BSc—the key reason being the focus on income generated from teaching into the BSc degree program. This was the most prominent thought bubble at the beginning, but over time, through a very inclusive process of consultation, it was marginalized. To cover the student life cycle we set up the following committees: (1) Steering Committee; (2) Structure of the BSc Committee; (3) Pedagogy Committee; (4) Student Experience Committee; (5) Honors and Careers Committee; (6) Government and Industry Committee; and (7) Alumni Committee. The Steering Committee was made up of executive deans and heads of school whose academics taught into the BSc and it set and drove the agenda. Like the Harvard review, it was asking: "Did the BSc degree convey a coherent vision of what the university was trying to achieve? Did the pieces mesh in sensible ways? How did the curriculum intersect other parts of the overall undergraduate experience?" To further assist the co-chairs of each committee, the Steering Committee developed a more explicit list of questions covering their specific charge. The purpose of these questions was to initiate the discussion but to in no way restrict the sphere of action, expression or influence of the respective committees. It was interesting to observe how different committees exercised these degrees of freedom.

Following a lead up in 2005, where open discussions were held around campus on global science education, each committee was invited to meet as frequently as possible during the first eight months of 2006, prior to the deadline for preparing the final submission. The Steering Committee was the first port of call for reports from the other committees, which were subsequently amalgamated and their recommendations delivered to two retreats. One of these retreats was held for two days, approximately 100 Km away from the St. Lucia campus, and along with the other retreat, was chaired by a professional facilitator. Although this may appear an expensive undertaking, it was apparent that we were trying to shape the future of the BSc and science education more generally in the State of Queensland, with faculty members from three faculties who did not know each other and had never shared a coffee or meal together. In forming the above committees, we gave a lot of attention to their membership and worked assiduously to ensure that younger faculty members were included. In shaping the discussion, we were cognizant of the fact that useful knowledge was doubling every two years and that the disciplines today may be very different in 2020. The futurologist Rodney Hill from Texas A&M University encapsulated this type of thinking when he asked, "Will any curriculum in any university in the top ten in the world in 2020 even resemble the curriculum of today's top ten universities?"

In addition to consulting faculty members, we also gave significant attention to seeking input from our students through surveys and inviting their representation on all the above committees. Initially, we reaffirmed Bok's (2006) principles that "the purpose of an undergraduate education is to foster generally accepted values and behaviors such as honesty and racial tolerance. Within this mandate, several aims are especially important: (1) ability to communicate; (2) critical thinking; (2) moral reasoning; (4) living with diversity; (5) living in a global society; (6) a breadth of interest; and (7) preparing global citizens." We fully appreciated that for advancement in today's world, students require more than the simple mastery of a body of knowledge. In reaffirming the above principles, we explicitly stated that our ambition is to instill in students wisdom and understanding, as well as knowledge and skills, the essence of a moral and ethical framework of education.

Students were given an opportunity to present the "student perspective" on the curriculum at each of the retreats. In all our discussions we emphasized that the BSc has at least three roles that are best explained by the three cohorts that exist within the student body: (1) those students who are present to obtain a general education; (2) those students who are positioning themselves to enter a professional degree (e.g., medicine, pharmacy, teaching, etc.); and (3) those students who see science as a worthwhile career and will proceed to Honors (4th year) and probably a PhD. The latter group comprised about 30% of students. Faculty members took very seriously their responsibility to show students, no matter their trajectory, the light on the hill in respect to all things science.

Like many other OECD countries Australia has been experiencing a downturn in its K–12 STEM performance. The most worrying sign coming from survey instruments like PISA, TIMSS and PIRLS is that Australia is not improving its performance; on just about every indicator it is slipping behind or remaining static compared to the top performing OECD countries (http://www.acer.edu.au). In this context it is important to see the BSc degree as part of a continuum spanning from K–12, university, and through to employment. If one considers the BSc degree in the light of a normal Australian lifespan, it is apparent that we only have a student in the undergraduate program for ~5% of their life, compared to the nearly 50% they spend in the workforce. This narrow window of opportunity highlighted the real responsibility to instill in students the knowledge, skills and habits of the mind for life-long learning. As the leading university in the state of Queensland it was apparent that most of the esteem in science education resided at the University of Queensland. Indeed it was clear from exchanges with high school representatives that the standards we set influenced the focus of the secondary school curriculum and the advice students were given about being university ready. During the review, we reaffirmed the entry requirements into the BSc being at least high school level English, intermediate mathematics (Maths B) and either chemistry or physics. We have also more recently awarded extra entrance points to students who have completed the highest level mathematics (Maths C). Regretfully, we are still beset by a problematic K–12 sector that is seeing expensive remedial teaching taking place.

The most significant outcome from the review was the enhanced collegial culture across the three faculties that brought different disciplines much closer together. This intangible gain was achieved through the hard work of the committees and its impact is often underestimated; it made the impossible possible and it is a great compliment to the faculty members involved in the review. There were many tangible gains: (1) reduction in the number of majors from 40 to 18; (2) stronger recognition that mathematics, physics, chemistry and biology are the enabling sciences that underpin our education philosophy; (3) introduction of two major first year courses entitled "Foundations of Science" and "Analysis of Scientific Data" to enhance the quantitative and information aspects of the degree; (4) a dramatic reduction in the number of first year courses (e.g. the number of biology courses was reduced from six to three); (5) a significant reduction in the number of second and third year courses to enable students to plan their program of study in a more rational manner and to simplify academic advising; (6) discipline-specific streaming to commence much later, in the second semester of second year; (7) a significantly enhanced focus on the undergraduate research experience; and (8) a renewed focus on teaching excellence, pedagogical advances and the science of learning. We also tabled a proposal during the review for an undergraduate research learning building to significantly enhance the quality of the environment for science students, which reflects current research and practice into how students learn.

The ultimate indicator of a successful review is whether the changes recommended have had any impact on student learning outcomes. The only way to do this is to foster a culture that is prepared to measure student learning in a realistic and meaningful manner that evokes general agreement (Bok, 2006; Matthews et al, 2012). The strong emphasis on quantitative skills in the 2006 review led to the development of an introductory mathematics-science interdisciplinary course, and the requirement that all BSc students complete a first level statistics course. To ascertain the impact of such an approach, the Faculty of Science (UQ changed its structure in 2008) funded teaching and learning grants under the priority area of building quantitative skills, which ultimately led to UQ leading a large national grant in the same area across science degree programs. As the trend data from 2008, 2011 and 2014 demonstrate, these efforts have significantly enhanced students' perceptions of quantitative skills (Figure 1). The data show that change can take years to impact students, even with sustained efforts and resources dedicated to this goal; 2014 students were still reporting levels of confidence and improvement lower than 75%. This is an important lesson to inform future BSc reviews; curriculum renewal to achieve aspirational graduate level learning outcomes is a long-term commitment. The need for similar data in other areas of the curriculum is axiomatic. Indeed, a faculty-wide mechanism is required to measure vertical and horizontal integration of recommended changes.

The word "curriculum" broadly defines the entirety of the student experience. The original Latin meaning of the word is "a race" or "the course of a race" that is derived from the verb *currere* meaning "to run/to proceed (http://en.wikipedia.org/wiki/Curriculum)." The responsibility for managing a curriculum review was as the meaning of the name conveys, like going to the races, as the variety of horses at the commencement was impressive. This is not necessarily unusual as all faculty members have experienced personally both the highs and lows of education, and higher education in particular. In other words, they

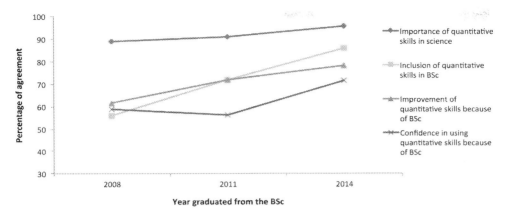

FIGURE 1. Percentage agreement from students for aspects of quantitative skills over time. Statistically significant differences existed between certain cohorts for quantitative skills within the following indicators: improvement ($F(2, 254.47) = 6.58$, $p = .002$); inclusion ($F(2, 251.80) = 15.13$, $p < .001$); and confidence ($F(2, 253.36) = 4.26$, $p = .015$).

have a healthy opinion on how the chips from such a race should fall. From ten years as executive dean at the University of Queensland, I (MEMc) would conclude that curriculum review is the most dangerous place for a dean to go. In pursuing the case for change, there was a burning need to manage up as much as down, as certain senior managers were often prone to overreact to noise in the change process. It was also a very enjoyable experience, especially when both young and old faculty members spoke passionately about their course and students. It was even more rewarding when students reminisced about their experiences and in the process lauded their teachers and the university. Similarly, positive feedback from representatives of government and industry about the quality of our graduates also gave reason for celebration. Our only wish is that all academics become blessed to see beyond their special interest course and see more clearly the universal: the whole student experience. The Belgian Nobel Laureate Maurice Maeterlinck once said, "At every crossroad on the road that leads to the future each progressive spirit is opposed by a thousand men appointed to guard the past (http://www.wisdomcommons.org/author/Maurice%20Maeterlinck)." Managing change in this digital age where knowledge can be delivered anywhere, any place, and at any time to remain competitive and relevant, will fundamentally influence curriculum. Healthy debate is ideal, but that debate must be future-looking.

REFERENCES:

Bok, D. (2006). *Our underachieving colleges: A candid look at how much students learn and why they should be learning more.* New Jersey: Princeton University Press.

Handelsman, J., Miller, S., and Pfund, C. (2007). *Scientific teaching.* New York: W. H. Freeman and Company.

Harvard University, Faculty of Arts and Sciences. (2004). *A report of the Harvard College Curriculum Review.* Cited and downloaded in mid-2005 from Harvard website; no longer available.

Matthews, K. E., & Hodgson, Y. (2012). The science students skills inventory: Capturing graduate perceptions of their learning outcomes. *International Journal of Innovation in Science and Mathematics Education, 20*[1].

National Research Council of The National Academies, Committee on Undergraduate Biology Education to Prepare Research Scientists for the 21st Century, Board of Life Sciences, Division of Earth and Life Studies. (2003). *Bio 2010: Transforming undergraduate education for future research biologists.* Washington DC: National Research Council of The National Academies.

Newman, F., Couturier, L., and Scurry, J. (2004). *The future of higher education: Rhetoric, reality and the risks of the marketplace.* San Francisco: John Wiley & Sons.

ABOUT THE AUTHORS

Michael E. McManus is an Emeritus Professor of the School of Biomedical Sciences and Centre for Advanced Imaging at the University of Queensland in Brisbane, Queensland, Australia.

Kelly E. Matthews is a Senior Lecturer of Higher Education in the Institute for Teaching and Learning Innovation and Faculty of Science at the University of Queensland in Brisbane, Queensland, Australia.

10

Key Elements to Create and Sustain Educational Innovation at a Research-Intensive University

Daniel Hastings and Lori Breslow

MIT is a research-focused university. As such, faculty are required to do cutting-edge research, and the internal reward system, centered on tenure, is built on that expectation. Nevertheless, MIT has introduced or modified a number of innovative ideas in education in the last several decades. This paper will focus on three of those initiatives. The first is the Undergraduate Research Opportunities Program (UROP), one of the Institute's oldest educational innovations. UROP began in 1969 as a way to involve undergraduates in faculty research; the vast majority of MIT undergraduates now enroll in at least one UROP by the time they graduate. The second innovation is Technology-Enabled Active Learning (TEAL), which changed the pedagogy used in MIT's first-year required physics courses from lecture/recitation to active learning. TEAL officially launched in 2003, and the majority of students currently take freshmen physics in the TEAL format. The third example is Conceive, Design, Implement and Operate (CDIO), an educational framework for teaching engineering that began at MIT in the late 1990s, and was established internationally as the CDIO Initiative in 2000.

The staying power of these three innovations is due to several elements they share in various combinations: a passionate faculty member, an initial pilot, support at the administrative level in the form of a department head or dean, substantial resources, the availability of expertise complementary to the faculty's domain knowledge, and, in some cases, the opportunity for formative assessment, particularly in the early stages of the project. We discuss each of these factors using the three examples cited above. We also describe the status of each project, and, finally, we offer some thoughts on why these innovations took root, their broader impact at MIT, and why institutional-level change has yet to take hold.

FACTORS NECESSARY FOR EDUCATIONAL CHANGE AT MIT

The first ingredient necessary to make substantial change to the existing educational paradigm at MIT is the leadership of a faculty member. The Institute has a strong faculty-driven culture, and the faculty feel they are the owners of the educational experience. In the three cases cited above, UROP, TEAL, and CDIO, there was an individual tenured faculty member who decided to devote extensive time and effort to start the innovation, implement it, and remain involved until it was institutionalized. Each of these faculty was passionate about the proposed effort and saw it as a way to significantly improve undergraduate education. That element of passion was necessary to convince others to buy into the change. It was also essential to provide the required energy and credibility to overcome the inevitable resistance from colleagues who were either unconvinced change was necessary, or who steadfastly supported the status quo.

For example, the physics faculty member who developed and championed TEAL, John Belcher, was frustrated by the relatively small number of students who were attending lectures, but, more importantly, he was troubled by the percentage of students who were failing freshmen physics (about 15% for first-semester mechanics and 10% for second-semester electricity and magnetism). He was aware of attempts at other universities to teach physics using active learning pedagogies, and along with physics senior lecturer Peter Dourmashkin, he was determined to experiment with so-called studio physics at MIT (Breslow, 2010).

For Margaret MacVicar, MIT's first dean for undergraduate education, the impetus to create UROP was the conviction that MIT's purpose was to "direct the best minds toward inquiries and enterprises concerned with the human condition" (Lerner, 1991). By providing students the opportunity to do research, UROP can set them on that path. And for Edward Crawley, then head of MIT's department of Aeronautics and Astronautics, and his colleagues who developed CDIO, motivation came from their view that engineering was a system and had to be taught as such. They believed that students who were to become successful engineers needed not only technical expertise, but also "social awareness and a bias toward innovation" (Crawley, Malmqvist, Östland, & Brodeur, 2007, p. 1).

In each of the three cases, the faculty member's initial step in moving toward full-scale implementation was to develop a pilot. Well-designed pilots teach something about what works and what does not so that the likelihood of success when the innovation is scaled up is increased. For UROP, one of the key questions was whether undergraduates could do any useful research with a faculty member. The initial pilot was small with only a handful of students and faculty. However, it showed faculty could use undergraduates in useful ways in their research and that undergraduates could learn something from the experience. It also suggested some boundary conditions that are still in place today: For example, a UROP has to be verified by the supervising faculty member as worthy of academic credit. This ensures that the UROP is not just "bottle wash-ing," but meets the same standards for academic credit as MIT courses.

Similarly, in designing TEAL, Belcher and Dourmashkin ran two "off-term" pilots in the fall semester 2001 and 2002, with two sections of about seventy students each. Those classes provided a number of useful lessons when TEAL went "on-term" with almost 800 students in the spring of 2003. The de-partment of Aeronautics and Astronautics experimented with several different pedagogies, means of student assessment, and changes in departmental norms around rewards for effective teaching, as Crawley and his colleagues developed CDIO.

The faculty champions, however, had to garner wider political and financial support if the reform was going to move beyond the pilot. As mentioned above, MIT faculty are responsible for the educational enterprise and a promising re-form can whither if the faculty champion steps away from it. At MIT, as in many other universities, power resides in the departments. Departmental level sup-port is absolutely essential over the life cycle of the innovation if it is going to be sustainable. That support can take several forms. For example, the vast majority of MIT faculty are so time poor that the passionate innovator can only succeed if time is made available for him/her to focus on the proposed transformation, which usually means relief from a regular teaching load. Or, the visible support of a department head or dean can offer some protection from critics. (This is what in the U.S. Air Force is called "top cover.") Thirdly, departmental support can ensure that space and administrative staff are made available.

In the case of TEAL, the department head and associate department head in the physics department backed the experiment and provided the necessary "top cover." TEAL faced significant opposition from a group of physics faculty who felt that only lectures could transmit the beauty of physics to students. (The students were not happy with the change to TEAL because they are required to come to class if they want to earn an A. Most other classes at MIT do not require attendance.) Yet departmental leadership, as well as the dean for undergradu-ate education, defended TEAL through its tumultuous first semester. Five years later, when the faculty champion wanted to move on, the department head made sure that the new faculty lead would continue TEAL in first-year physics.

In the case of CDIO, department leadership saw to it that the pedagogy was adopted across numerous undergraduate courses in the department. For UROP, the support was at the level of deans, as the home department was not that supportive. However, this had the positive effect of getting UROP adopted across the Institute.

Institute-level leadership has the power to provide funding. Academic units often operate with much of their resources fully committed. This means that any new—and often controversial—idea has to compete for resources against the established way of doing business. For an educational reform to even begin in a pilot phase, resources are necessary, particularly for additional staff, such as teaching assistants, instructors, experts in learning and assessment, and/or educational technologists (more on this below). As significant work on these initiatives is often done over summers, it is also important to underwrite the summer salary of the faculty who are involved, or they may be tempted to spend that time on the research for which they do have funding.

In the three cases cited, substantial resources were made available in their initial phase. TEAL benefited from two sources of funding that came to MIT in 2000: the d'Arbeloff Fund for Excellence in Education (named after the donor, the former chair of the MIT Corporation, and his wife), and iCampus, which was funded by Microsoft Research to support innovation in education technology. These two funding sources gave grants to TEAL via an internal proposal process that was not too onerous. The grants were used for summer salary for faculty and teaching assistants, as well as the salaries of some of the institutional level staff who collaborated on TEAL. Both the UROP pilot and the CDIO initiative were funded initially by large grants from outside MIT. The gift to launch UROP came from Edwin Land, the founder of Polaroid, while a foundation in Sweden initially sponsored the research that led to CDIO.

Each of the three reforms also benefited from a number of specialists employed at MIT. Faculty in an institution like MIT are chosen for their scientific, technical and humanistic scholarship; it is no secret that the vast majority of them have no scholarly background in education or educational technology. Ideally, then, the faculty, who have domain expertise, will have the opportunity to collaborate with educationalists, whose expertise is in the science of learning; educational technologists, who can develop effective software; instructional designers, particularly those with knowledge of online learning; and educational researchers, who can design and implement studies to assess the extent to which the innovation is succeeding in meeting the educational goals it was designed to achieve.

At MIT this kind of expertise resides in the Teaching and Learning Laboratory (TLL), the Office of Digital Learning (ODL), and the Office of Strategic Educational Initiatives (SEI), among other units. For example, ODL employs project managers who are typically supported partially with institutional funds and partially with funding associated with faculty-led projects. This model ensures their availability to work on different projects is flexible, and that they are responsive to faculty needs. TEAL and CDIO were helped substantially by collaboration with educationalists in TLL, who are themselves PhDs in STEM fields with additional expertise in teaching and learning. Educational technologists/

programmers were available through the Office of Educational Innovation and Technology (OEIT), now SEI.

In addition, TEAL and CDIO benefited from collaboration with educational researchers with specific expertise in the assessment of learning. A member of the initial TEAL project team, Technion Professor Yehudit Dori, is an internationally recognized expert in science education. She was involved in a study to compare TEAL with the lecture/recitation model of teaching during the semester that electricity and magnetism was taught in both formats (Dori & Belcher, 2005). Dori then collaborated with TLL researchers to follow a subset of these students to understand the long-term retention of E&M concepts for students in each condition (Dori, Hult, Breslow, & Belcher, 2007).

The d'Arbeloff and iCampus grants went over several years, which enabled the long-term development of TEAL over a period of several years. It is important that the resources were continued to be made available to the lead faculty so he could focus on gathering data to assess the impact of TEAL, and to use those data for continuous improvement.

CURRENT STATUS OF THE CHANGE INITIATIVES

The Undergraduate Research Opportunities Program is now more than forty years old. It has become institutionalized in that routinely over 85% of undergraduates participate in the program at some point in their undergraduate career. Many of the faculty see UROP as a way to interact more deeply with undergraduates (as compared to lectures in a class, for example), and it has earned wide support among the faculty. A number of policies have been developed to ensure both the quality of the program and its smooth functioning. For example, as mentioned above, all UROPs should be eligible for academic credit as certified by the supervising faculty member. This helps guarantee the academic validity of the UROP project, and it allows students to move from a paid UROP to one for academic credit and vice versa. Many faculty now insist on a semester of academic credit for a student doing a UROP before he/she gets paid. In other words, the faculty are using academic credit as a gatekeeper to ensure the student is serious.

Secondly, UROP projects can be either faculty initiated or student initiated. In the former, faculty members advertise for UROP students through an internal MIT website and by word of mouth among the students. They decide whether they want to offer funding (usually from a research grant) or academic credit. In a student-initiated project, the student needs to a find faculty member willing to supervise her/him; this is done both by providing all students with a list of faculty expertise and by word of mouth. If a faculty member is willing to sign a proposal to verify the project is credit worthy, but the student wants to be paid, then the student can apply for central MIT funding. Students are most likely to get credit during the semester, but get paid during the summer. For many students on financial aid, the rate of pay in the summer is sufficient to meet MIT's "summer earnings expectation." Thus a paid summer UROP both provides necessary resources for students and connects them personally with a faculty member.

Surveys show that by doing UROPs, students learn about the nature of research, and, in particular, how to handle ambiguity. We believe this is the case because in research, unlike most problem sets, there are not clear-cut answers. Students also learn about disciplines and areas that they do *not* wish to pursue as a result of doing a UROP. Thus, students can explore different fields without having a grade attached to a course.

TEAL is now firmly embedded in the physics department and has survived several transitions to different faculty leads. This is a tribute to the department leadership over the years who had withstood both faculty criticism and student resistance. The department expects that new TEAL faculty will be trained in active-learning pedagogy since the skill to stand up at a blackboard and lecture is quite different from the ability to work with student teams and respond to the difficulties they may be having at any given moment. New faculty are asked to go to one or more classes the term before they teach, and their own class is scheduled after that of an experienced instructor so they can watch him/her teach the class before they do. As positive affirmation of the effect of TEAL, the failure rate in freshmen physics has been more than halved.

Even though there have been several Aeronautics and Astronautics Department heads in the last few years, CDIO has been adopted and institutionalized in the department. The undergraduate curriculum is now oriented around a set of classes that help students to conceive, design, implement and operate aerospace engineering products. All faculty teaching undergraduate classes structure their learning outcomes around CDIO and write reflective memos each term to the associate department head in which they outline what they will do to improve their teaching in the coming year.

The department has also retained a full-time communications and learning specialist to help the faculty integrate communications skills into classes. The specialist collaborates with the faculty evaluate the learning that is taking place in the courses. Another example of the effect of CDIO is that for many years the department administered its own student course evaluation, developed by a department-level assessment specialist along with the Teaching and Learning Lab. The evaluation included questions that gave the instructor a sense of how and what the students were learning. In recent years, the questions were taken over by the institute course evaluation system as being best practice and the department system was supplanted by the Institute system.

THE IMPACT OF CHANGES ON MIT UNDERGRADUATE EDUCATION

Has MIT undergraduate education been transformed as the result of these re-forms and other pedagogical innovations at MIT? If the meaning of the word "transformed" is that all courses have adopted educational practices that have been shown to strengthen teaching and learning, then the answer must be "hardly." But that does not mean that the undergraduate educational enterprise has not been affected beyond the particular courses or programs that under-took a reform.

CDIO and TEAL caused other departments at the Institute to examine their own curriculum and pedagogical methods. The Department of Mechanical En-gineering, for example, has undertaken a large-scale curricula review that has resulted in more flexible requirements for the students, allowing them more latitude in how they structure their undergraduate degree. The Department of Electrical Engineering and Computer Science (EECS), the largest undergradu-ate major, has introduced much more active learning and peer instruction into its introductory courses. Material Science and Engineering rethought its un-dergraduate curriculum, embracing many of the core ideas of CDIO, if not its specific form. Courses in EECS and math now teach in the TEAL classroom, adopting a number of its pedagogical innovations. As mentioned above, UROPs are offered in every department in MIT, including those in the humanities and the social sciences.

Why were these efforts not adopted across the board? To answer that question, we draw upon a framework developed by our colleagues at the MIT Sloan School of Management called "The Three Lenses" (Ancona, Kochan, Van Maanen, & Westney, 2005). It provides different perspectives through which organizations and organizational change can be understood, which Ancona and her colleagues define as the strategic design lens, the political lens and the cultural lens. The strategic design lens posits that an organization is engineered to achieve agreed-upon goals based on the environment in which it sits, as well as its own strengths and weakness. Change comes about when a threat or op-portunity arises in the environment, or when parts of the organization are not aligned to achieve defined goals. The political lens sees organizations as contests for power, and the driver of change is a shift in power among stakeholders. The cultural lens defines organizations as entities in which members share symbols, identities, norms and assumptions, and the driver for change is new interpreta-tions of meaning within the organization (Carroll, 2006).

How does this framework help to explain why we have not seen educational change Institute-wide? We believe that the structure, the politics and the culture of MIT (and, we would argue, at many research-intensive universities) were not conducive to wide-scale transformation. Nor are drivers of change yet sufficiently strong to move the entire educational enterprise. For example, although MIT's strategic goals include both research and learning, it is no surprise that in the wider academic environment, prestige in research is still the pre-eminent goal for universities like MIT. (Whether the balance is moving toward education with shifts in both government policy and popular opinion toward higher education remains to be seen.) We can also find an explanation for the absence of institutional transformation through the cultural lens. At MIT, competition, the survival of the fittest, and the superiority of quantitative analysis over other kinds of work are strong cultural values. But they do not necessarily lend themselves to educational methods whose goals are to guarantee that all students master a pivotal set of concepts and skills.

The political lens probably allows the most straightforward explanation. The locus of power at MIT resides in the departments, and there is no strong centralized control to balance it. But departments are reluctant to adopt ideas from other departments wholesale; change is impeded by the Not Invented Here syndrome. As other MIT colleagues of ours so eloquently described NIH in a 1977(!) paper, "If we have not invented the innovation we cannot claim credit for it and thus fail to gain the prestige that accompanies something new" (Halfman, MacVicar, Martin, Taylor, & Zacharias, p. 3). NIH is very much operational among departments at MIT, and faculty in one department are loathe to simply replicate what faculty in another department have done. We have observed that reforms are best spread at MIT when faculty champions describe their innovation and are content with the fact that it will be implemented differently and on a different scale in other departments. In fact, this mirrors results from an international study of successful reforms in engineering education that found that "departments appear to be the engine of change" (Graham, 2012, p. 2).

In October 2014, MIT released the final report of the Institute-wide Task Force on the Future of MIT Education (see http://future.mit.edu/final-report). It was the product of three working groups of faculty and staff, which had examined residential education and facilities, global opportunities, and new financial models for higher education. The working groups solicited input from the entire community, both by organizing face-to-face meetings and by requesting comments on a preliminary draft. The final report outlines a number of recommendations for improving both undergraduate and graduate education at MIT, including increasing flexibility through modularity, transforming physical spaces to support learning, and defining new educational opportunities through service. It may yet provide a blueprint for institutional educational change.

REFERENCES

Ancona, D., Kochan, T., Van Maanen, J., & Westney, E. (2005). *Managing for the future: Organizational processes and behavior.* 3rd edition. Nashville, TN: Southwest Publishing.

Breslow, L. (2010). Wrestling with pedagogical change: The TEAL initiative at MIT. *Change: The Magazine of Higher Learning, 42*(5), 23–29.

Carroll, J. S. (2006). The three lenses. Retrieved from http://ocw.mit.edu/courses/sloan-school-of-management/15-301-managerial-psychology-fall-2006/lecture-notes/lec2.pdf, February 28, 2015.

Crawley, E. , Malmqvist, J., Östlund, S., & Brodeur, D. (2007). *Rethinking engineering education: The CDIO approach.* New York, NY: Springer.

Dori, Y. J. & Belcher, J. (2005). How does technology-enabled active learning affect undergraduate students' understanding of electromagnetism concepts? *The Journal of the Learning Sciences, 14*(2), 243–279.

Dori, Y. J., Hult, E., Breslow, L., & Belcher, J. W. (2007). How much have they retained? Making unseen concepts seen in a freshman electromagnetism course at MIT. *Journal of Science Education and Technology, 16*(4), 299–323.

Graham, R. (2012). *Achieving excellence in engineering education: The ingredients of successful change.* London, UK: The Royal Academy of Engineering.

Halfman, R. L., MacVicar, M. L. A., Martin, W. T., Taylor, E. F., & Zacharias, J. R. (1977). *Tactics for change.* Unpublished manuscript.

Lerner, R. M. (1991). Undergraduate dean MacVicar dies at 47. *The Tech.* Retrieved from http://tech.mit.edu/V111/N40/macvicar.40n.html, February 22, 2015.

ABOUT THE AUTHORS

Daniel Hastings is the Cecil and Ida Green Education Professor of Aeronautics and Astronautics at Massachusetts Institute of Technology in Cambridge, Massachusetts.

Lori Breslow is the Founding Director Emeritus of the Teaching and Learning Laboratory at Massachusetts Institute of Technology in Cambridge, Massachusetts.

11

Changing Practice Towards Inquiry-Oriented Learning

Les Kirkup

The Office for Learning and Teaching (OLT) is part of the Australian Government's Department of Education with a mandate to promote and support change in higher education institutions for the enhancement of teaching and learning. This chapter describes the goals, processes and outcomes of a one-year OLT National Teaching Fellowship[1] awarded to the author in 2011 to transform institutional practice by mainstreaming inquiry-oriented learning (IOL) in science in Australian universities.

Inquiry plays a critical role in the professional lives of scientists. By comparison, until recently, inquiry has assumed a modest role in the undergraduate science curriculum (Alkaher & Dolan, 2011). IOL activities have the potential to enhance students' problem-solving skills, stimulate creativity and foster innovation within students (Hanif, Sneddon, Al-Ahmadi, & Reid, 2009; Lee, 2012). These are essential attributes for students who complete a degree in science (LTAS, 2011). Through IOL activities, students: engage with scientific questions that have no predetermined answer; develop and implement approaches to address those questions; refine their approaches in order to enhance the quality of their data; gather evidence, and; communicate explanations and conclusions based on that evidence (adapted from Olson & Loucks-Horsley, 2000). As such, IOL reflects processes employed by scientists in their discipline-based research.

Evidence has steadily accumulated of the effectiveness of IOL to enhance student engagement and learning in science (see for example, Casotti, Rieser-Danner, & Knabb, 2008). The question arises as to why few science degrees programs in Australia have embedded IOL or similar approaches in their curriculum. Part of the answer lies in the absence of a critical mass of stakeholders able to drive curriculum change on a large scale. This situation has altered

---

1. Strictly, the author was awarded a National Teaching Fellowship of the Australian Learning and Teaching Council (ALTC), which is the direct antecedent of the Office for Learning and Teaching, and which had the same responsibilities. This chapter draws on the report of the OLT Fellowship which can be found at http://www.olt.gov.au/resource-kirkup-les-uts-altc-national-teaching-fellowship-final-report-2013

recently with drivers at national and institutional levels uniting to bring impetus to enhancing student learning of science at universities through participating in the processes of inquiry.

These drivers include:

- Australia's chief scientist who provides high-level advice to the Australian Federal Government, is determinedly advocating students be given insights into processes by which scientific knowledge is created and challenged, through engaging in inquiry (Office of the Chief Scientist, 2012).
- Evidence accrued over many years that active learning strategies, such as IOL, increase student performance (Cobern, Schuster, Adams, Applegate, Skjold, Undreiu, Loving, & Gobert, 2010; Freeman, Eddy, McDonough, Smith , Okoroafor, Jordt, & Wenderoth, 2014).
- The potential of inquiry to engage students, thereby arresting student attrition prevalent in science courses (Pitkethly & Prosser, 2001).
- Increased awareness that undergraduate inquiry emulates research activities of scientists and consequently is viewed favorably by research-focussed academics (Healey & Jenkins, 2009).

The OLT Fellowship awarded to the author was a response to these drivers. A diversity of approaches was adopted to engage and support stakeholders in mainstreaming IOL in science in Australian universities, including: funding and mentoring teams in Australian universities to develop IOL activities; running hands-on IOL workshops in universities, enabling IOL to be experienced and critiqued by discipline-based academics, teaching and learning specialists, and senior academic administrators; developing a partnership with Australia's premier scientific organization, the Commonwealth Scientific and Industrial Research Organisation (CSIRO), in order to co-develop laboratory-based IOL activities of broad appeal linked to the national science agenda, and; exploring the student perspective of IOL through student surveys and focus groups. The fellowship program described in this chapter promoted the national drive towards embedding more inquiry in undergraduate science degrees in Australian universities.

STRATEGIES FOR CHANGE

Elton (2003) remarked: "The appropriate collaboration of relevant agencies, both inside and outside universities may be able to use certain systematic strategies to achieve positive systemic change."

In order to promote changes in practice, Elton argued that "education and reason," i.e., the unidirectional exposition of innovations to audiences through papers, seminars, or similar and supported by a clear evidence-base of the value of the innovations are necessary but not sufficient for change to ensue. With due regard to Elton's insight, the author took the position that the participation of relevant agencies, allied to other strategies could achieve what education and reason alone could not. These agencies included the Office for Learning and Teaching, students, technical staff, academics and senior administrators, including deans and deputy vice chancellors, at universities that participated in the fellowship program. Elton (2003) also emphasized that the probability of change and innovation in higher education is enhanced when there is a confluence of top-down and bottom-up pressure (top-down being facilitative, and bottom-up being innovative). A goal of the fellowship was to galvanize individuals and groups within universities to build, and sustain, that pressure.

The strategies adopted to facilitate change in Australian universities towards IOL in science will now be described.

FELLOWSHIP-FUNDED ACTIVITIES (FFAS)

Recognizing that a "climate of readiness" (Southwell, Gannaway, Orrell, Chalmers, & Abraham, 2005) is key to successful propagation of innovations, and that the innovations must be owned from the beginning by the participants, requests for expressions of interest (EOI) were advertised nationally from academics intent on developing, trialling and embedding inquiry-oriented activities in their curriculum. Each successful EOI was supported with modest funding of $2,000. The author acted as an external agent who assumed several roles, including facilitator and mentor (Vilkinas & Cartan, 2006).

Academics were given the opportunity to be part of a national, multi-disciplinary cluster and to share their experiences and progress with others engaged in similar activities at several universities across Australia. Nine successful applications for funding originated in science faculties at the universities across Australia. The core science disciplines of physics, chemistry and biology were equally represented amongst the FFAs.

The requirements for FFAs encouraged teams to form with diverse backgrounds and capabilities to develop, trial and embed IOL activities within the curriculum. The initiative engaged institutional leaders, senior academics and educational developers in the IOL development. A goal of this strand of the fellowship program was to enhance recognition for the work being done by academics in developing inquiry activities within their own institution by being involved with a national program of activities and to act as a seed to attract more funding within their institution and/or externally.

Support from the fellowship for FFA recipients came in several forms, including running focus groups with students to explore their experience of, and attitudes towards, IOL activities. The FFA recipients wrote reports on the activities they developed which appear on the fellowship website: http://www .iolinscience.com.au/our-iol-activities/new-partnerships-and-networks/. Several recipients have since attracted money from within their universities to continue their innovation and published their work in peer-reviewed journals (see, as examples: Rayner, Charlton-Robb, Thompson, & Hughes, 2013; Creagh & Parlevliet, 2014).

STUDENT FOCUS GROUPS

For an informed insight into how students view learning through inquiry, the student experience was explored by means of focus groups at five universities in four states that received FFA funding. Examination of attitudes and experiences permitted students' understandings of inquiry to surface and issues to emerge that had not been anticipated.

Focus groups were conducted with students who had recently completed an inquiry-type activity. To bring emphasis to institutional issues, questions relating to the student experience were developed in conjunction with academics in each institution. While some questions specifically addressed the activity students had completed, others focused on more general themes, such as: How did students respond to the activity and how were enjoyment, relevance and learning viewed by students; what were students' expectations of an inquiry activity, and; what were students' recommendations for the future development of the IOL activities.

A confidential report outlining the research findings was provided to academics at each university to support the development and evaluation of their IOL activities.

INSTITUTIONAL WORKSHOPS

Inquiry-oriented activities require students to be imaginative, inventive and "think outside the box." The student perspective of IOL activities and the challenges facing teaching assistants intent on supporting students carrying out such activities were explored in institutionally-based workshops in 2011/2012 in Australia, New Zealand and the United Kingdom. By placing academics and others in the role of students, the perspective of those participating in the workshops was shifted to more closely align with that of students. This allowed for an exploration of the value and challenges of IOL activities from the student viewpoint. It also introduced an amount of uncertainty and anxiety in participants, mirroring experiences of undergraduates required to engage in inquiry.

Workshop participants included full-time and casual academics, technical-support staff, educational developers and senior academic managers such as associate deans. The participants, working usually in pairs, carried out, then critiqued, an IOL activity. Participants were encouraged to explore the: activity from a student perspective; scaffolding of activities which would best support students; challenges faced by students carrying out the activity experiment, and; challenges faced by academics in supporting students in inquiry-oriented activities.

To maximize the value of the workshop to the participants and to their institution, a report summarizing the workshop and its outcomes was sent to each workshop participant and senior academic manager (such as the dean of the faculty). A short anonymous survey was administered at the end of the workshop and results of the survey were included in the report.

The workshops were influential, with several universities indicating they had impacted on plans for introducing IOL into their curriculum. The following is an extract of a letter received from academics at Monash University:

> The workshop, together with further discussions with you about our IOL initiatives, galvanised our IOL program and laid the foundations for development and later implementation of IOL-based practicals in first year units in biology, chemistry and physics. Your workshop also provided a catalyst for us to consider the teaching associate perspective, given their importance in student learning in science teaching laboratories.
>
> —Gerry Rayner (Biology), Chris Thompson (Chemistry)
> and Theo Hughes (Physics), Monash University

NATIONAL FORUM

A one-day national forum entitled *Enhancing Learning in Science Through Inquiry and Technology* attended by full-time and casual academics, educational developers and teachers from 18 universities was held at the University of Technology, Sydney (UTS), to bring prominence to learning through inquiry. In addition to academics developing, or on the cusp of developing IOL activities, the forum was intended to appeal to students wanting to engage with, or contribute to, the conversation on learning through inquiry, as well as educational developers working with science academics on inquiry.

The forum comprised international keynotes, topic-driven presentations and a panel discussion. Goals of the forum included bringing national prominence to IOL, intensifying the conversation on IOL and providing forum participants with the ideas, tools and techniques to support student learning through inquiry and technology. Details of the forum can be found at: <www.iolinscience.com.au/wp-content/uploads/2012/05/program_booklet.pdf>

To emphasize the importance of inquiry, not only to the undergraduate curriculum, but to supporting a university's research agenda, the forum was formally opened by UTS' Deputy Vice-Chancellor (Research), Professor Attila Brungs, who observed:

> As the Deputy Vice-Chancellor (Research), I recognise the role of inquiry in inspiring the next generation of talented researchers, which is so critical for creating a strong, vibrant and sustainable research culture.

An outcome of the raising of the profile of IOL within UTS, partly as a result of the forum, was the creation a community of practice (CoP) entitled "Inquiry and Research Integrated Learning" designed to explore and disseminate IOL across all faculties at UTS. A forum was held in 2013 showcasing many non-science IOL innovations at UTS: http://www.uts.edu.au/sites/default/files/UTS%20IRIL%20Showcase%202013%20Program.pdf

THE CSIRO-UNIVERSITY UNDERGRADUATE INQUIRY INITIATIVE

To extend the influence of IOL beyond the confines of the university sector, a partnership was formed with the Commonwealth Scientific and Industrial Research Organisation (CSIRO) to promote learning through inquiry in the undergraduate curriculum. This became known as the CSIRO-University Undergraduate Inquiry Initiative. The goal of the initiative was to harness the complementary skills and energies of CSIRO and Australian universities to co-develop adaptable inquiry-rich learning resources. These resources were intended to enhance undergraduates' capacities to explore contemporary scientific issues of strategic importance to Australia and Australians.

The co-developed resources were designed to enhance student engagement, especially in the first year at university. It was anticipated that connecting undergraduates at universities with CSIRO would raise the profile and visibility of CSIRO within a large and influential group of citizens, namely undergraduate students who will be the next generation of Australian scientists, science policy makers and scientifically literate members of society; for universities, the resources developed would offer their undergraduates context-rich opportunities to engage in practice-oriented and research-integrated IOL activities.

A prototype activity was co-developed with the CSIRO based on research into organic solar cell technology: http://www.csiro.au/en/Research/MF/Areas/Innovation/Systems-and-devices/Flex-Electronics/Printed-Solar-Cells. The activity offered significant scope for students to design and carry out an investigation into technology that taps into a renewable energy source. The activity has been developed, trialled, reviewed and refined at UTS. A paper detailing the background to the initiative and the activity itself has been submitted for publication.

OTHER OUTCOMES THAT MAINTAINED THE MOMENTUM FOR CHANGE TOWARDS IOL IN SCIENCE

OLT grant focusing on teaching assistants: The influence of teaching assistants on student learning and engagement in IOL activities emerged throughout the fellowship. Funding was granted to the author by the OLT in 2013 for a cross-institutional, international project examining the impact of alignment between the background, ambitions, and views on teaching and learning of students and their teaching assistants on student engagement and satisfaction in first-year laboratories.

A special issue of the International Journal of Innovation in Science and Mathematics: In order to maintain the national prominence of IOL, a special issue on "Inquiry and Problem Solving in the Undergraduate Curriculum" is to be published in the journal in 2015.

Threshold Learning Outcomes (TLOs): The author was a member of the advisory group to discipline scholars given a mandate to develop a national set of TLOs for use in Australian universities. A key TLO concerns inquiry and problem-solving and which closely mirrors the goals of IOL.

Good-Practice Guide: The author and Liz Johnson of Deakin University were commissioned to write a good-practice guide for national distribution on "Inquiry and Problem Solving." It is available as a good-practice guide from: http://www.olt.gov.au/resource-learning-and-teaching-academic-standards -science-2011. The publication assisted in maintaining the prominence of IOL at a national level.

Extension grant: Academics from Monash applied successfully for funding from the OLT to develop initiatives emerging from the fellowship. An account of their work, and a good-practice guide, which was an outcome, can be found at:

http://www.iolinscience.com.au/wp-content/uploads/2014/07/Good-Practice -Booklet_FINAL.pdf.

DISCUSSION

The following lessons emerged as a result of the work described in this chapter:

- Elton's view of the conditions necessary for change, i.e., that a confluence of agencies within and outside a university heightens the probability of curriculum change, was well supported by outcomes of this fellowship program. As facilitator, innovator and mentor, the author was able to support and promote change within several universities. Being an outside agent also freed the author of association with issues that might adversely influence the progress of the initiative at institutional or faculty levels.
- As Southwell et al. (2005) pointed out: "A climate of readiness is important if successful innovation and dissemination are to take place. Such a climate recognises the need for change, engages in reflective critique, supports risk-taking, . . . and recognises and rewards those engaged in enhancing teaching and learning, and builds capability." That climate of readiness to move towards learning through inquiry characterized the FFA recipients and played a large part in the success of their innovative IOL activities.
- Face-to face visits, focus groups with students, as well as hands-on workshops, successfully built engagement as well as forged links between academics and others to promote institutional change through engaging with stakeholders at all levels.
- The standing that came with being awarded a national teaching fellowship opened many doors, for example those of the senior executives within university. This allowed both the national agenda towards IOL to be communicated at the highest levels within an institution, while at the same time bringing attention to work happening within that institution.
- Working closely with academics and others in several universities meant the fellowship program could connect academics together facing similar challenges, allowing good ideas to be quickly disseminated. The award of a small amount of funding to progress the development of IOL activities acted as a catalyst for the work and also as a seed for further funding, and raised the profile of the recipients within their own institutions.

- Workshops and seminars were held within many institutions during the fellowship period in order to engage practitioners and disseminate the findings of the fellowship. Keynotes and plenary presentations were additional opportunities to communicate the messages of the fellowship, and to engage academics in consideration of IOL.
- As a discipline-based academic with contemporary experience developing, delivering and evaluating IOL activities, the author understood the challenges facing academics and students created by open-ended inquiry activities. As a consequence, the confidence of stakeholders was quickly gained, which is necessary when promoting change.

The impact of the fellowship on changing practice towards inquiry-oriented learning nationally and institutionally has been uneven. The scale and sustainability of change of institutional practice toward IOL, based on the experience of the fellowship program described in this chapter, is correlated with:

- A significant and growing number of academics within an institution committed to curriculum reform towards IOL.
- Demonstrable support for IOL innovations from senior academics and policy makers.
- Regular conversations between innovators and the author as external agent, including face-to-face discussions to explore local issues and anticipate potential road blocks.
- Genuine recognition and buy-in within an institution that national initiatives promoting IOL were (or would become) a priority at institutional and program levels.
- The existence of IOL innovations already progressing within an institution that would benefit from external validation and influence.
- The determination of academics to publish their work on IOL and disseminate it nationally and internationally.

Some outcomes and deliverables of the fellowship were predictable (for example, a website devoted to the fellowship, http://www.iolinscience.com.au/. Others were less predictable, but no less welcome. As an example, the author received a communication from Peter Coolbear, director of New Zealand's National Centre for Tertiary Teaching Excellence, Ako Aotearoa:

> Of considerable interest to us was that some of the attendees at Les' workshop at the Victoria University of Wellington were not science teachers. Our staff member, Ian Rowe, who hosted the workshop at Victoria provided the following comment: "For me as the organiser, the

most revealing comment came from two tutors from Whitireia [Community Polytechnic] who taught on a social work degree. . . . They were certain the method could be used in their non-science subject and they were excited to plan how they would alter some parts of their work to include inquiry methods."

—Dr. Peter Coolbear, February 2013

FINAL REFLECTIONS

National teaching fellowships, supported by the OLT and its antecedents, began in 2006. To date, there are in excess of 80 scholars from 29 institutions that have undertaken a fellowship program spanning the whole range of disciplines offered by Australian universities. Many of the scholars are drawn from disciplines such as science and engineering, and are not education research specialists. As such, they bring an amount of "street credibility" when they embark on a fellowship. This is advantageous when the goal is to promote change or exert influence in higher education, especially with academics who work directly with students. This, coupled with the fact that many fellows occupy (or move into as a result of the fellowship) senior positions within their institutions and are well respected by their professions, means they are able to promote and grow communities that have the critical mass to effectively stimulate change at institutional and national levels.

ACKNOWLEDGMENTS

I would like to give special thanks to Andrea Mears. Andrea was the project officer who supported the fellowship and its activities through 2011/2012. Her contributions to the work described here were vital to its success. Many other people assisted in the endeavor to promote and embed change toward IOL. I would like to thank: The Office for Learning and Teaching, Siobhan Lenihan, Manju Sharma, Stephen Billett, Kelly Matthews, John Rice, Liz Johnson, Stephanie Beames, Nicole Eng, Bruce Milthorpe, Attila Brungs, Shirley Alexander, Linda Foley and Nirmala Maharaj.

REFERENCES

Alkaher, A., & Dolan, W. (2011). Instructors' decisions that integrate inquiry teaching into undergraduate courses: How do I make this fit? *International Journal for the Scholarship of Teaching and Learning, 5* (2), 1–24.

Casotti, G., Rieser-Danner, L., & Knabb, M. T. (2008). Successful implementation of inquiry-based physiology laboratories in undergraduate major and non-major courses. *Advances in Physiology Education, 32*, 286–296.

Cobern, W.W., Schuster, D., Adams, B., Applegate, B., Skjold, B., Undreiu, A., Loving, C.C., & Gobert, J.D. (2010). Experimental comparison of inquiry and direct instruction in science. *Research in Science & Technological Education, 28*(1), 81–96.

Creagh, C., & Parlevliet D. (2014). Enhancing student engagement in physics using inquiry oriented learning activities. *International Journal of Innovation in Science and Mathematics Education, 22*(1), 43–56.

Elton L. (2003). Dissemination of innovations in higher education: A change theory approach. *Tertiary Education and Management, 9* (3), 199–214.

Freeman, S., Eddy S. L., McDonough, M., Smith M. K., Okoroafor, N., Jordt, H., & Wenderoth, M. P. (2014). Active learning increases student performance in science, engineering and mathematics. *Proceedings of the National Academy of Sciences, 111* (23), 8410–8415.

Hanif, M., Sneddon, P. H., Al-Ahmadi, F. M. & Reid, N. (2009). The perceptions, views and opinions of university students about physics learning during undergraduate laboratory work. *European Journal of Physics, 30*, 85–96.

Healey, M., & Jenkins, A. (2009). *Developing undergraduate research and inquiry.* Accessed 1 Jan. 2015, from https://www.heacademy.ac.uk/sites/default/files/developingundergraduate_final.pdf

Lee, V. S. (2012). What is inquiry-guided learning? *New Directions for Teaching and Learning*, 2012: 5–14. doi: 10.1002/tl.20002.

LTAS Project. (2011). *Learning and teaching academic standards statement.* Accessed 1 Jan. 2015 <www.olt.gov.au/resource-learning-and-teaching-academic-standards-science-2011>.

Office of the Chief Scientist. (2012). *Mathematics, engineering & science in the national interest.* Accessed 1 Jan 2015, from www.chiefscientist.gov.au/wp-content/uploads/Office-of-the-Chief-Scientist-MES-Report-8-May-2012.pdf

Olson S & Loucks-Horsley S (Eds.). (2000). *Inquiry and the National Science Education Standards: A guide for teaching and learning.* (Available online at: http://www.nap.edu/books/0309064767/html/

Pitkethly, A., & Prosser, M. (2001). The first year experience project: A model for university change. *Higher Education Research and Development, 20*, 185–198.

Rayner, G., Charlton-Robb, K., Thompson, C. D. & Hughes, T. (2013). Interdisciplinary collaboration to integrate inquiry-oriented learning in undergraduate science practicals. *International Journal of Innovation in Science and Mathematics Education, 21*(5), 1–11.

Southwell, D., Gannaway, D., Orrell, J., Chalmers, D., & Abraham, C. (2005). *Strategies for effective dissemination of project outcomes.* Sydney, Australia: Carrick Institute for Learning and Teaching in Higher Education.

Vilkinas, T. & Cartan, G. (2006). The integrated competing values framework: Its spatial configuration. *Journal of Management Development, 25* (6), 505–521.

ABOUT THE AUTHOR

Les Kirkup is a Professor of Faculty of Science at the University of Technology, Sydney, in Sydney, NSW Australia.

SECTION C

Case Studies—Projects at the Course and Department Level

Colleges and universities are often heterogeneous collections of departments, schools, or other units—each with a distinct culture. This variability even within institutions makes large-scale transformation exceedingly challenging. Innovation begins often at the course, department, or school level and spreads. The collection of chapters in this section provides case studies for a range of course or curricular innovations. The first three papers describe innovations in engineering education. While all have a similar goal—more broadly educate engineers to possess both the technical knowledge and the problem-solving skills necessary for addressing 21st century challenges—the approaches are quite different. The Vertically Integrated Project model, described by Coyle, builds vertically and horizontally articulated teams. This model, which connects teaching and learning directly to faculty research, started as a course in a single department and has been adopted by several other universities. The iFoundry model, described in the chapter by Sheets, is envisioned as a cross-disciplinary curriculum incubator for testing new courses and programs across an engineering college at a large research university. The third engineering transformation chronicles the creation of a new multi and interdisciplinary engineering program that combines a liberal education with technical knowledge. In contrast to the blank slate approach chronicled by Challah, the PPI program (Mili) presents a purposefully disruptive model for radically changing an entire existing college of technology. The Squires chapter describes a mature five-year redesign of a mathematics department built on faculty teamwork and data-driven decision making for continuous improvement. Finally, Smith's chapter communicates a model of curricular transformation built on collaboration fostered through Faculty Learning Communities.

1

The Vertically Integrated Projects (VIP) Program: Leveraging Faculty Research Interests to Transform Undergraduate STEM Education

Edward J. Coyle, James V. Krogmeier, Randal T. Abler,

Amos Johnson, Stephen Marshall, and Brian E. Gilchrist

Modern universities, especially those classified as research-intensive institutions, are highly fragmented:

- *By Discipline*—With the exception of a handful of faculty with joint appointments in two or more disciplines and a small percentage of students with dual majors, students and faculty are sorted by discipline. There are few opportunities for students to interact in a meaningful way with those in other disciplines. For faculty, there are few incentives or rewards to initiate multidisciplinary collaborations.
- *By Time*—The semester or quarter and the academic year are the fundamental units of time on campuses. Within a semester, the fundamental unit is a credit hour that is associated with a fixed amount of time in a classroom or a laboratory.
- *By Mission*—Universities typically define their missions to be research, education, and service. Commercialization, economic development, and globalization are recent additions to this list. In very few cases are these missions integrated in a way that the boundary between them is not obvious. In fact, most faculty partition each day between their research activities and assigned lecturing in undergraduate or graduate classes.

Efforts to transform higher education must account for this fragmentation. If they do not, any success they achieve will likely be temporary or affect only a small number of students and faculty.

In this paper we report on our effort, the Vertically Integrated Projects (VIP) Program (Coyle, Allebach, & Garton-Krueger, 2006), to address and potentially overcome fragmentation in higher education. The lessons learned when initiating and growing VIP within several institutions have helped to improve the program as measured by the learning outcomes for students, research outcomes for faculty, and new opportunities for partnering with other organizations. The key has been to leverage faculty interests in and Institutions' reward structures for research, especially when research is defined in a discipline-independent way as *innovation*. VIP provides a mechanism that enables undergraduates to participate in and contribute to innovative efforts led by faculty and their graduate students.

THE VIP PROGRAM

Sustaining and accelerating the pace of innovation in society requires a continuous stream of graduates in all disciplines who understand how the processes of research and technology advancement can be integrated to enable innovation. Current approaches to the education of undergraduates and graduate students are simply not up to this challenge:

- Undergraduates rarely achieve a deep understanding of, or have an opportunity to contribute to any aspect of, their chosen discipline.
- Master's students are typically not involved in either research or the development of new technology/techniques.
- PhD students rarely see their innovative ideas/discoveries have an impact beyond the publication of their theses, conference papers, and journal papers.

We have thus developed a new curricular approach that integrates education and research: the Vertically Integrated Projects (VIP) Program. It creates and supports teams of faculty, graduate students, and undergraduate students that work together on long-term, large-scale projects. The teams are: *multidisciplinary*—drawing students from all disciplines on campus; *vertically-integrated*—maintaining a mix of sophomores through PhD students each semester; *large-scale*—with 10 to 20 undergraduates per team; and *long-term*—each undergraduate may participate in a VIP project for up to three years and projects may last for years, even decades. The continuity, technical depth, and disciplinary breadth of these teams provide:

- The time necessary for students to: learn and practice many different professional skills; make substantial contributions to the project; experience many different roles on a large design/discovery team; and work effectively in a multidisciplinary environment.
- A compelling context: The research efforts of the faculty and graduate students ensures they are engaged because they benefit from the efforts of the undergraduates. It also enables the undergraduates to understand complex issues in their field of interest.

- The mentoring necessary for students to learn about, contribute to, and lead the parts of the project on which they are focused. The mentoring crosses all boundaries, enabling faculty, graduate students, and sophomores through seniors to collaborate successfully.

GOALS OF THE VIP PROGRAM

The goal of the VIP Program is to achieve systemic reform of higher education by:

- Unifying the missions of research and education by enabling undergraduate *teams* to work together with faculty and graduate students in a way that benefits everyone.
- Overcoming the fragmentation by discipline of the university through its focus on large-scale projects that are, almost by necessity, multidisciplinary.
- Eliminating the fragmentation of time in the standard curriculum by enabling projects that last for many years, even decades.

CURRENT STATUS OF VIP AT GEORGIA TECH

There are 300 undergraduates currently enrolled in 27 VIP teams at GT: http://vip.gatech.edu/teams. The teams' advisers are researchers from the following colleges and other organizations on campus: College of Architecture (CoA), College of Computing (CoC), College of Engineering (CoE), College of Science (CoS), Center for Education, Teaching and Learning (CETL), Georgia Tech Research Institute (GTRI), and the Ivan Allen College (IAC).

The primary task so far has been to establish how the VIP courses count toward undergraduate degrees in *each* discipline on campus. Once this is in place for all disciplines, the creation and maintenance of long-term, large-scale, multidisciplinary teams will be possible for almost any ambitious effort that can be envisioned. In the CoC and in ECE and ISyE within the CoE, mechanisms that enable VIP credits to "count" for students' junior and senior design projects have been developed. For the current status of this long-term, multidisciplinary curricular effort, please see: http://vip.gatech.edu/how-vip-credits-count.

Mature VIP teams typically have 12 to 20 students/team, while new teams have six to eight students/team. Students from more than 20 different disciplines are represented on the current set of 27 VIP teams. These students are in some cases from disciplines that have not yet determined how VIP credits count other than as free electives; e.g., psychology and economics.

Faculty request to have teams of their own; they are not required to have VIP teams. To initiate a team, they must demonstrate that the proposed project would be: challenging to undergraduates; last at least four years, preferably more; and have a broad enough goal that the project can evolve based on contributions from everyone on the project, including the undergraduates.

The evaluation of the VIP program is being led by Julia Melkers of Public Policy. With NSF support and support from a GT Global-FIRE grant, she has been characterizing knowledge exchange amongst the students both within and between VIP teams. She has also been evaluating the learning outcomes for students who participate in the VIP Program [Melkers, et al, 2012]. These results build on earlier efforts to evaluate the effect of vertically integrated project teams on students' development of both disciplinary and professional skills (Coyle, Jamieson, & Oakes, 2005). Melkers has recently received IRB approval

FIGURE 1. The Fall 2010 eStadium VIP team. They design, develop, and deploy systems that enable studies of wireless network traffic during football games. Fans in the stands go to: http://estadium.gatech.edu/iphone to access stats and on-demand video clips of plays for GT football games. Students from CEE, ECE, CS, and ISyE participate in the team. They also design social network apps and wireless sensor networks.

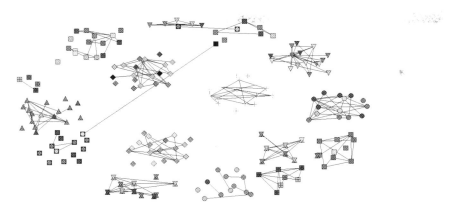

FIGURE 2. A network showing the exchange of technical information within and between VIP teams for the Fall 2011 semester at Georgia Tech. Students on the same team are grouped together and represented by a common shape (square, triangle, plus-sign, etc.). The colors represent the disciplines of the students. The mostly disconnected subgraph in the lower left is a team with a hub-and-spoke advising model, with the adviser meeting and working one-on-one with each student. New teams tend to be smaller in size (4-8 students) than more mature teams (12+ students).

to use, after obtaining appropriate written permission from them, students' grades within VIP and in the curriculum in general to determine the effect of VIP on student learning. This will enable the evaluation of VIP to progress well beyond the students' self-reported data from surveys and interviews that are currently used.

BENEFITS OF VIP FOR STUDENTS

The VIP Program provides many benefits to undergraduate students, graduate students and faculty. The most important of the benefits is the way it enables members of all of three groups to form a *community* in which they work together over an extended period of time on a project of common interest. This experience builds mutual respect, cultivates curiosity and creativity, enables meaningful collaborations on both a disciplinary and personal basis, and provides benefits to all participants in the project.

The benefits to the undergraduates who participate in VIP include, but are not limited to:

- The opportunity to apply what they are learning in their regular coursework to real, challenging problems that are of current interest in their field.

- Learning to work with and eventually take a leadership role in a sophisticated team that is working on a challenging project.
- Development and regular honing of professional skills, including: oral and written communication ability, making presentations to a variety of audiences, collaborative brainstorming and problem solving, developing resilience in the face of failure, etc.
- Becoming a truly active/independent/life-long learner and a creative problem solver.
- Learning about cutting-edge issues in their discipline.
- Achieving significant depth of expertise in some aspect of their home discipline.
- Learning how to communicate and work productively with people from other disciplines and widely varying backgrounds.
- Learning how to "come up to speed" on and then contribute to a large-scale, intellectually ambitious project.
- Learning that challenging problems may have many possible solutions and even many approaches to defining them.
- Experiencing the great satisfaction of solving a challenging problem, designing a complex system and seeing it work as intended, or discovering a new way of looking at an issue or problem.
- Learning to be a creative, resourceful, self-motivated, and responsible member of a team.
- Learning to adapt as situations change and to recover/restart after setbacks.

Many students have already reported that when they interview for jobs after participating in VIP, that their experiences and accomplishments on their VIP team quickly become the focus of the interview. A number of innovations produced by VIP students have been patented/copyrighted and licensed. We also believe that it is only a matter of time before there are a number of successful commercialization efforts based on VIP projects.

BENEFITS OF VIP TO THE UNIVERSITY

The benefits of VIP to students and faculty are measurable, as are benefits to the university. We believe that there are additional benefits to universities, although they are perhaps harder to measure:

- Provides students with a compelling reason to be on campus, even in an age of MOOCs and other distance education approaches.

- Enhances innovation on campus by enabling everyone to participate.
- Enables projects of large scope and duration to be attempted.
- Enables new partnerships with organizations both on and off campus.
- Opens up and deepens multidisciplinary opportunities across campus.
- Deepens and broadens the university community by providing an environment in which faculty, graduate students, and undergraduate students can get to know each other very well.

CHARACTERISTICS OF FIVE VIP PROGRAMS

In this section, we describe five VIP Programs—those at Georgia Tech, Morehouse College, Purdue University, University of Michigan, and the University of Strathclyde—in terms of origin and type of implementation strategy; number of disciplines involved; type of institution; implementation in the curriculum; resources and support available; growth of the program; grading/assessment strategy and tools; relationship with other discovery and design programs; software tools for program administration; and number of students and faculty involved. In some cases, fewer than five programs will be described, as appropriate.

Origins, Institutions, and Numbers of Disciplines, Faculty and Students Involved

The VIP Program was launched at Purdue University in 2001 with the creation of the first VIP team, the eStadium team. VIP was designed to improve upon its predecessor, the Engineering Projects in Community Service (EPICS) Program (Coyle, Jamieson, & Oakes, 2005). Both of these programs at Purdue used large-scale, long-term teams composed of undergraduates that could register for credit for up to three years. The difference between them was the context of the projects. In EPICS, the teams designed, developed and deployed products and systems for non-profit organizations in the local community. In VIP, the teams design, develop and deploy products and systems that assist faculty and their graduate students with their research. The hypothesis was that more faculty would be willing to advise teams for many years if doing so would benefit their research activity, for which they are regularly rewarded.

The choice of VIP as the name of the new program focused on the mechanism that enabled the teams to function well, as opposed to the discipline(s) on which the program was focused, which was engineering in the case of EPICS. The result is that VIP has spread to many disciplines besides engineering, including architecture, business, computing, the humanities, public policy, and science.

At Purdue, VIP grew primarily within Electrical and Computer Engineering, but has also spread to other disciplines over time, including Aeronautics and Astronautics, Human Development and Family Studies, and Earth, Atmospheric and Planetary Sciences. There are currently 14 VIP teams at Purdue.

VIP spread to Georgia Tech in 2009 and was founded with the intent of being a campus-wide program and on developing processes and tools that would enable it to grow to support many teams. There are currently 300 undergraduates enrolled in 27 VIP teams. The teams' advisers and students involved come from four colleges—Architecture, Computing, Engineering, and Science—and such organizations as the Georgia Tech Research Institute; the Center for Education, Teaching, and Learning; and the Centers for Disease Control.

In 2010, a VIP team was created at Morehouse College. The team collaborated with the eStadium VIP team at Georgia Tech. This enabled a research focus and access to a large stadium that were not available at Morehouse. The team functioned for three semesters. Efforts are underway to develop an effective way for VIP to function at Morehouse, where faculty teaching loads are higher than at R-I universities like Georgia Tech.

In 2011, VIP was implemented at the University of Strathclyde in Glasgow, Scotland. It started in many different disciplines simultaneously, including Biology, Computing, Engineering, English, and Management. There are currently 10 VIP teams at Strathclyde.

Other universities have started VIP-like programs, including projects within the University of Michigan's Multidisciplinary Design Program (Daly, Bell, Gilchrist, Hohner, & Paul Holloway, 2011) http://mdp.engin.umich.edu and Texas A&M's AggiE-Challenge, which currently fields 13 teams:

http://engineering.tamu.edu/easa/areas/enrichment/aggie-challenge

Implementation in the Curriculum

It is critical that curricular implementation of VIP provide incentives for students to participate for at least two years. The School of Electrical and Computer Engineering at Georgia Tech approved guidelines for VIP credits, described below, that provide an example of how VIP can be integrated into senior design. To encourage long-term student participation, VIP credits cannot be used as ECE electives unless students take six or more credits of VIP, as per the following guidelines.

NOTE: There is a six-credit limit on the following types of courses: Independent Research, Special Problems and VIP. Students can take all three types of courses, but only six credits can be applied toward the degree. Students interested in doing more than six credits of VIP should consider using VIP for Senior Design, which allows for an additional semester of VIP participation through the Senior Design course. To do this, students must follow a specific timeline.

For students not using VIP projects for Senior Design:

If five or fewer VIP credits are earned:
Five credits can be used as approved electives.
During the junior year (or after ECE 2031), VIP can be used to fulfill the ECE 3006 Professional Communications requirement.
If six VIP credits are earned with the same team:
Three credits can be used as ECE electives.
Three credits can be used as approved electives.
During the junior year (or after ECE 2031), VIP can be used to fulfill the ECE 3006 Professional Communications requirement.

For students using VIP for Senior Design, at least five VIP credits will be earned with the same team prior to Senior Design (per the required timeline):

Three credits can be used as ECE electives.
Two to three credits can be used as approved electives.
During the junior year (or after ECE 2031), VIP can be used to fulfill the ECE 3006 Professional Communications requirement.
Three Senior Design Credits: The student will register for the VIP section of Senior Design, with the technical portion of the students' design experience completed as a member of the student's VIP team.

An explicit timeline for this process is available at:

http://www.vip.gatech.edu/how-credits-count-electrical-computer -engineering

Resources and Support

Resources available for VIP programs depend upon the size and age of the program. The most common element in a VIP Program is a research-active faculty member who serves as the director and is also the adviser for a team. Once a program has at least 10 to 12 teams, a program manager is necessary to assist with operations. With the addition of software tools for processing permits to participate in the program, for grading and peer evaluations, and the unique IT needs of the program, it can expand significantly. We believe at this time that with these resources, a VIP Program that fields hundreds of teams with thousands of students is feasible.

One unique resource available to the Georgia Tech VIP Program is a co-director for technology, Randal Abler, who in addition to other tasks, defines and develops technology solutions to assist with both program and team operations.

Tools for Program Administration and Growth

Tools that have been developed to enable both scaling and evaluation of student outcomes include:

- On-line team advertising combined with a process for students to apply to join a team of their choice.
- Course-permit administration tools to assign students to teams.
- Grading and peer evaluation tools and a database used by all advisers/teams. IRB approval has been obtained to enable program evaluators to access grades in these databases.

THE VIP CONSORTIUM

Sixteen universities have recently created the *VIP Consortium*. It consists of schools that have or plan to adopt the VIP model and have committed to collaborate to improve and expand it. Schools in the U.S. that have already fielded VIP teams are Georgia Tech, Purdue University, and Morehouse College. Please see the website of the Purdue VIP Program: https://engineering.purdue.edu/vip. Schools in the U.S. that have programs very similar to VIP include the University of Michigan and Texas A&M University. Internationally, the University of Strathclyde and National Ilan University in Taiwan have created VIP Programs.

The universities that have formed the VIP Consortium are: Boise State University, Colorado State University, Florida International University, Georgia Tech, Howard University, Morehouse College, National Ilan University, Purdue University, Rice University, Texas A&M, University of Hawaii—Manoa, University of Michigan, University of Strathclyde, University of Washington, and Virginia Commonwealth University. The rapid dissemination of the VIP model to other universities is confirmation of the compelling nature of the educational *and* research benefits of the program. It is also a unique opportunity to achieve systemic reform on a national scale.

The University of Strathclyde in Glasgow, Scotland has an active VIP Program, currently fielding 10 VIP teams: http://www.strath.ac.uk/viprojects. They were the first non-U.S. university to participate in the consortium. Their experiences and challenges in creating and sustaining VIP have been quite different than those of U.S. institutions. Their insights have already been of significant value to the consortium's effort.

The creation of the VIP Consortium is being supported by a $5M grant from the Leona M. and Harry B. Helmsley Charitable Trust. The grant, which started Jan 1, 2015, supports U.S. institutions' participation in consortium-wide efforts to develop and share ideas, processes and software tools that enhance the operation, growth and evaluation of all VIP Programs. Georgia Tech is the lead institution of the consortium, and the University of Michigan is the co-lead.

REFERENCES

Coyle, E. J., Allebach, J. P., and Garton-Krueger, J. (2006). The Vertically-Integrated Projects (VIP) Program in ECE at Purdue: Fully integrating undergraduate education and graduate research. *Proceedings of the 2006 ASEE Annual Conference and Exposition*, Chicago, IL.

Coyle, E. J., Jamieson, L. H., and Oakes, W. C. (2005). EPICS: Engineering Projects in Community Service. *International Journal of Engineering Education*, Vol. 21, No. 1, pp. 139–150.

Daly, S., Bell, H., Gilchrist, B. E., Hohner, G., Paul Holloway, J. P. (2011). Making a college-level multidisciplinary design program effective and understanding the outcomes. *Proceedings of the 2011 ASEE Annual Conference and Exposition*, Vancouver, B.C.

Melkers, J., Kiopa, A., Abler, R. T., Coyle, E. J, Ernst, J. M., Krogmeier, J. V., and Johnson, A. (2012). The social web of engineering education: Knowledge exchange in integrated project teams. *Proceedings of the 2012 ASEE Annual Conference and Exposition,* San Antonio, TX.

ABOUT THE AUTHORS

Edward J. Coyle is the John B. Peatman Distinguished Professor and a GRA Eminent Scholar in the School of Electrical and Computer Engineering at Georgia Tech in Atlanta, Georgia.

James V. Krogmeier is the Associate Head and Professor of Electrical and Computer Engineering at Purdue University in West Lafayette, Indiana.

Randal T. Abler is a Principal Research Engineer in the School of Electrical and Computer Engineering at Georgia Tech in Atlanta, Georgia.

Amos Johnson is an Associate Professor in the School of the Computer Science Department at Morehouse College in Atlanta, Georgia.

Stephen Marshall is a Professor of the Department of Electronic and Electrical Engineering at the University of Strathclyde in Glasgow, United Kingdom.

Brian E. Gilchrist is a Professor of Electrical Engineering and Computer Science and Professor of Atmospheric, Oceanic and Space Sciences in the College of Engineering at the University of Michigan in Ann Arbor, Michigan.

2

Transformative Initiatives: How iFoundry Reimagines STEM Education for the 21st Century

Diana E. Sheets

THE FOUNDATIONAL BEGINNINGS OF IFOUNDRY

The origins of iFoundry can be traced back to "The Engineer of 2020 Project," which forecast the needs of civilization, as well as the education, training, and perspective necessary for the next generation of engineers to succeed. The first phase of the project, *The Engineer of 2020: Visions of Engineering in the New Century,* published by the National Academy of Engineering (NAE), set forth "a vision for engineering" and the scope of work to be done (2004, p. xi). The second phase, *Educating the Engineer of 2020: Adapting Engineering Education to the New Century,* examined how "to enrich the education of engineers who will practice in 2020" (2005, p. xii).

As the project noted, the frontiers of science are on the cusp of life-altering advances in "nanotechnology, logistics, biotechnology, and high-performance computing" (*The Engineer of 2020*, 2004, p. 1). Meanwhile, technological advancements in developing nations present both opportunities and challenges.

These circumstances warrant initiatives to fostering transformative innovation. Here in the College of Engineering at the University of Illinois, we seek to address these issues in an effort to anticipate the educational needs of the next generation of technological leaders to invent "the new new thing" (Lewis, 1999).

iFoundry was born in the summer of 2007. It was co-founded by Andreas Cangellaris and David Goldberg, faculty in the College of Engineering.[1] It was conceived as a "cross-disciplinary curriculum incubator" ("Who We Are", n.d.). Initially called the Illinois Foundry for Tech Vision and Leadership, it was soon renamed the Illinois Foundry for Innovation in Engineering Education with the notable tagline "Transforming Engineering Education for the 21st Century" (iFoundry website, 2012).

---

1. Cangellaris was promoted to department head of the Electrical and Computer Engineering Department at the University of Illinois in 2008 before becoming dean of the College of Engineering in 2013. Goldberg is the Jerry S. Dobrovolny Professor Emeritus in Entrepreneurial Engineering.

Because of concerns that traditional engineering and associated computer science curricula that dated back to the cold war might be ill-equipped for the "radical changes in transportation, communication, and computer technology" that rendered today's society "a very different world", Cangellaris and Goldberg wrote a white paper ("Whitepaper for an Illinois Foundry for Tech Vision and Leadership", 2007). Their report highlighted some of the driving influences prompting the creation of iFoundry: enhancing engineering diversity, emphasizing excellence and analytical insight, strengthening the curriculum to nurture creativity while promoting professionalism and leadership acumen critical for success in the "civic arena"—what in an earlier era C. P. Snow would have referred to as the "corridors of power" (Snow, 1964).

> If the United States is to continue its leadership in technological innovation and the creation, regulation and management of new technologies and new markets, its engineering force needs to be rejuvenated by a more diverse talent pool, where excellence in scientific education and analytical skills is complemented by a broader curriculum that inspires creativity and innovation and includes training in the professionalism and leadership traits needed for successful participation in the civic arena. ("Whitepaper", 2007)

What capabilities does *The Engineer of 2020* suggest that engineers should possess? They should not be circumscribed by disciplinary boundaries. Engineers should embrace "creativity, invention, and cross-disciplinary fertilization to create and accommodate new fields of endeavor" including the "nonengineering disciplines" (2004, p. 50). They should be responsive to global trends. Their education should equip them with the skills and know-how "to address the technology and societal challenges and opportunities of the future" (2004, p. 51). The engineering curriculum needs to be responsive "to the disparate learning styles" characteristic of our increasingly diverse student populations (2004, p. 52).

The training and attributes of engineers will determine their success. If they aspire to leadership and management positions, they need analytical capabilities, as well as ingenuity and creativity. They must communicate and possess business and management acumen (*The Engineer of 2020*, 2004). Their education should include "interdisciplinary learning" and "case studies of engineering successes and failures" (*Educating the Engineer of 2020*, 2005 pp. 2–3).

iFoundry was conceived as a pilot program, an educational enrichment environment where new courses and programs could be tested and evaluated before introducing them into departments within the College of Engineering.

Modeling its mandate from "The Engineer of 2020 Project", iFoundry fosters courses and programs that nurture "a variety of cross-cutting skills and disciplines: communications; leadership; teamwork; arts & design cross-fertilization; better utilization of humanities and social science hours; more general understanding of the societal and human contexts of engineering and technology" ("Whitepaper", 2007).

PROGRAMS IMPLEMENTED BY IFOUNDRY

iFoundry began its spiritual awakening at the "Workshop on the Engineer of the Future" held at the University of Illinois in September of 2007 (Goldberg, Cangellaris, Loui, Price, Litchfield, 2008b). Keynote speakers were William Wulf, retiring president of the NAE, and Sherra Kerns, founding vice president for innovation and research at Franklin W. Olin College of Engineering (Goldberg, Cangellaris, Loui, Price, & Litchfield, 2008a).

During the 2007–2008 academic year, a team of faculty and students at iFoundry reviewed the undergraduate programs and courses internally in the College of Engineering and with respect to other programs offered throughout the country. The first incubator course, ENG 498, was introduced in Spring 2008. Its objective was investigating innovation initiatives in America's engineering programs ("History of iFoundry", n.d.).

iFoundry was formally integrated into the undergraduate program of the College of Engineering in August that year. The following month iFoundry formed a partnership with the Olin College. This led three years later to a formal Olin-Illinois Exchange program where students can spend up to two semesters in residence. The program enables Illinois engineering undergraduates to develop entrepreneurial design concepts at Olin in "a practical engineering environment where students attempt to solve real engineering problems" in an entrepreneurial setting with approximately 400 like-minded students (Lamb, n.d.).

In the fall of 2009, 75 entering students in the College of Engineering joined iFoundry. They participated in iLaunch, a pilot program to familiarize them with the opportunities available and jumpstart their first-year experience. iLaunch, held at a university retreat center known as Allerton Park, encouraged students to take the initiative. They connected, worked on projects, and took David Goldberg's workshop ENG 198, "The Missing Basics: What Engineers Don't Learn & Why They Need to Learn It" (Goldberg video, 2008). The workshop fostered the "critical and creative thinking skills of engineering" with the express purpose of creating "a whole new engineer" (Goldberg and Somerville, 2014, p. 52).

"The Missing Basics" exposed students to critical thinking and creative solutions. They learned the seven essentials to every successfully executed project: (1) Ask the right questions, (2) Label and categorize, (3) Model the problem, (4) Break down a project to its manageable parts, (5) Collect data and analyze it, (6) Visualize solutions, and (7) Communicate outcomes (Goldberg video, 2008).

iLaunch was a success. It was subsequently renamed the Illinois Engineering First-Year Experience (IEFX). The following year 300 iFoundry students participated. By 2011 the pilot program was rolled out to all 1,500 entering first-year students in the College of Engineering ("Introduction to: iFoundry and IEFX", 2013).

Today, IEFX is administrated through the College of Engineering Undergraduate Programs Office. Corporate involvement with entering students is nurtured through the IEFX NETWORK. Freshmen are offered one-credit "mini" elective courses to showcase their opportunities. IEFX collaborates with the NAE to feature "Grand Challenges" to encourage students to collaborate and design instructional approaches to solving important global issues ("IEFX electives", n.d.).

The iFoundry offerings continued. In 2009, Olin College and the University of Illinois jointly sponsored a conference, "The Engineer of the Future 2.0". Woodie Flowers, an MIT professor, gave the keynote address. He discussed student research suggesting that recent graduates in mechanical engineering at MIT felt that "soft skills" were more critical to their professional development than many of their required courses. This confirmed the findings of a recent NAE report in which soft skills were characterized as "teamwork, leadership, creativity and design, entrepreneurial thinking, ethical reasoning, and global contextual awareness" (Miller, 2010).

That conference was followed in November of 2010 with "The Engineer of the Future 3.0" held at the University of Illinois. Daniel H. Pink, the celebrated management and motivational expert, was the keynote speaker ("iFoundry Encourages Creativity in Education", 2010). Some 300 students, faculty, and educators attended the meeting, which emphasized the creative potential of "student-centered learning" ("History of iFoundry", n.d.).

Three years later "The Engineer of the Future 4.0" Conference was held. The focus was "Community and how it matters in engineering education". In each of the four conferences the focus was "to probe radical new ways to transform the student experience in engineering education" ("Engineer of the Future 4.0 Conference", n.d.).

By Fall 2011, iFoundry was empowered. IEFX had been introduced to all first-year students in the College of Engineering. Several new courses were rolled out including "Aspirations to Leadership" and "Interdisciplinary Senior

Design." The following year Intrinsic Motivation (IM) Course Conversion was unveiled, which fosters student self-taught mastery of selective courses in the core curriculum.

In November of 2013, an Innovation Certificate was initiated in conjunction with the iFoundry Technology Entrepreneur Center (TEC). The admission process for this program is partially based on a creative essay and an interview. Enrollment is limited to 25 students who apply during their first year in what amounts to a three-year program with potentially 75 students in all. The focus: "entrepreneurship, innovative product design, and transformative technical products and services" in order to create "breakthrough new products" ("Innovation Certificate", n.d.). Specially designed courses—12 credits in all—are featured, as well as coaching and support. Students learn how to understand customer problems, create solutions, meet timetables for deliverables, and work in real-world situations with entrepreneurially-driven companies.

In the Campus Honors Program, which accepts less than one percent of all undergraduates, engineers account for approximately one third of its students. In addition, the College of Engineering offers other qualified students an opportunity to enroll in the James Scholar Honors Program. iFoundry introduced the James Scholar Quest in 2013. It customizes the James Scholar Honors program to specific engineering departments ("James Scholar Quest", n.d.).

At some point the list of associated programs and opportunities in iFoundry become almost too numerous to mention. Nevertheless, CUBE Consulting, a Junior Enterprise organization, merits attention. It is a student-initiated engineering consulting group pioneered in 2013. It offers engineering and business undergraduates the opportunity to work on project teams for businesses, including startups, nonprofits, and research organizations.

It is these kinds of programs and initiatives within iFoundry that offer the students within the College of Engineering the possibility to obtain not only one of the best technical educations imaginable, but also to maximize personal opportunities for enrichment.

"HEROIC SYSTEMS" BLOCKBUSTER COURSE CREATED BY IFOUNDRY

In many respects, the Blockbuster Course[2] "Heroic Systems: Pushing the Boundaries of Greatness, Past, Present and Future" (ENG 298), introduced in Fall 2014, encapsulates the mission of iFoundry. It is a cross-disciplinary

---

2. The concept of Blockbuster courses is "multidimensional", a general education class approaching a theme from several disciplines including the "humanities, social science, and science to address real-world problems" (http://cee.illinois.edu/node/2967).

offering taught by faculty in the College of Engineering, the humanities, and business. The opening lecture, with its overarching perspective, is given by Andreas Cangellaris, dean of the College of Engineering.

"Heroic Systems" represents a new type of undergraduate course, one that consciously reinvents STEM education as a transformative initiative for the 21st century. Its purpose is to reimagine the engineering experience so that our graduates become the movers and shakers of strategic global policy, the very essence of "a whole new engineer", rather than narrow technical experts typically relegated to the backroom.

The faculty consisted of a culturally diverse group of scientists and humanists. They discussed a variety of "heroic systems" from Roman antiquity to the creation of the Midwest, to the rejection of a craft society in favor of an engineered outcome. The course examined the telecommunications, the electric grid, the space program, business analytics and its intersection with sports, and current developments in bioengineering and biomedicine, as well as the possibilities for heroic systems in developing marketplaces. The outcome: Students were exposed to big ideas, strategic thinkers, and creative intersections linking the science and the humanities. Recent results of a survey questionnaire indicate the course's strategic value ("ENG 298 Heroic Systems Survey Results", 2015).

THE IFOUNDRY TEAM: THE FINANCIAL AND ADMINISTRATE CHALLENGES TO IMPLEMENTING THE PROGRAM, PAST, PRESENT, AND FUTURE

The iFoundry team is small and dynamic. The dean of the College of Engineering, Andreas Cangellaris, remains involved, as does David Goldberg. Ray Price, who along with Goldberg serves as co-director, keeps a watchful eye on iFoundry's operations. Price, who works on creativity and innovation, collaborated with Abbie Griffin and Bruce Vojak in a study on how "tech visionaries" sustain their innovation in mature firms (*Serial Innovators*, 2012). Goldberg's book, *A Whole New Engineer*, examines the creative impetus at iFoundry and Olin (2014).

Karen Hyman serves as associate director. Bruce Litchfield, an assistant dean in the College of Engineering, is an iFoundry Fellow, as are approximately twenty other academics who participate. Geoffrey Herman, a visiting assistant professor, is engaged in programs related to engineering curriculum reform including the Intrinsic Motivation Course Conversion.

The administrative mandate for iFoundry is simple. Keep costs minimal. iFoundry funding is about $250,000 annually. It has been maintained through the College of Engineering with support from corporations and foundations. Some of these generous sponsors have included the following: Advanced Micro Devices (AMD), Autodesk, The Boeing Company, Hewlett-Packard Company (HP), International Business Machines Corporation (IBM), National Collegiate Inventors and Innovators Alliance (NCIIA), Procter and Gamble Company (P&G), Shell Oil Company, and the Severns Family Foundation (R. Price, personal communication, February 10, 2015).

Administrative appointments have been mostly part-time, as in the case of David Goldberg and Karen Hyman. Ray Price's position is largely supported through his academic appointment as the William H. Severns Chair of Human Behavior. One full-time administrative staff position was created in 2009. iFoundry Fellows are selected based on their expertise, interest, and support for the iFoundry mission. Their compensation is minimal (R. Price, personal communication, February 10, 2015).

Much of the financial support for the programs developed within iFoundry is borne by the academic units, whether TEC or specific departments, or even by the College of Engineering, where these innovative courses, when successful, become embedded into a reinvigorated curriculum.

Thus, in the case of the Innovation Certificate, iFoundry coordinates efforts with TEC, which sponsors most entrepreneurial courses in the college. One creative outcome is that engineering students not enrolled in the Innovation Certificate program can, nevertheless, take many of these course offerings and pursue entrepreneurial opportunities independently through TEC.

The same principle applies to the educational opportunities at Olin. Students from the College of Engineering at the University of Illinois can study for one or two semesters there and have the opportunity of working closely with Olin's faculty and students in an interdisciplinary "hands-on" environment where student ideas are formalized into conceptual designs for potential implementation into the marketplace. These exchange opportunities are reciprocal, allowing Olin students to participate in programs offered by the College of Engineering at the University of Illinois.

The benefit for students in our College of Engineering at the University of Illinois is that they obtain "the Olin effect" without the university having to "change the whole curriculum" or "build new buildings" or "remake the classrooms" or "overhaul the teaching or teachers" ("Introduction to: iFoundry and IEFX", 2013). Thus, academic excellence is maintained while students have the opportunity and entrepreneurial freedom to participate in Olin College's pathbreaking design and development environment.

The largest cost associated with the iFoundry initiatives thus far was the development of IEFX. The first year 75 students participated and the budget was

$120,000. When 300 students participated, $275,000 was allocated. When IEFX had 1,500 students participating, the budget was $500,000. Since 2012 IEFX has been incorporated into the College of Engineering where the costs are directly born by the college (R. Price, personal communication, February 10, 2015).

What has made the iFoundry experience work is its pilot stature as a curriculum incubator. It is small and nimble. It can be responsive to student interests and needs. Their voices can be heard and acted upon. Programs and courses and be tried without great expense or protracted discussion about the long-term implications for the curriculum. If successful, they can become part of the curriculum. iFoundry provides an intimate environment where individual student needs can be addressed. The experimental and innovative nature of the iFoundry incubator nurtures the development of "a whole new engineer", one uniquely equipped with the skills and outlook to address the "grand challenges" of the 21st century.

CONCLUSION: THE IMPACT OF IFOUNDRY

What makes the iFoundry experience so transformative? Upon arrival to the College of Engineering, students are immersed into IEFX. Along the way they weigh the entrepreneurial potential of the iFoundry Innovation Certificate or the entrepreneurial TEC programs or the Olin experience. They pursue internships, personalized courses of study, consulting opportunities, and work with faculty advisors on research or design projects. They consider overseas opportunities. They take courses recommended or developed, as in the case of "Heroic Systems", through iFoundry. All told, the sum of these experiences— some with the college, some with the departments, some with TEC, and some specifically created in the iFoundry incubator—is transformative. The outcome dramatically reshapes the perception students have of what it means to be an engineer in the 21st century and empowers them to invent the future.

Consult the iFoundry website for more information, (http://ifoundry.illinois.edu/).

REFERENCES

Goldberg, D. E. (2008). *The missing basics: What engineers don't learn & why they don't learn it.* YouTube video retrieved August 29, 2014, http://ifoundry.engineering.illinois.edu/media/readings/missing-basics.

Goldberg, D., Cangellaris, A., Loui, M., Price, R., & Litchfield, B. (2008a). AC 2008-1667: iFoundry: Engineering curriculum reform without tears. *ASEE*

Conference Proceedings. Retrieved February 10, 2015, https://uofi.app.box.com /s/bc1nkjit0hw9rs5rt9h0.

Goldberg, D., Cangellaris, A., Loui, M., Price, R., & Litchfield. (2008b). Preparing for substantial change: The iFoundry initiative and collective learning. *Proceedings of the 2008 ASEE IL/IN Section Conference*. Retrieved February 11, 2015, http://ilin.asee.org/Conference2008/SESSIONS/Preparing%20for%20 Substantial%20Change%20The%20iFoundry%20Initiativ.pdf.

Goldberg, D. E. & Somerville, Mark. (2014). *A whole new engineer: The coming revolution in engineering education*. Douglas: ThreeJoy Associates, Inc.

Griffin, A., Price, R., & Vojak, B. (2012). *Serial innovators: How individuals create and deliver breakthrough innovations in mature firms*. Redwood City: Stanford University Press.

iFoundry. (2007). Whitepaper for an Illinois foundry for tech vision and leadership. (July 18, 2007). *iFoundry website*. Retrieved August 29, 2014, http://threejoy .deg511.com/wp-content/uploads/2011/12/iFoundry_concept.pdf.

iFoundry. (2013). *Engineer of the Future 4.0 Conference*. iFoundry website. Retrieved February 10, 2015, http://ifoundry.illinois.edu/engineer-future-40-conference.

iFoundry. (2013). Introduction to: iFoundry and IEFX: Student engagement *iFoundry PowerPoint presentation*.

iFoundry. (2015). *ENG 298 Heroic Systems survey results*. Unpublished report. University of Illinois.

iFoundry. (n.d.). History of iFoundry. *iFoundry website*. Retrieved September 8, 2014, http://ifoundry.illinois.edu/who-we-are/history-ifoundry.

iFoundry. (n.d.). IEFX electives. *IEFX website*. Retrieved February 14, 2014), http:// www.iefx.engineering.illinois.edu/#!iefx-electives/c1xqq.

iFoundry. (n.d.). Innovation certificate. *iFoundry website*. Retrieved August 29, 2014, http://ifoundry.illinois.edu/student-opportunities/innovation-certificate.

iFoundry. (n.d.). James scholar quest. *iFoundry website*. Retrieved September 8, 2014, http://ifoundry.engineering.illinois.edu/student-opportunities/james-scholar -quest.

iFoundry. (n.d.). Who we are. *iFoundry website*. Retrieved August 29, 2014, http:// ifoundry.illinois.edu/who-we-are.

Lamb, K.(n.d.). Engineering at Illinois Student and OIX Participant. *iFoundry website*. Retrieved August 29, 2014, http://ifoundry.illinois.edu/student -opportunities/olin-illinois-exchange-oix/karen-lamb-0.

Lewis, M. (1999). *The new new thing: A silicon valley story*. New York: W. W. Norton & Company.

Miller, R. K. (2010). From the ground up: Rethinking engineering education for the 21st century. Keynote address given at the *2010 Symposium on Engineering and Liberal Education*, Union College, N.Y., June 4–5, pp. 1–20. Retrieved February 10, 2015, http://www.olin.edu/sites/default/files/ union_college_from_the_group_up.pdf.

Snow, C. P. (1964). *Corridors of power*. London: Macmillan Publishing Company.

The Daily Illini. (2010, November 15. Updated 2012, October 17). iFoundry encourages creativity in education. Retrieved August 29, 2014 from http://www.dailyillini.com/article_326d6a2f-14cf-56d3-8fc0-edee056f72e2.html?mode=jqm.

The National Academy of Engineering. (2004). *The engineer of 2020: Visions of engineering in the new century*. Washington, D.C.: The National Academies Press. [Electronic version available for download]. Retrieved July 20, 2014, http://www.nap.edu/catalog.php?record_id=10999.

The National Academy of Engineering. (2005). *Educating the engineer of 2020: Adapting engineering education to the new century*. Washington, D.C.: The National Academies Press. [Electronic version available for download]. Retrieved September 8, 2014, http://www.nap.edu/openbook.php?isbn=0309096499.

ABOUT THE AUTHOR

Dr. Diana E. Sheets is an iFoundry Fellow in the College of Engineering and a Research Scholar in the English and History Departments from the University of Illinois, Urbana-Champaign in Urbana, Illinois.

3

Current Directions in Modern Undergraduate Engineering Education

Anas Chalah, David Hwang and Fawwaz Habbal

ENGINEERING IN A LIBERAL ARTS CONTEXT

The School of Engineering and Applied Sciences (SEAS) is one of 12 schools at Harvard University. It offers to Harvard College students a full undergraduate curriculum, as well as Master's and PhD programs. Established in 2007, SEAS is the newest school in America's oldest university, and is transforming undergraduate engineering education. The school has no departments; most research is interdisciplinary, and the curriculum includes significant cross-disciplinary and system-level courses.

SEAS is embedded in a liberal arts school. Harvard undergraduates are first admitted to Harvard College. Unlike some programs in engineering and applied sciences, students choose a major ("concentration" in Harvard parlance) midway through their sophomore year. Some choose more than one area. In addition to a technical area, such as mechanical engineering or applied math, they are simultaneously immersed in a liberal arts environment. This mixing provides students a foundation for understanding the societal context for their technical problem solving.

In addition to SEAS' mission to educate engineers and applied scientists, SEAS aims to provide some level of technology understanding to all Harvard College students, by providing introductory courses that emphasize active learning.

An understanding of rapidly changing technology is essential for devising solutions to the world's most wicked problems. The SEAS curriculum is designed to educate students so they can respond to these societal challenges. Engineering has become essential core knowledge for every broadly educated person and an indispensable background for leaders. In addition, critical thinking skills, derived from broad exposure to the arts, humanities, and social sciences, provide students with tools to find the root cause of a problem before employing technology to solve it.

The SEAS strategy is to create the "21st century engineer" by educating students who excel in engineering and applied sciences, but who also have a broad knowledge of other disciplines. These "T-shaped" individuals[1]—possessing depth in one engineering discipline and educated broadly in other liberal arts disciplines—will be capable of collaborating seamlessly across multiple fields that are required to solve complex problems. Engineering with its mathematical language can impose challenges to many students. Thus to create this 21st century engineer, SEAS is finding new ways to engage students, deliver content, collaborate across the university, and connect classroom experiences to the wider world.

By investing in innovative new instruction techniques and making engineering more accessible to all students, enrollment in engineering and applied sciences courses has increased steadily since the establishment of the school in 2007, and the number of concentrators has also increased significantly (see Figure 1).

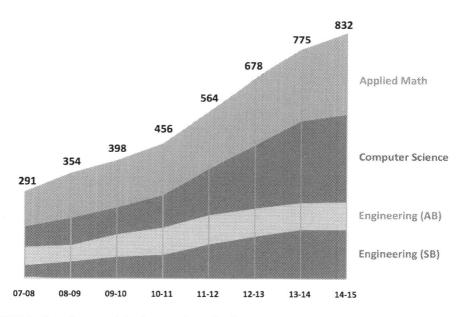

FIGURE 1. Steady growth in the number of college concentrators enrolled in Engineering and Applied Sciences

---

1. This term is used to indicate students who have a broad background, yet possess deep knowledge in a discipline. See for example: Tranquillo, (2013).

EDUCATIONAL MISSION—ACTIVE LEARNING AND DESIGN

The undergraduate curriculum at SEAS is organized around the premise that engineering and the applied sciences are both *multi-* and *inter-*disciplinary. This philosophy leads to a curriculum with a balance of theory and critical thinking skills, as well as deeply integrated hands-on design projects that provide active learning opportunities throughout the curriculum. By emphasizing the *skills of solving problems* through applying iterative feedback to a creative idea, the SEAS curriculum provides every student an understanding of the design process and the tools needed to solve complex problems.

Harvard is among only a few programs in the U.S. to offer both a Bachelor of Arts (A.B.) degree and an ABET[2]-accredited Bachelor of Science (S.B.) degree in Engineering Sciences. The *A.B. degree* requires a minimum of 14 to 16 courses. This degree provides solid preparation for the practice of engineering and for graduate study in engineering, and also is an excellent preparation for careers in other professions (business, law, medicine, etc.). The *S.B. degree* program requires a minimum of 20 courses, and the level of technical concentration is comparable to engineering programs at other major universities and technical institutions. In addition to the flexible Engineering Sciences A.B. and S.B. degrees, SEAS offers a rigorous S.B. degree in Electrical Engineering and Mechanical Engineering, and an A.B. degree in Biomedical Engineering. It also offers A.B. degrees in Applied Mathematics and Computer Science. The curriculum has a multitude of project-based design courses that teach engineering principles in a multi-disciplinary context.

The emphasis on design thinking, experiential learning, as well as peer-to-peer learning (Mazur, 2013; Bruff, 2009) has permeated across most courses. These elements are integrated within the curriculum and supported by teaching staff and appropriate infrastructure.

The Role of the Undergraduate Teaching Laboratories

A multitude of state-of-the-art rapid prototyping, fabrication and testing resources are placed in the SEAS Teaching Labs[3]. These labs are staffed by professionals with higher degrees in electrical engineering, environmental engineering, bioengineering, chemical engineering and mechanical engineering. The mission of the Teaching Labs is to provide students with infrastructure and

---

2. The S.B. program in Engineering Sciences is recognized by the national accreditation agency for engineering programs in the United States: "Engineering Accreditation Commission of the Accreditation Board for Engineering and Technology, Inc. (ABET)."

3. Visit: http://www.seas.harvard.edu/teaching-labs

hands-on learning experiences and tools for problem-solving across multiple disciplines.

Faculty work closely with the Teaching Labs staff in designing experiments and activities appropriate for their courses and educational outcomes. The teaching staff is responsible for preparing the required infrastructure, whether it is hardware or simulations. Faculty are frequently present during active Teaching Lab periods and work closely with the teaching assistants and staff to ensure that students gain maximum educational benefit from the engagements.

The Teaching Labs are also the place where visiting students, from high schools, universities, and other countries work with SEAS students to conduct a variety of projects, some of which are open-ended research projects. These projects vary in depth and breadth, but they all require multidisciplinary problem solving skills. Examples include dealing with water and air pollution mitigation, generating green energy, designing medical devices, and developing different types of software.

Undergraduate students use the Teaching Labs for executing ideas they have, either as individuals or as part of groups and students clubs. In most cases, students have mentors from the Teaching Labs or faculty. A long list of student organizations at SEAS provide additional opportunities for SEAS concentrators as well as other Harvard College students to collaborate on real-world problem solving.

Students show their ideas and projects through an annual *SEAS Design and Project Fair,* organized by faculty, SEAS teaching staff, and other staff. The fair attracts SEAS concentrators as well as student projects from across Harvard College (for example[4]). The range of projects displayed every spring is incredibly broad, with dozens of SEAS courses with project components represented.

Extra-Curricular Projects Are a Complementary Part of a Student's Education

Many SEAS engineering students choose to increase the depth and breadth of their technical knowledge by working on extracurricular design projects, either individually or in teams. The goal of these projects is often to implement or disseminate a solution to a problem in the real-world context, outside of the classroom. SEAS encourages students to come up with their ideas and projects that may have commercial value. Students' inventions and related IP are owned by the inventing student(s). SEAS and its faculty do not share or participate in the ownership of such IP.

---

4. http://www.seas.harvard.edu/news/2010/12/es-51-drives-home-principles-engineering -design

Furthermore, SEAS offers small financial support for these extracurricular projects through *the Nectar process*[5]. Nectar is the official funding process at SEAS to support undergraduate co-curricular initiatives, defined as extra-curricular initiatives with curricular (technical) content. Students or groups of students working on co-curricular projects are eligible to apply for a semester of funding or for longer-term funding. Grants for semester projects are typically $2,000 or less, while long-term projects are eligible for a higher funding amount. All students engaging in Nectar projects are required to work with a faculty advisor, and those that require physical prototyping space are often supported by the Teaching Labs. Posters from Nectar-sponsored projects are displayed at the end of each funding period.

"HOLISTIC LEADERS" WITH SYSTEMS THINKING, DESIGN AND INNOVATION SKILLS

As the world grows in complexity and societal problems shift to broad multidisciplinary concerns, a new type of engineering leadership is needed. This leadership requires not only deep technical expertise, but also the ability to examine issues broadly and work and communicate collaboratively. At Harvard SEAS, one of the critical areas of transformation has involved a shift to train such future "holistic leaders", who are skilled particularly in (1) systems thinking, (2) design capability, and (3) innovation.

One platform used at Harvard to foster the above skills is Engineering Sciences 96 (ES 96), which is a junior-level core/required course for Bachelor of Science engineering students. Thirty years ago, two faculty members with EE and ME backgrounds established ES 96 as *Engineering Problem Solving and Design Projects*. The goal was to train students to work on open-ended problem solving. This course evolved over the years, and incorporated complex, multi-dimensional problems with a client who becomes the recipient and an evaluator of the project outcome. As the SEAS engineering program expanded, the course was revised to utilize the design thinking process, multi-disciplinary thinking, and computer simulations[6]. Normally, the class is composed of 10–15 students from all SEAS engineering disciplines (Bio, Electrical, Mechanical, Environmental) who, for one semester, consider an open-ended problem, often provided by an outside "client." Recent clients and areas of study have included

---

5. http://www.seas.harvard.edu/nectar
6. Fred Abernathy, Victor Jones, Rob Howe, Kit Parker, David Mooney, Woody Yang, Fawwaz Habbal, Jim Anderson and Karena McKinney made significant contributions to ES 96 in recent years.

using technology to combat gang-related violence with the Springfield, Massachusetts Police Department; addressing patient-doctor challenges in managing non-healing wounds in diabetic patients with the Harvard Medical Center; and improving environmental mitigation strategies for the Fukushima nuclear disaster. The course continues to train future holistic engineering leaders. Students often cite this course as pivotal in their learning and career direction.

Importance of Posing an Open-Ended Problem in Undergraduate Education

In courses such as ES 96, students are posed with a large-scale, open-ended problem. Students are asked to determine collectively a specific area(s) that can result in an outcome that is significant and is feasible to address in a single semester. Students are guided by faculty and provided principles of open-ended design, but they must determine the final plan of action themselves. They must consider the problem at hand, the varying skill sets and technical expertise of the team, and determine a course of action and project plan. Spending too little time scoping the problem may result in an ineffective solution; while spending too much time scoping would result in time pressures to complete a full study.

Faculty encourage the development of milestones both internally and with the client, but students must manage the project from end to end to attain a successful outcome. This type of project requires a significant faculty-student interaction and often results in rewarding outcomes.

Broad "Systems" Thinking and Communication Ability are Required to be Successful in Creating Effective Engineering Systems and Large-Scale Solutions

Students are encouraged to not only understand the technological issues at stake, but also the broad impacts to the environment, society, and the economy. They are trained to understand that technical solutions must also be implementable in order to have a lasting impact. Thus the questions that are posed to students probe technical knowledge, as well the broad thinking and ability to create an impact.

For example, in today's world, often the decision-makers for addressing large technological projects are not technical leads, but business leaders or policy makers. Therefore, students are trained to communicate their results to a non-technical audience. A surprise quiz where students must provide an "elevator pitch" to a simulated external party is often used for training. At the end of the course, students produce a final written, collective document, as well as provide an hour-long final presentation (where all students participate) to the client.

Design Thinking Process is a Good Tool for Problem Solving

Students are also taught the design cycle shown in Figure 2. In this design cycle, students are taught to first work with their client/stakeholders to empathize and define the problem at hand. They are then tasked to brainstorm and ideate, create prototypes, and test the various prototypes. Due to the length of the semester, the number of design iterations are typically limited in number; although students are required to execute at least one complete iteration in a semester.

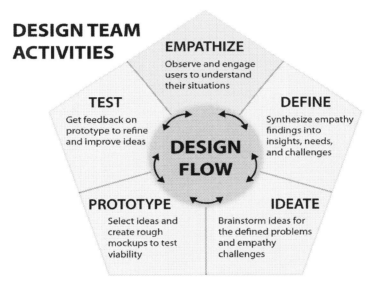

FIGURE 2. A simplified design flow with interactive team design exercises used to teach students to work within a team and with clients.

Innovation is Part of Finding Solutions

Students are encouraged to consider approaches to real-world problems using innovative thinking. Faculty, many of whom have entrepreneurial experiences, train students to consider approaches that are transdisciplinary, applying knowledge from one domain or discipline into another. For example, a knowledge of analytics for databases can be applied to crime patterns and tracking. Another example is a critical understanding of biological fluid flow applied to understand failure modes in environmental contamination. Since students in a course such as ES 96 come from multiple disciplines and have a range of backgrounds and experiences, team-based innovation is fostered and encouraged.

After experimenting with the revised version of ES 96 for approximately three years (and assessing students based on detailed rubrics across multiple outcomes), the course has become a pillar of the SEAS engineering curriculum. ES 96 is one example of the changes and transformations that SEAS is developing to train engineers to be holistic leaders in the 21st century.

LESSONS LEARNED: HOW CHANGE WAS BROUGHT ABOUT AND APPLICABILITY TO OTHER INSTITUTIONS

This paper has discussed changes made to Harvard's engineering programs in the last few years, with particular focus on active learning, design, and creating holistic leaders. The origin of this institutional change dates back to 2006, when the Harvard provost and dean of the Faculty of Arts and Sciences (FAS) called for the re-examination of the educational curriculum in all Harvard schools in an effort to promote interdisciplinary education and research. The provost created a university-wide committee with regular progress reports by each school.

The dean of the School of Engineering and Applied Sciences (at that time, the Division of Engineering and Applied Sciences) convened a faculty group to assess the engineering research agenda and curricula in response to the provost and FAS dean's initiative. The faculty concluded that while Harvard's curriculum was strong and broad in engineering sciences, newly hired faculty could create specialized courses in mechanical, electrical, and biomedical engineering that could lead to new degrees in these areas.

Around this same time, in recognition of the critical importance of technology and engineering to the university, the president, the Harvard governing board, and the dean of Engineering and Applied Sciences led efforts to elevate the Division of Engineering and Applied Sciences to become a separate School of Engineering and Applied Sciences. The School of Engineering and Applied Sciences (SEAS) was thus established in 2007. The establishment of the school, combined with the provost's initiative, as well as the faculty's desire to establish new degrees, led the founding academic dean and his successor to re-evaluate the engineering curriculum.

Faculty committees appointed by the SEAS dean initiated the discussion about new pedagogy in engineering education, continuing the work begun by the provost and the FAS dean's initiative. The SEAS dean and faculty determined that a 21st century engineering curriculum should emphasize active learning, design, interdisciplinary education, and leadership in a societal context. The two deans invested resources to make this change happen, setting aside both space and funding needed to increase active learning content in courses. The recommendations of the committees, with strong backing of the dean, thus led to the hiring of design faculty, establishment of modern Teaching Labs supported by high-caliber engineers and technicians, and a new era of support for innovation and entrepreneurship discussed in the previous sections of the paper. It also led to the establishment of new concentrations (i.e., majors) in Biomedical Engineering, Electrical Engineering, and Mechanical Engineering.

A core group of faculty members were on the initial committees and were strong proponents of the increase in active learning. As more faculty experimented with active learning in their courses, momentum built in this new direction. Junior faculty (i.e., tenure-track faculty), who were passionate about this type of learning, were also hired. College students responded enthusiastically to these curricular improvements, with undergraduate enrollment in engineering programs increasing dramatically, continuing the momentum. With a strong core of administrative leadership, senior faculty, junior faculty, and student body excited about the pedagogy, the change has been embedded in the culture and we believe it is now an integral part of SEAS.

There are a number of lessons learned in our experience that may be applicable to other institutions considering such a change. Most importantly, for the change to be successful, the following factors should be in place:

- The right institutional environment and leadership. Extensive initiatives that require investment and support require the leadership of the institution. In our case, this meant the strong leadership of the president, provost, the dean of the Faculty of Arts and Sciences, and the dean of SEAS to rally together to create the appropriate educational support.
- Faculty who will champion the initiative. A group of engaged senior faculty with diverse technical backgrounds realized the need for active learning and a more innovative pedagogy and gave credibility and strength to the effort.
- Appropriate resources/infrastructure. Our particular initiative needed resources and infrastructure to be successful. The initiative gained traction because the SEAS dean made it a priority.
- Ability to leverage the existing strengths in an institution. The initiative was successful because it leveraged the existing strengths of Harvard. For example, Harvard is primarily a research university; the initiative used this fact as a strength by setting up some labs and experiments that were undergraduate-appropriate versions of experiments conducted in a professor's research lab. This allowed faculty to impart their research in a way that was accessible and appropriate for undergraduates, giving students access to cutting-edge technologies.

Our experience provides evidence that making a dramatic difference in curriculum and pedagogy across a university requires leadership, vision and commitment at the senior administrative level (provost and president tier), as well as at the course level (instructors engaged in the new pedagogies). The described pedagogical change at Harvard SEAS was successfully accomplished due to strong dean leadership, and a core of passionate faculty pushing for this change.

Author Notes

achalah@seas.harvard.edu, dhwang@seas.harvard.edu, habbalf@seas.harvard.edu
School of Engineering and Applied Sciences, Harvard University
29 Oxford Street, Cambridge, MA 02138, USA

REFERENCES

Bruff, D., (2009). *Teaching with classroom response systems: Creating active learning environments.* San Francisco, CA: Jossey-Bass Publishing- Wiley.

Mazur, E.,(2013). *Peer instruction: A user's manual.* Boston, MA: Addison-Wesley.

Tranquillo, J., (2013). The T-shaped engineer: Connecting the STEM to the TOP. *Proceedings of the Annual Conference of ASEE.* Available from http://www.asee .org/public/conferences/20/papers/7610/view

ABOUT THE AUTHORS

Anas Chalah is the Executive Director of Active Learning and the Director of Lab Safety in the John A. Paulson School of Engineering and Applied Sciences at Harvard University in Cambridge, Massachusetts.

Fawwaz Habbal is the Executive Dean for Education and Research in the John A. Paulson School of Engineering and Applied Sciences at Harvard University in Cambridge, Massachusetts.

David Hwang is the Assistant Dean for Education in the John A. Paulson School of Engineering and Applied Sciences at Harvard University in Cambridge, Massachusetts.

4

How an R-1 University Rallies Around Transforming Education: Opportunities and Challenges

Fatma Mili, Robert Herrick, Tom Frooninckx

The Purdue Polytechnic Institute is an initiative to transform the Purdue University College of Technology while setting a general exemplar useful for others. It was selected as a *Purdue Moves* initiative, one of 10 pillars that define the university's strategic direction, and has received tremendous resources and attention. Now in our second year, we reflect on where we are, what we have accomplished, and how we make sense of the opportunities and challenges within the general framework of social science scholarship in the field of transformational change.

THE PURDUE POLYTECHNIC INITIATIVE

The last decade has seen a growing desire to "do something" about higher education in general, and higher education in STEM in particular. The voices for action were driven by three categories of concerns:

1. The graduates we are producing are not well prepared for the economy and the world in which they are called to function. Surveys of employers and reports by professional organizations and federal agencies are consistent in their findings (AACU, 2011; Levy & Murnane, 2013; Wagner, 2010) that students are not developing key skills and habits of mind.
2. Thanks to advances in research in education, motivation, and learning, we now know much more about effective learning environments and methods (Sokol, Grouzet, & Muller, 2013; Deci & Ryan, 2013; Pink, 2006). We also know that we are serving a generation of students with different aspirations, different anxieties, and different modes of learning and collaborating (Brown & Seely, 2011; Levine, 2012).
3. Department of Labor statistics show that higher education is becoming almost a prerequisite to earning a living; surveys of students and

parents show that higher education is seen as a key part of the education of a fulfilled citizen (Bowen et al., 2009; Sahlberg, 2011). Yet access to and acceptance by institutions of higher education remain difficult or unattainable for a significant percentage of our youth.

The Purdue Polytechnic Institute aims to address these three shortcomings by starting from a *tabula rasa* position, to question all assumptions and to rebuild education for the 21st century. We knew that this complete re-acculturation process would require time and resources, as well as a "safe space" in which to work. Extensive funding was given to this initiative to allow it to emerge and thrive.

A call was issued in the summer of 2013 to Purdue faculty interested in designing an education that accounts for the most recent findings on human motivation and human learning, and for the future needs of graduates and employers. A group of 16 faculty was selected; they are affiliated primarily with the College of Technology with representation from the Colleges of Liberal Arts (English, Communications, and Performing Arts), Education, and the Libraries. The group was given the following guidelines and expectations:

1. *Create a pilot learning experience to begin in the Fall of 2014*: Part of the experiment is to break away from the traditional "production studio" mode of operation where extensive preparation and very close review take place upfront. Once something is approved, its offering is routinized. By contrast, we were aiming for a "research laboratory" mode of operation where teaching and learning are research, never routinized. We preferred therefore to start sooner rather than later with the understanding that everything is subject to questioning and change.

2. *Spend one semester creating the right mindset*: Paradoxically, while we wanted action fast, we also wanted to avoid falling into the traditional, deeply ingrained reflexes. It was very important for the faculty to spend the semester creating a common baseline of research, adopting an agreed upon charter, and especially creating the right mindset to take risks and ask hard questions.

3. *Think outside of existing constraints*: Along with slowing down decisions, part of creating the right mindset required that we do not start with existing rules, regulations, and processes.

In light of the above expectations, the team of faculty fellows did the following:

August 2013–December 2013: Creating the right mindset; community building; learning together: The team met on average twice a week for a few hours at a time. They invited speakers, attended seminars and workshops, and read and discussed books, research papers, and blogs. By the end of the first semester, they adopted a set of values (Mili, 2013 c) and agreed upon a small number of guiding principles. In particular, the faculty subscribed to the idea that learning happens best when it is driven by student's intrinsic motivation (Deci & Ryan, 2013) and that a growth mindset is a more important and more lasting asset to learning than talent alone (Dweck, 2006). The faculty also agreed on the need to cultivate more than technical skills, especially in light of the findings that graduates often do not have what it takes to succeed past their first job (AACU, 2011; Wagner, 2010).

January 2014–May 2014: Self organizing, self-governing, and designing: The faculty started the year with a two-day retreat in which they agreed to a foundation for the learning experience to be designed. There would be two pillar settings: Design studio and Socratic seminar. Both of these settings would be transdisciplinary, and problem-based. Topics and skills would be iterated using a spiral approach. The focus now was to design the Fall 2014 offering.

Although the proposal approved in June 2013 was to create a relatively autonomous institute within which faculty fellows would operate, the structure remained virtual. This was in part a deliberate choice because we considered the governance model as an integral part of the innovation. We were aware of the interplay between governance and culture (Kezar, 2004) and wanted to take the time to design it "right." In the process, we experimented with governance by hierarchy, which led to resentment; no governance, which led to chaos; and decisions by consensus, which led to paralysis. We finally settled on Holacracy (Robertson, 2015) which we are still using. Some of the key characteristics of Holacracy that we found appealing: It is a distributed governance system in which authority and accountability are tightly coupled; governance is dynamic and responsive to changing needs and situations in a very transparent and codified manner; everyone is empowered to make proposals to make changes; and it is strongly biased towards action and against paralysis. Faculty took to it very quickly and adopted it wholeheartedly. It allowed us to organize in "circles" with well-defined domains and authority that were operating with a high level of autonomy and full transparency. This has contributed to strengthening the level of trust among the faculty. The circles include: Design Studio, Seminar, Assessment, Space, Credentialing, and Degree Architecture.

April 2014–August 2014: Recruiting and welcoming Fall 14 cohort; Finalizing 1st-year design; Designing Bachelor of Science Degree in Transdisciplinary Studies: During the summer the first cohort of students was recruited and welcomed. Because we did not have our own degree, the first year offering was designed so that it could serve students in all of the College of Technology majors (7 majors) and students in Exploratory studies (undecided). We invited students accepted in the College of Technology and in Exploratory Studies to apply. A group of 36 students was accepted. We prepared the space and ordered equipment and supplies for the design studio and the seminar class. As we proceeded through the first two semesters, it became very clear that unless we had our own degree, we would have very little curricular decision space. We invested considerable time in defining the Purdue Polytechnic degree with the intention of taking it through the requisite approval processes so that it is in place for fall 2015. The degree requirements are defined in terms of learning experiences and milestones, and competencies the student must demonstrate. The implementation of the competency-based approach touches upon almost every aspect of university management and requires collaboration from all: registrar, bursar, financial aid, academic units, etc. The adoption of the competency-based approach was rewarded with $5,000 and full support from all administrative offices.

August 2014–December 2014: Running the first offering; Learning with the students; and Managing the mixed reactions: Nothing was as exhilarating as welcoming the students, putting in practice what we had been designing for a year. Very often the outcome surpassed our expectations, and several times we had to backtrack and fix the plane while flying. We feel a tremendous gratitude towards the 36 students who enrolled with us. They exhibited a level of energy and optimism that is very inspiring. We ended the semester with many research questions that need further exploration, but with the strengthened conviction that this is the right direction. Students benefitted from the experience much more than from a traditional setting, even (and maybe especially) those who were not ready for the level of autonomy and self-direction we gave them. Even for those things that did not work, we find solace in David Price's 2013 statement: "No student ever had his entire education ruined because of a learning innovation that didn't come off. But I can show you plenty of students whose curiosity and imagination were strangled by being trapped in a repetitive, uninspiring, unimaginative learning enclosure."

As exhilarating as it was this period is also the one when we experienced the most intense and persistent resistance. We managed to remain under the radar as long as our activities were abstract designs with no visible manifestations. With the classrooms buzzing with students and activities, scrutiny and objections multiplied. The rationale we used for creating the Purdue Polytechnic Institute was to create an autonomous, safe space for faculty to innovate and experiment with a different model of teaching, learning, and governance. We have largely succeeded in creating that different culture and setting; but the Purdue Polytechnic does not exist in a vacuum. It is a transformative unit within a larger system, and not all of the system components were receptive to its activities.

INSTITUTIONAL TRANSFORMATION, GENERAL FRAMEWORK

The topic of institutional transformation has been extensively studied (see Westley, et al., 2013 for a recent survey). This topic has been studied from two perspectives: resilience and transformation. Once established, resilient systems create mechanisms by which they self-perpetuate. Institutional transformation in an established system takes deliberate resources and actions to overcome the perpetuating resilience processes.

Institutional changes or transformations "arise when organizational actors with sufficient resources (institutional entrepreneurs) see in them an opportunity to realize an interest that they value highly" DiMaggio (1988). This formulation is consistent with the consensus in the literature (e.g,. Kezar & Eckel, 2002; Garud, Hardy, & Maguire, 2007) that the three most critical elements in an institutional transformation are: actors (agents), resources, and opportunities.

The motivation and creativity of agents constitute the most important trigger and sustainer of transformations. The power of the vision driving them and their passion for its realization lead them to break away from scripted thoughts and actions and to take considerable risks. Emirbayer and Mische (1998) argue that actors differ by their temporal embeddedness (i.e., whether they are focused on the past that they are trying to preserve and reproduce, the present to which they are trying to react, or the future that they envision and want to create). The temporal embeddedness captures the predominant orientation, mode of thinking and acting, and preferred form of agency: routine, sense-making, and strategic (Dorado, 2005). Agents oriented towards the past lean towards routine thinking through the selective reactivation of past patterns of thought and action. Agents oriented towards the present lean towards sense-making agency through the continuous scanning of the environment in an effort to frame it in a coherent way and react to it. Agents oriented towards the future lean towards strategic agency through the imaginative generation of possible future trajectories of action defined by their hopes, fears, and desires. Strategic agency is the most clearly related with transformational change; it is goal-oriented and focused on the long term, thus abstracting away daily distractions and derailment. This is especially necessary because periods of change create high levels of uncertainty and rapidly changing environments (Emirbayer & Mische, 1998). Sense-making agency is important in framing the situation, explaining it, and communicating it so as to keep the focus on the strategic change (Weick, 1998). Successful transformation requires a combination of sense-making and strategic agency.

An institutional transformation is by definition an effort to depart *significantly* from the usual patterns of behaviors and actions. Because departures from the usual norms "threaten dominant positions and break away legitimate order" (DiMaggio, 1988), they carry a high risk to the actors. Mobilizing resources enables them to take these actions and buffers them from some of the resistance and backlash. The literature identifies three distinct resource mobilization processes: leveraging, accumulating, and convening (Dorado, 2005).

- *Leveraging* is the purposeful fundraising and grant seeking process. Actors define a project, then seek support from backers.
- *Accumulation* is the process by which independently initiated actions and interactions over extended periods of time converge and create a momentum in support of an idea (Van de Ven & Garud, 2004).
- *Convening* is social activism driven by the desire to see an issue get due attention and eventually be addressed.

Dorado (Dorado, 2005) defines opportunities as "the likelihood that an organizational field will permit actors to identify and introduce a novel institutional combination and facilitate the mobilization of the resources required to make it enduring." Opportunities depend on at least two characteristics: multiplicity and the degree of institutionalization.

Institutional entrepreneurship scholars (Battilana, Leca, & Boxenbaum, 2009; Dorado, 2005) define multiplicity in terms of the number and overlap between the different institutional referents experienced by actors. The multiplicity of referents creates tensions that trigger actions; it also broadens the agents' minds and cultural toolboxes (Swidler, 1986) and enhances their ability to frame new institutional arrangements in ways that make them acceptable to all parties (Douglas, 1986).

An environment with little or no institutionalization is characterized by unpredictability and high levels of uncertainty; this inhibits strategic agency. A very high level of institutionalization codifies patterns of thinking and behavior, and punishes deviations from the norm. This also inhibits strategic agency. A medium level of institutionalization is more conducive to innovation.

Based on these two dimensions of multiplicity and institutionalization, three categories of environments, known as predominant forms, are identified. When multiplicity is low and institutionalization is high, there is little room for deviation from the norm and actors have few reasons to do so; the environment is *Opportunity Opaque*. This is the most traditional setting for mature organizations. When there is a moderate level of institutionalization to provide structure and stability, but not too high to the point of stifling innovation, and when there are several institutional referents, the environment promotes innovation, it is *Opportunity Transparent*. Innovative organizations deliberately create this transparency through their organizational choices, hiring, and professional development. When there is a high multiplicity and low institutionalization, the level of uncertainty and unpredictability is too high. The environment is *Opportunity Hazy* (Dorado, 2005). This characterizes social structures in transition; for example, a country right after a revolution.

TRANSFORMATIONAL PROCESS

Institutional transformation is the process by which agents in the system create something new to realize a vision or correct a situation deemed untenable. Holling captured the process by which ecological (or social) systems evolve and persist or transform in a seminal paper (Holling, 1986) as an infinite looping process between four phases defined below:

Launch and Growth: This is the successful launch of new ideas. It requires the availability of free (non-attached) resources and a high level of diversity of ideas and referents (multiplicity). The successful launch leads to the growth of the innovation. This is a slow and lengthy process. As long as all conditions are favorable, the innovation continues to consume resources and introduce a level of uniformity and conformity. As we proceed through the growth, a higher percentage of the resources becomes attached and the level of sameness and conformity increases, thus multiplicity decreases.

Conservation and Institutionalization: When innovations reach their level of maturity, growth slows down, most resources tend to be locked in the system, and sameness and uniformity are at the highest level. This phase is characterized by a high level of order and institutionalization of processes. There is a definite norm and tradition for everything. This phase is both very robust (resilient) but also vulnerable to major changes or disruptions.

Release and Creative Destruction: Large institutions are at risk from major disruptions that can be triggered by internal or external agents or events. Disruptions break existing structures and release resources. At this stage, the level of uniformity remains high, growth stops, but resources are available. Creative destruction is a short lived phase that prepares the ground for exploration and reorganization.

Reorganization and Exploration: The release of resources from the creative destruction triggers an exploration of ideas and a high multiplicity of views and options. When one of these emerges as the dominant idea and resources are secured, the system reorganizes to support its launch and exploitation; and a new growth phase starts.

The same loop captures incremental changes, slow evolution, resilience, and innovation. Resilient systems adopt changes and reorganize in a way that minimizes disruptions and allows them to resume growth in a minimally modified manner. In other words, a resilient system can absorb a perturbation and adapt to it with minimal change. Innovation and transformation on the other hand follow the same cycle but *reorganize* in more fundamental ways redistributing resources.

GOVERNANCE AND CULTURE

The description of the transformational prerequisites and process refers to several features that can aid or hamper transformation. There is a direct relationship between institutionalization and multiplicity on the one hand and the opportunity for innovation on the other. What active interventions can one make to affect these parameters? The two main candidates are administrative structures (i.e., governance systems), and social structures (i.e., culture, leadership, and human relationships).

In an article titled "What is More Important to Effective Governance: Relationships, Trust, and Leadership, or Structures and Formal Processes?" Adrianna Kezar (2004) examines the literature on the inadequacy of academic governance. Academic structures are predominantly hierarchical, slow, inefficient, and ineffective in using expertise-based decision making and responding to the needs and concerns of internal and external constituencies. All of this breeds mistrust and disengagement from important players and potential innovators.

Campuses who have responded to the call to adopt different and more participatory processes did not necessarily show significant improvements in efficiency or effectiveness. Kezar (2004) examined several case studies and concluded that culture, able leadership, and human relationships are much more important than the formal processes used. With the right leadership and climate of trust, the exact processes used are almost irrelevant. If trust is absent, even when shared governance processes are used, they remain suspect.

THE PURDUE POLYTECHNIC TRANSFORMATION—AGENCY, RESOURCES, INSTITUTIONAL OPPORTUNITIES

The Purdue Polytechnic Institute was triggered through an initial strategic definition of goals and aspirations captured in foundational white papers by Mili and Bertoline (2013 a,b,c). The faculty fellows and staff brought an additional mix of strategic and sense-making agency, and they each bring something unique in terms of their temporal inclination, as well as motivation, skill, work, and educational experience. In particular, sense-making has proved to be critical as we navigated the many backlashes and sources of resistance.

Because the transformative innovation was taking place primarily within the College of Technology, most of the questions, anxiety, and resistance came from there. We capitalized on the standing of one of the faculty (second author) in the college to help communicate with all the department heads and advisors and to listen to their concerns and address them. His recognition (distinguished professor), his track record with national and professional organizations (ASEE, ABET), and his length of service in the college contributed much of the sense making as we proceeded through the design and implementation.

Because the initiative required extensive collaboration with several other offices on campus, we used a team approach to sense-making on campus. The fact that one of the faculty fellows is the vice chair of the Purdue Faculty Senate proved to be an asset in navigating much of the faculty governance. The first author is a member of the Purdue ADVANCE steering committee, as a result, even though relatively new at Purdue, she created many trust relationships with women department heads, associate deans, and faculty, which proved instrumental. Support from the highest office on campus, combined with sense-making and familiarity with the institution on the ground were instrumental.

The resource mobilization for the Purdue Polytechnic project falls more closely under leveraging, although the research and scholarship aspects of it are more closely related to accumulation and convening. A proposal was developed in the summer of 2013 and submitted to the president and the Board of Trustees. Upon approval, we started the further development and the implementation.

The resources from the president came in three different forms:

- *Financial Resources:* Significant funding was allocated to this initiative in the form of a start-up budget and a recurring budget conditioned by enrollment. Two aspects of this funding proved to be significant: The dean of the College of Technology (Bertoline) is a co-PI on this initiative, and the support was given by the president so that Purdue Polytechnic incubates the transformation of the whole college.

- *Decision Authority Resources:* The support and directives from the president erased many potential barriers. For example, because credit hours are embedded in everything we do, the adoption of the competency-based approach required a significant investment and participation from almost every administrative structure on campus (e.g., the registrar, financial aid, and enrollment management). Implementing competencies without the president's support would have been very difficult.

- *Promotion:* The president has repeatedly praised this initiative and its potential in many public settings inside and outside Purdue. This backing lends us credibility with the Purdue community, students, parents, and employers.

INSTITUTIONAL OPPORTUNITIES

The Purdue Polytechnic Institute is an innovation taking place within a larger context: Purdue University, and the College of Technology. Although the initial proposal and subsequent efforts define it as a transdisciplinary, university-wide effort, many pressures reduced its scope so that it is primarily a College of Technology initiative. Most of the resistance we have faced came from within the College of Technology.

Institutionalization: The College of Technology at Purdue is not unlike any other college in an R-1 university. The history, the size, and the complexity of the institution by definition necessitate the creation and following of highly codified processes, rules, and deadlines, very often to a fault. As has been observed in most higher education institutions (Birnbaum, 2000), these lead to inefficient, slow, and ineffective processes that stifle quick actions and experimentation. It is because of our experience with these processes that we proposed (and failed so far) to have the Purdue Polytechnic a relatively autonomous entity with new, more agile, and faster processes. In summary, institutionalization in our context is high.

Multiplicity: Because of its roots and history, the College of Technology attracts faculty who are passionate about teaching and teaching by doing, and who have prior industrial work experience. The latter used to be a requirement. As such, most of them have been exposed to multiple referents, the industrial, and the academic; but these are non-overlapping referents that do not influence each other; an industrial process for example is seen as irrelevant in an academic setting. Furthermore, in part because of the specificity of the discipline (there are few colleges of technology that graduate PhDs.) and because of the geographic location of Purdue, for most faculty, this is the only academic position they have held, and for some, this is also the college from which they graduated. This greatly reduces the multiplicity of referents.

On the one hand, we have a self-selected extremely diverse team of faculty fellows ready to create a new transparent environment in the form of Purdue Polytechnic Institute. On the other hand, they need to operate under the opaque environment of the traditional College of Technology. Giving the institute some level of autonomy creates a somewhat transparent environment, whereas putting it fully under the college jurisdiction makes the environment opaque. This is the nature of the debate held for two years now. The hesitancy stems from the fear that an autonomous institute will have little impact on the college it is created to transform. There is also an important resource aspect to the setting. Transformations happen when "free" resources that are not locked into existing processes and structures become available. In this case, there is a significant injection of new resources allocated to this innovation. This has highlighted another rationale for creating an autonomous unit to house the innovation: As long as the initiative is fully embedded within the traditional college structure, the new resources get automatically locked in old processes and hamper rather than help the transformational efforts.

TRANSFORMATIONAL STATUS

The cycle capturing the Purdue Polytechnic transformation is summarized in the following timeline:

1982–2012: 30 years of Growth, Institutionalization, Conservation, Opaque

This period captures the slow and steady growth of the College of Technology. The college outgrew its focus on associate degrees and started offering world-class accredited bachelor's degrees (e.g., the CIT degree is one of the first ABET-CAC IT degrees accredited). Several graduate degrees were added, including a PhD in technology. The growth of the faculty shifted from an almost exclusive focus on undergraduate teaching to a focus on research and graduate teaching. Through its 50-year history, the college has developed and established a well-recognized culture, and a comprehensive set of norms, processes, and traditions.

2012–2013: 1.5 years of Disruption, Opportunities, Uncertainty, Haziness

No disruption is completely sudden or instantaneous, but usually it takes time to come to the surface and truly disrupt the orderly working of a well-institutionalized system. Several factors conflated towards this disruption: At the national level, the expressions of dissatisfaction with some aspects of higher education kept getting louder and more frequent. At Purdue, a new president was hired followed by a complete change of personnel in upper administration. These factors happened to align all in the same direction of a desire to change the way we design and deliver undergraduate education. In contrast with other crises, we did not really have anything broken; it was more of a unique opportunity than an outside disruption. Tremendous resources have been invested in the creation of the Purdue Polytechnic Institute. This created a high level of uncertainty surrounding the nature and the role of the institute, as well as the expectations from the rest of the college and the expectations from the different constituencies. The resources that have been released are sufficiently significant to disrupt the routine functioning of the college.

2014: Reorganization Haziness?

Since January 2014, efforts in the Purdue Polytechnic Institute have focused on creating an innovative student experience, curriculum, and culture that are as authentic and faithful to the scientific and experiential finding as possible. These efforts included adopting a distributed leadership model that empowers all participants and builds on the strength of the team (Harris, 2013).

We are now at a critical crossroads. This effort can take off as a truly transformative effort, or as a more modest incremental change. The real innovation requires time, space, and resources to grow into a coherent project that is given freedom to evolve and to incorporate lessons learned. A safer but less transformative path would consist of weaving the changes and the resources into the system immediately. This will result in some incremental change with the main focus on minimizing the disruption of the bigger institution and preserving its smooth operation and power structure. The former requires autonomy, time, and a high level of risk-taking, assessment, and authentic questioning and experimentation. The latter demands scalability, predictability, and guaranteed timely results.

GOVERNANCE AND CULTURE

The conclusion reached by Kezar's (2004) study that "Relationships, Trust, and Leadership" are more important than formal administrative structures is sobering and inspiring at the same time. One of the key values underlying the Purdue Polytechnic approach is that of trust. We as faculty must trust the students' intrinsic curiosity and desire to learn. We empower students to take risks and trust that they can learn from failure and recover from it. We trust the students by making ourselves vulnerable with them and learning with them, rather than acting as the penultimate experts. We trust our colleagues and open our classrooms for collaboration, critique, and suggestions. As we debated and subscribed intellectually to these principles, we also realized how much unlearning and retraining we needed to do. There is an important faculty development component on which we barely touched the surface.

In the process of creating this new model of education, we also realized the magnitude of the interdependence between what happens in the classroom and the remaining governance that surrounds it. We have researched different governance models for our team and adopted Holacracy (Robertson, 2015). As Kezar (2004) notes, changing administrative structures alone is not sufficient. We are aware of the need to do both: Working on nurturing a climate of trust, empowerment, and shared leadership and adopting a governance structure consistent with it.

SUMMARY, ASSESSMENT

In his book *Change Leader: Learning to do What Matters Most,* Michael Fullan, (2011), argues that "the creative premise for the change leader is not 'to think outside the box' but to get outside the box, taking your intelligent memory to other practical boxes to see what you can discover."

We are standing at a crossroads. We understand the stark difference between radical innovation and incremental innovation. The latter is safe; it has almost guaranteed results and little risk, but it also has commensurate impact and rewards. At the other end, radical innovation involves the adoption of approaches that challenge and disrupt the broader context with its cultural norms and power structures. It requires larger leaps of understanding and often new ways of seeing a problem. The chances of success are difficult to estimate, and there is initially often considerable opposition to such ideas. We are just beginning the second year and still aiming for and hoping for the true transformation with its rewards to our students and many students in the future.

REFERENCES

AACU. (2011). *The LEAP vision for learning: Outcomes, practices, impact and employers' views.* Washington, DC: Association of American Colleges and Universities.

Augustine, N. (2005). *Rising above the gathering storm: Energizing and employing America for a brighter economic future.* Washington, DC: National Academies Press.

Battilana, J., Leca, B., & Boxenbaum, E. (2009). How actors change institutions: Towards a theory of institutional entrepreneurship. *Academy of Management Annals, 3*(1), 65–107.

Birnbaum, R. (2000). *Management fads in higher education.* San Francisco: Jossey-Bass.

Bowen, W., Chingos, M. M., & McPherson, M. S. (2009). *Crossing the finish line: Completing college at America's public universities.* Princeton, NJ: Princeton University Press.

Brown, D., & Seely, J. (2011). *A New culture of learning: Cultivating the imagination for a world of constant change.* Lexington, KY: CreateSpace Independent Publishing Platform.

Deci, E., & Ryan, R. (2013). *The handbook of self-determination research.* Rochester, NY: University of Rochester Press.

DiMaggio, P. (1988). Interest and agency in institutional theory. In L. Zucker (Ed.), *Institutional patterns and organizations: Culture and environment,* (pp. 3–22). Cambridge, MA: Ballinger.

Dorado, S. (2005). Institutional entrepreneurship, partaking, and convening. *Organization Studies 26(3), 385–414.*

Douglas, M. (1986). *How institutions think.* Syracuse, NY: Syracuse University Press.

Dweck, C. (2006). *Mindset: The new psychology of success.* New York: Balantine Books.

Emirbayer, M., & Mische, A. (1998). What is agency? *Amercian Journal of Sociology,* 962–1023.

Eyring, C. C. (2011). *The innovative university: Changing the DNA of higher education from the inside out.* Hoboken, NJ: John Wiley and Sons.

Fullan, M. (2011). *Change leader: Learning to do what matters most.* Hoboken, NJ: John Wiley & Sons.

Garud, R., Hardy, C., & Maguire, S. (2007). Institutional entrepreneurship as embedded agency: An introduction to the special issue. *Organization Studies* 28(7): 957.

Harris, A. (2013). *Distributed leadership matters: Perspectives, practicalities, and potential.* New York, NY: SAGE Publications.

Holling, C. (1986). The resilience of terrestrial ecosystems: Local surprise and global change. In *Sustainable development of the biosphere,* (pp. 292–317). Cambridge, UK: Cambridge University Press.

Kezar, A. (2004). What is more important for effective governance: Relationships, trust, and leadership or administrative structures and formal processes? In *Restructuring shared governance in higher education.* Wiley Periodicals Inc. New Directions in Higher Education.

Kezar, A., & Eckel, P. (2002). Examining the institutional transformation process: The importance of sensemaking, interrelated strategies, and balance. *Research in Higher Education,* 43(3), 295–328.

Levine, A. (2012). *Generation on a tightrope: A portrait of today's college student.* San Francisco: CA: Jossey-Bass.

Levy, F., & Murnane, R. J. (2013). *Dancing with robots: Human skills for computerized work.* Third Way: Fresh Thinking. Retrieved from http://content.thirdway.org/publications/714/Dancing-With-Robots.pdf

Mili, F. (2014, July). *Purdue Polytechnic as an entrepreneurial endeavor with strategic partnerships.* Retrieved from Purdue Polytechnic Blog: https://polytechhub.org/blogs/2014/08/purdue-polytechnic-as-an-entrepreneurial-endeavor-with-strategic-partnerships

Mili, F., & Bertoline, G. (2013 a). *Marrying the red bricks and the free bits of learning.* Purdue University: Polytechhub.org. Retrieved from https://polytechhub.org/resources/7

Mili, F., & Bertoline, G. (2013 b). *The overdue revolution in higher education.* Purdue University: polytechhub.org. Retrieved from https://polytechhub.org/resources/3

Mili, F., & Bertoline, G. (2013, c). *The PPI vision: Values, beliefs, and signature.* Retrieved from Purdue Polytechnic Institute: https://tech.purdue.edu/sites/default/files/files/Incubator/PPIsignature.pdf

Pink, D. (2006). *A whole new mind: Why right-brainers will rule the future.* New York, NY: Riverhead, Rep Upd Edition.

Price, D. (2013). *Open: How we'll work, live and learn in the future.* London, UK: Crux Publishing.

Robertson, B. (2015). *Holacracy: The new management system for a rapidly changing world.* New York, NY: Henry Holt and Co.

Sahlberg, P. (2011). *Finnish lessons: What can the world learn from educational change in Finland?* New York, NY: Teachers College Press, Columbia University.

Sokol, B., Grouzet, F., & Muller, U. (2013). *Self-regulation and autonomy: Social developmental dimensions of human conduct.* Cambridge, UK: Cambridge University Press.

Swidler, A. (1986). Culture in action: Symbols and strategies. *American Sociological Review,* 52(2), 273–286.

Van de Ven, A., & Garud, R. (1994). The coevolution of technical and institutional events in the development of an innovation. In J. Baum, & J. S. (Eds.), *Evolutionary dynamics of organizations* (pp. 425–443). New York, NY: Oxford University Press.

Wagner, T. (2010). *The global achievement gap: Why even our best schools don't teach the new survival skills our children need—and what we can do about it.* New York, NY: Basic Books, First Trade Paper Edition.

Weick, K. (1998). Sensemaking as an organizational dimension of global change. In D. Copperider, & J. Dutton (Eds.), *Organizational dimension of global change* (pp. 39–56). Thousand Oaks, CA: Sage.

Westley, F., Tjornbo, O., Schultz, L., Olsson, P., Folke, C., Crona, B., & Bodin, O. (2013). A theory of transformative agency in linked social-ecological systems. *Ecology and Society, 18*(3), 27.

ABOUT THE AUTHORS

Fatma Mili is Professor of Information Technology in the Purdue Polytechnic Institute; the co-principal investigator of the Purdue Polytechnic Institute and the inaugural lead of the Educational Research and Development arm. She is now heading the Center for Trans-Institutional Capacity Building at Purdue University in West Lafayette, Indiana.

Robert Herrick is the Robert A. Hoffer Distinguished Professor of Electrical Engineering Technology and Faculty Fellow Liaison in the Purdue Polytechnic Institute at Purdue University in West Lafayette, Indiana.

Tom Frooninckx is the Managing Director of the College of Technology and Purdue Polytechnic Institute at Purdue University in West Lafayette, Indiana.

5

Departmental Redesign:
Transforming the Chattanooga State Math Program

John Squires

In higher education, teachers have been looking for innovative ways to improve instruction and student learning in their classrooms. As society becomes technologically saturated, many of these innovations involve utilizing online resources. Terms like course redesign, blended learning, hybrid classrooms, asynchronous instruction, and flipped classes have become prevalent. These new approaches are aimed at the individual classroom with the goal of improving student engagement. However, as long as they are done as stand-alone initiatives, their impact is limited to a handful of students who perform better under these alternative delivery methods. In order to maximize the impact, a programmatic approach is needed, a concept we will call departmental redesign.

In fall 2009, the Chattanooga State Math Department embarked on a comprehensive, long-term project in order to improve student engagement and success. The project took five years to complete and had a significant impact on both the department and the college. We start with an overview of how this was accomplished by the department through an atmosphere of teamwork and continuous improvement. Then we examine the major aspects of this initiative, including course redesign, the emporium model, online instruction, working with high schools and innovative approaches to scheduling.

TEAMWORK AND CONTINUOUS IMPROVEMENT

Transforming a department is not possible without teamwork at all levels, which requires faculty buy-in. The key to faculty buy-in is straightforward: involvement. Once faculty members take ownership of a project, they can work together to improve the program over time. This concept of teamwork stands in stark contrast to the traditional model of faculty members working autonomously in silos of "hollow collegiality" (Massey, Wilger, and Colbeck, 1994).

For the redesign of the math department at Chattanooga State, faculty teams were formed for the overhaul of each course. Lead teachers were assigned to monitor the outcomes of a course and work to improve its quality over time.

Feedback was sought from a number of stakeholders, including students, faculty, program directors, and deans at the college. Student focus groups were formed to glean suggestions for improvement, and faculty members from other disciplines were invited to review the courses. Courses were reviewed regularly and input from teachers was sought. Working together, the math department at Chattanooga State was able to improve the quality of the program.

Once an atmosphere of teamwork had been established and course redesign had been implemented, continuous improvement became possible. The key to improving steadily over time is the appropriate use of data. This does not mean that improvement always occurs. Rather, it means that when problems arise, they are addressed programmatically. Likewise, when improvement occurs it impacts the entire program. By paying attention to data, having faculty work together, following a proven process in course redesign and making incremental changes over time, it is possible to dramatically improve the outcomes of a program over a period of time. The math program at Chattanooga State is an example of this.

COURSE REDESIGN

The five principles of course redesign are: redesign the whole course, encourage active learning, provide students with individualized assistance, build in ongoing assessment with prompt feedback, and ensure time on task while monitoring student progress (Five, 2015). If realized, these principles will create an ideal learning environment for students in which they can flourish.

Redesigning the whole course means including all of the sections for that course in the project (Five, 2015). Since all of the sections of a course are included instead of a few pilot sections, teamwork and efficiency are maximized, and continuous improvement becomes possible. Redesigning the whole course also means starting from scratch and building both activities and assessments, which will provide the student with a rich learning experience.

Most faculty members know that student engagement is the most important key to learning. Students who are disengaged do not learn anything, and as a result they do not succeed. If they do happen to survive, it is either because they enter the course with a set of skills and knowledge sufficient to carry them through the course, or because they happen to be good at taking tests. Sadly, any students who skate by in spite of not being engaged learn a fraction of what they could have. By building a system which not only encourages, but actually requires, students to be fully engaged in learning, course redesign can improve the quality of the educational system for all students (Five, 2015).

If student engagement is essential to learning, providing students with individual assistance is just as important (Five, 2015). Engaged students will read the content, do the work, perform the activity, and think about what they are doing. However, in the process they will have questions, make mistakes, miss a point, and not understand something. It is important at those times that there is someone to help them; if they can receive individual help in a friendly learning environment, they can overcome the hurdles they face and then conquer the material.

Ongoing assessment and prompt feedback minimize down time for students, creating a more efficient learning system than the traditional classroom (Five, 2015). In many college classes, student homework is not graded simply because of the burden that grading papers represents. Many students have little or no feedback until the time of the exam, leaving them without a sense of their progress in the course material. This problem can be solved by utilizing online learning systems to provide continuous assessment and instant feedback. Programs such as Pearson's MyMathLab, which was used throughout the math program at Chattanooga State, provide instant feedback to students whenever they are using the software, which might be at midnight or on the weekends.

By restructuring the classroom setting and designing the course in a way that encourages students to keep working, time on task can be greatly increased (Five, 2015). In course redesign, the role of the instructor often shifts towards that of a mentor or coach, freeing the teachers from grading stacks of papers so they can concentrate on monitoring student progress. Students who are working and getting help from teachers tend to perform better and learn more.

The Chattanooga State Math Department embraced course redesign in fall 2009 and redesigned 12 courses in a span of three years. The National Center of Academic Transformation has developed a proven concept and framework for improving instruction through the use of technology, and the work of the Chattanooga State Math Department shows the potential of adopting that instructional model and implementing it on a wide scale.

EMPORIUM MODEL

The emporium model eliminates class meetings and replaces them with a large computer lab where students access online learning materials and receive individual assistance (Emporium, 2015). A combination of faculty, professional tutors, and peer tutors are used to staff the lab, which is open weekdays, weeknights, and often weekends, for students to utilize at their convenience. The emporium model is one of the most successful models because of its ability to reduce costs while increasing student engagement and providing personal assistance in the learning process.

The Chattanooga State Math Department utilized the emporium model throughout its curriculum, from Basic Math to Calculus III. Every student in every class was expected to attend the math lab as part of the class. The college hired two full-time staff members to manage the lab, which expanded from 60 computers in fall 2008 to 185 computers in fall 2012. Additionally, computer labs of 32 computers each were installed at each of the remote sites, and computer classrooms of 24 computers were installed at the main campus. Overall, the college invested in almost 500 computers at six locations. The labs were staffed by a combination of full-time and part-time faculty, as well as professional and student tutors.

The result of shifting to the emporium model was clear; the math lab became the most utilized resource on campus. A proper implementation of the emporium model results in increased student engagement combined with individualized assistance, which leads to a friendly environment where students can work on math and receive the help they need. A Wall of Fame in the math lab highlighted student success stories, promoting a culture of success in math. Faculty offered mini-lectures to help students with specific topics, and students could form study groups to work with each other at designated times.

ONLINE INSTRUCTION

In fall 2009, the Chattanooga State Community College math department faced a problem not uncommon to colleges around the nation; online course offerings had high failure rates and were not a quality experience for the students. When the department leadership examined the state of the online math courses in fall 2009, the data revealed these courses had unacceptably high failure rates. Some classes had failure rates as high as 90%, where only a few students passed an on-line math course. It was clear that the courses did not incorporate online best practices—in fact, the "online courses" consisted of little more than a traditional course that happened to be offered online. After examining the data and the courses, a decision was made to put a moratorium on offering all online math courses, effective spring 2010. This critical first step provided a fresh start; while the decision was highly controversial and raised eyebrows among the campus community, it proved to be the correct decision in the long term.

Once the moratorium was put into place, the department began working to redesign its college math courses. The redesign of online courses focused on four best practices—course organization, quality resources, proactive faculty, and student engagement—and each of these is outlined below. For faculty wishing to dig deeper, there are a number of resources outlining best practices for online instruction; the University of Maryland's Quality Matters rubric is one example (Quality, 2014).

Course layout makes a huge difference in student engagement and is very important in online courses. The National Center of Academic Transformation has five principles of successful course redesign, and two of these principles, "Continuous Assessment" and "Time on Task", are directly linked to course structure (Five, 2015). Too often, online courses simply list a series of two or three big exams dates, often with very little else in the course in terms of assessment. Continuous assessment means regular homework, quizzes and exams throughout the course. Constant quizzing is listed as one of the characteristics of a great online course (Williams, 2013). This stands in stark contrast to the typical two to three big exams approach, which often leads to the night-before-the-test syndrome in which students do nothing until the night before the big exam. The course should be organized into bite-size topics, with the appropriate resources for each topic organized in a manner that makes them easily accessible. Weekly deadlines must be established and student progress must be monitored throughout the course. Proper course organization can go a long ways towards keeping students on task and on track.

For online classes, the resources should be fully integrated into the course in a seamless manner, providing the students with a guided learning path. While there may be some flexibility in the organization of the course that reflects different learning styles, access to learning resources is critical to providing the student with a quality online experience. Students should not be provided with a plethora of resources with little or no instruction on how to use them. Faculty should locate quality resources to integrate into their courses and develop resources locally when needed. These resources must follow best practices; for example, videos should be 5–10 minutes in length, and care must be taken to avoid color combinations that will not be discernible to color-blind students. When making videos, instructors should follow the 90-second rule for media, not spending more than 90 seconds on the same screen. This attention to detail makes a difference in the overall quality of the resources, and if a number of these practices are not followed, then the experience for the student will be frustrating, thereby negatively impacting course success.

Faculty engagement is crucial to the success of any course, traditional or online. The faculty member's approach in an online course should be active, not passive. Faculty must be vigilant in both monitoring student progress and providing individual assistance to each student, one of NCAT's Principles of Course Redesign (Five, 2015). If a faculty member only responds to e-mails from students, the online course will most likely fail because of a lack of faculty engagement. The faculty member teaching an online course should be consistently trying to keep students on task, while also offering them assistance as needed. While some of this assistance can be done at scheduled times, most of it will need to be provided on an ad hoc basis due to the asynchronous nature of an online course. Student work must be reviewed promptly by the faculty and strengths and weaknesses discussed with the students. Faculty engagement is also directly linked to student engagement. The more proactive the faculty member is the more engaged the students will be. Rita Sowell, a faculty member at Volunteer State Community College, has taught online courses successfully for many years and was consulted as a resource for Chattanooga State's online math program. When asked about the key to having a successful online program, her response was straight forward, stating that "the key was that the faculty members have to be willing to do a lot of hard work—period" (Sowell, 2012).

In any course, traditional or online, student engagement is the key to student success. Students who stay engaged, working in the course regularly throughout the semester, tend to succeed at much higher levels than students who procrastinate and are not engaged. Given this, the question becomes, "What can be done to increase student engagement?" The answer is that course layout, quality resources, and proactive faculty members can all contribute to increasing student engagement. When students understand how they are to go about learning the concepts, when they are provided with quality resources, and when they perceive that the faculty member cares about their success and is willing to assist them and teach them as needed, then they tend to be more engaged in the courses. Also, there are a number of strategies that can be used to increase student engagement, such as giving points for posts on discussion boards and course activities, making involvement a requisite part of completing the course, and encouraging peer-to-peer assistance and cooperation. These strategies can make a huge difference in increasing student engagement, which ultimately leads to student success.

Prior to the moratorium in fall 2009, the department offered three college math courses online. Over a period of five years, these courses served a total of 704 students, with a 36% success rate and a 1.29 GPA. Since the courses were redesigned and reintroduced in an online format beginning in fall 2011, the department offers seven college math courses online. Over a period of three years, these courses have served a total of 1,494 students, with a 61% success rate and a 2.15 GPA. Before, online math courses had success rates below 50%, much lower than on-ground classes. Now, online math courses have success rates of 50–70% and are comparable to on-ground classes. By giving the program a fresh start, the Chattanooga State Math Department has seen a significant increase in student success in online courses.

WORKING WITH HIGH SCHOOLS

In fall 2009, Chattanooga State was working with only four high schools in offering dual-enrollment math classes. In order to expand this program, the department had to overcome staffing obstacles associated with dual-enrollment classes. Simply put, the challenge in expanding dual enrollment is that most high schools do not have qualified faculty to teach college math classes. In order to meet this challenge, the department created the Early College Hybrid Online (ECHO) model of instruction, which started at Red Bank High School and quickly spread throughout the college's service district (ECHO, 2015). With ECHO, a college faculty serves as the teacher for an online class, and the high school teacher serves as the on-ground facilitator for that class. This model of hybrid instruction and blended learning, combined with the teamwork between the high school and college faculty, expanded opportunities for dual enrollment while maintaining quality. The program expanded to 22 high schools in fall 2014, and students in the program had a 97% success rate with a 3.6 GPA in the college math classes. Also, by utilizing the one-room schoolhouse strategy, several small high schools were able to expand their course offerings, including Calculus I and II. Some of these schools were unable to offer these courses prior to the ECHO program, but their students were now able to take these advanced courses.

In January, 2012, John Squires of Chattanooga State Community College worked with Deb Weiss of Red Bank High School to create the SAILS program, whereby students needing remediation in math are able to complete it during their senior year in high school (SAILS, 2015). The SAILS program is a highly strategic response to one of the state's greatest educational challenges: the high cost of remedial education and its effect on certificate and degree rates. In May 2012, the Chattanooga State Math Department received a grant form the Tennessee Board of Regents to pilot the program at 10 high schools in the Chattanooga area. At the same time, he worked with three other community colleges in Tennessee to introduce the program at their colleges. During the 2012–2013 academic year, the program involved four community colleges working with ten high schools and served 600 students. The results were impressive. For example, in the Chattanooga State pilot, 83% of the students completed the program, meaning they entered college ready to take a college math class. In May 2013, Governor Haslam and the Tennessee Higher Education Commission granted Chattanooga State $1.1 million to scale-up the project statewide.

In Year 1 of the statewide scale-up (2013–2014 AY), the project served over 8,000 students and had a 67% success rate, saving these students over 10,000 semesters in college and over $6.4 million in tuition and books. Because of the project's phenomenal success, $2.45 million was approved for continued state-wide scale-up. In Year 2, (2014–2015 AY), the program is expected to serve over 13,000 students in 184 high schools in 79 school districts. The SAILS program is attracting national attention for its innovative, yet practical, solution to the college readiness problem, and has been featured in *Inside Higher Education* (Going, 2013).

INNOVATIVE APPROACHES TO SCHEDULING

Because the teacher is not lecturing in the mini-lab version of the emporium model, it is possible to combine classes, with several students from one class in the same room as students from another class. This strategy is known as the one-room schoolhouse. In this setting, all that is required from the teacher is the ability to shift gears between courses. The ability to combine classes offers a solution to low-enrollment classes, further reducing the scheduling roadblocks and meeting the needs of students.

The Continuous Enrollment Plan allows students who can display mastery of the topics to move quickly through a course, completing multiple courses in one semester. This option accelerates students to graduation, saving them both time and money. Since implementing this innovative approach to class enrollment, over 1,000 students at Chattanooga State Community College have taken advantage of this option and completed two or more courses in the same semester. In addition to the time they have saved, these students have saved over $500,000 in tuition and books. It should be noted that implementing this strategy required the cooperation of offices across the college—from Financial Aid to Veterans Affairs to Academic Affairs; these separate areas of the college worked together in order to find solutions to challenges and benefit students.

CONCLUSION

In five years, the Chattanooga State Math Department was able to transform its program, positively impacting students in the program through teamwork and innovative approaches. The results were dramatic, with success rates increasing in both developmental math and college math courses. During a five year period, the enrollment at the college decreased by 5%, while the enrollment in college math classes increased by over 60%, with students succeeding in college math skyrocketing by over 70%. The college invested over $1 million in the project, but that investment was paid back several times over due to these amazing results and the hard work of the Chattanooga State Math Department faculty.

REFERENCES

Chattanooga State Community College. (Accessed February 2015). *ECHO*. Retrieved from https://www.chattanoogastate.edu/mathematics-sciences/echo

Chattanooga State Community College. (Accessed February 2015). *SAILS*. Retrieved from https://www.chattanoogastate.edu/high-school/sails

Fain, P. (September 2013). Going to the root of the problem. *Inside Higher Education*. Retrieved from https://www.insidehighered.com/news/2013/09/13/promising-remedial-math-reform-tennessee-expands

Massey, W. F., Wilger, A. K., & Colbeck C. (1994 Jul/Aug). Overcoming "hollowed" collegiality, *Change*, 26(4), pp. 10–20.

National Center for Academic Transformation. (Accessed February 2015). Five principles of successful course redesign, National Center for Academic Transformation (NCAT). Retrieved from http://www.thencat.org/PlanRes/R2R_PrinCR.htm

National Center for Academic Transformation. (Accessed February 2015). The emporium model, National Center for Academic Transformation (NCAT). Retrieved from http://www.thencat.org/PlanRes/R2R_Model_Emp.htm

Sowell, R. (2012). *Personal communication*. Volunteer State Community College, Tennessee.

University of Maryland. (2014). *Quality matters*. University of Maryland. Retrieved from https://www.qualitymatters.org/rubric

Williams, C. (accessed February 2015). 5 best practices for designing online courses. Retrieved from http://blog.heatspring.com/designing-online-courses/

ABOUT THE AUTHOR

John Squires served as Associate Professor of Mathematics and Math Department Head at Chattanooga State from 2009 to 2014. He retired in December 2014 and was named Professor Emeritus of the college. He now serves as Director of High School to College Readiness at the Southern Regional Education Board in Atlanta, Georgia.

6

Successful Model for Professional Development: Creating and Sustaining Faculty Learning Communities

Ann C. Smith, Gili Marbach-Ad, Ann M. Stevens, Sarah A. Balcom,
John Buchner, Sandra L. Daniel, Jeffrey J. DeStefano,
Najib M. El-Sayed, Kenneth Frauwirth, Vincent T. Lee, Kevin S. McIver,
Stephen B. Melville, David M. Mosser, David L. Popham,
Birgit E. Scharf, Florian D. Schubot, Richard W. Selyer, Jr.,
Patricia Ann Shields, Wenxia Song, Daniel C. Stein, Richard C. Stewart,
Katerina V. Thompson, Zhaomin Yang, and Stephanie A. Yarwood

BACKGROUND

Improving undergraduate science, technology, engineering, and mathematics (STEM) education is both an urgent national need and a long-term challenge (PCAST, 2012; AAU, 2011).The STEM fields are critical to generating new ideas, companies, and industry that drive our nation's competitiveness, and will become even more important in the future (Arum & Roksa, 2011). Nevertheless, there has been a steep decline in both the number and persistence of students in STEM majors. The decline in popularity of STEM programs is particularly marked among freshmen, who often leave the major soon after completing introductory science courses (Green, 1989; Seymour & Hewitt, 1997). Evidence is mounting that introductory coursework fails to inspire students and provide them with the foundational knowledge they need to persist and excel in STEM degree programs (Hurtado et al., 2010; Wood, 2009; Handelsman, 2004). Students leaving STEM majors express dissatisfaction with both the curriculum and the instruction, often perceiving that professors care more about research than student learning (Johnson, 1996; Marbach-Ad & Arviv-Elyashiv, 2005; Seymour, 1995; Seymour & Hewitt, 1997; Sorensen, 1999).

Faculty members clearly play a pivotal role in undergraduate STEM education reform. Through their enthusiasm and expertise, they shape the attitudes and aspirations of their students (Cole & Barber, 2003; Gaff & Lambert, 1966).

However, faculty members have essentially no formal preparation for their university teaching responsibilities (Tanner & Allen, 2006). While they are generally aware that prior knowledge plays an important role in the ability to acquire new concepts, they lack expertise in evaluating their students' prior knowledge and adjusting their teaching practices to frame their course as part of a learning progression (Marbach-Ad, Ribke & Gershoni, 2006; Duschl, Maeng & Sezen, 2011).

A recent Association of American Universities Report (AAU, 2011) urges a cultural change in how faculty members approach teaching. The traditional mode of undergraduate STEM instruction, characterized by long lectures where students take a passive role, emphasizes content coverage over conceptual mastery and leaves students deeply dissatisfied (Henderson, Beach & Finkelstein, 2011; Henderson & Dancy, 2008; Henderson et al., 2008; Seymour & Hewitt, 1997). Faculty members must move away from teaching a "sea of facts" and instead help students develop a meaningful conceptual understanding. The American Association for Advancement in Science report, *Vision and Change: A Call to* Action (AAAS, 2009) provides a consensus list of the major concepts that students in the biological sciences should understand deeply. Disciplinary societies such as the American Society for Microbiology (ASM) have voiced strong support for this approach and have developed curriculum recommendations that are grounded in a focused list of concepts aligned with those proposed in *Vision and Change* (Merkel, 2012). Using the process of scientific teaching (Handelsman et al., 2004), these curriculum guidelines can serve as the basis for designing courses that achieve specific learning outcomes using best practices for student learning. However, before meaningful change can be implemented in the classroom, it is imperative that we have a thorough understanding of the knowledge base of the incoming students, with an appreciation of their conceptual understanding about science and the world around them.

CONCEPT INVENTORIES AS A TOOL TO PROBE STUDENTS' CONCEPTUAL UNDERSTANDING

Well-designed concept inventories (CIs) are important tools for assessing the extent of student concept mastery. CIs generally consist of a series of multiple-choice questions that are informed by research into students' prior knowledge of a topic. Distractors for the multiple choice questions are developed with awareness of naive ideas, misconceptions, and faulty reasoning commonly shared by students (D'Avanzo, 2008; Fisher 2004). Misconceptions are ideas that differ from valid scientific explanations and also (1) tend to be shared by a significant proportion of the population; (2) cut across age, ability, gender, and cultural boundaries; (3) produce consistent error patterns (Osborne & Freyberg, 1985); and (4) are highly resistant to instruction (Fisher, 1983; Thijs & van den Berg, 1993). We are aware that some consider the term "misconception" to have a negative connotation and suggest instead using the terms "alternative conception" or "naïve conception," however for simplicity we hereafter use the term "misconception."

The first of the CIs to have widespread influence on undergraduate instruction was the Force Concept Inventory (FCI), which was developed in the '80s by the physics community to assess student understanding of fundamental Newtonian concepts (Hestenes, Wells & Swackhamer, 1992). The FCI has provided powerful evidence of the effectiveness of active-learning teaching methods over traditional, lecture-based methods (Crouch & Mazur, 2001; Hake, 1998; Mulford & Robinson, 2002). Following this lead, CIs have been developed across a range of STEM disciplines, including chemistry (Mulford & Robinson, 2002), geosciences and astronomy (Libarkin, 2008), engineering (Evans et al., 2003) and in biology and its subdisciplines (Smith & Marbach-Ad, 2010).

While CIs are widely used to assess student learning of targeted concepts, questions have been raised about how well multiple choice questions can measure deep learning (Smith & Tanner, 2010). This issue has been at the heart of a series of national, NSF-funded workshops on Conceptual Assessments in Biology (CAB)(DBI-0957363). The consensus of attendees at the most recent conference (CAB-III, 2011) was that there is value in CIs that call for students to provide open-ended responses in addition to selecting multiple-choice responses (Smith & Marbach-Ad, 2010). These types of instruments provide both quantitative and qualitative evidence of student learning, giving them great utility as faculty professional development tools. For example, CAB-III participants recognized the power of using CI data to create a state of cognitive dissonance in faculty members who declare that they have "covered" a concept in class, but learn from CI data that students poorly understand the concept.

OUR FACULTY LEARNING COMMUNITIES

Faculty Learning Communities (Cox, 2004) have emerged as a powerful mechanism for teaching reform and faculty professional development. Communities inspire faculty members to develop shared vision and expertise, and they provide motivation and support for those seeking to adopt new teaching practices. There are various types and models for faculty learning communities (see Chapter D2). We built our communities along the lines of Wenger's (1998) theory of community of practice where we focus on collaborative projects.

Here, we describe our two communities: the UMD Host Pathogen Interactions (HPI) FLC and the VT Microbiology (MICB) FLC. We will introduce the

UMD HPI FLC and then explain how the model for this community spurred the development of the Virginia Tech (VT) community, now an active and vibrant force for course transformation at VT.

The Host Pathogen Interactions Faculty Learning Community

In 2004, as part of a college-wide effort to reinvigorate the undergraduate biology curriculum, UMD faculty members with research expertise in the area of HPI formed a teaching community with the expressed purpose of creating a research-intensive undergraduate curriculum informed by best practices in teaching and learning. Collectively, these faculty members share responsibility for teaching nine undergraduate courses in the undergraduate microbiology curriculum, including a large introductory course in general microbiology. Prior to the establishment of the HPI FLC, the UMD faculty had operated as individuals, each of us teaching the way that we had been taught and rarely assessing our learning outcomes. With the increasing body of knowledge on how students learn science, we felt that it was time for a more collaborative and forward-thinking approach to teaching. The HPI teaching community was founded on shared research and teaching interests, and it mirrors the classic research group, where science faculty members gather regularly to share ideas, review data, and discuss current findings. We have detailed the history and initiatives of our FLC in a series of publications (Marbach-Ad et al., 2007, 2009, 2010).

Over the last ten years the number of members in the HPI FLC has varied due to new hires and retirement. The HPI Teaching Community now includes 14 members who represent all faculty ranks, including those with primarily teaching responsibilities (lecturers and instructors), as well as tenured/tenure-track faculty members who have done research in the area of host pathogen interactions. Gili Marbach-Ad, the director of the College Teaching and Learning Center (http://www.life.umd.edu/tlc/), is also an integral part of the group, providing expertise in science pedagogy and assessment. During our time as a community, we have developed thirteen HPI concepts, an assessment tool (HPI Concept Inventory) and transformed our courses according to current research in student learning in STEM courses (Cathcart et al., 2010; Injaian et al., 2011; Quimby et al., 2011; Senkevitch et al., 2011). Members of our group have become active in campus-wide and national STEM educational initiatives, including Vision and Change, and ASM curriculum reform.

The HPI Concept Inventory, Our FLC Tool

The HPI Concept Inventory was developed by the UMD HPI FLC as a way of measuring the success of various curricular initiatives (Marbach-Ad et al., 2010). We give the HPI CI as a pre-test and post-test to provide insight into student gains in understanding of HPI concepts within each of our courses and across the full program of nine courses. It consists of 18 multiple choice questions validated through an iterative process (Marbach-Ad et al., 2009, 2010). The multiple choice nature of the inventory allows for quantitative analyses with large samples of students. Students complete the CI online and provide their student ID to enable matching of pre-test and post-test scores, and allow for retrieval of demographic information (e.g., gender, major) from institutional records. After students answer each question, they are asked to provide an explanation for the answer they chose. These open-ended explanations provide a rich source of data for qualitative analysis. Since 2006, at UMD we have implemented the HPI CI in four to six courses each semester.

At the conclusion of every semester, our team meets for an extended work session to review the data, according to a specified protocol (Table 1).

Through this systematic analysis of our data, we have gained insights into our program and, as a result, have made substantial changes to our curriculum, including the development of a new introductory course for students majoring in microbiology (Marbach-Ad et al., 2010). Further, the data analysis has served to spur rich conversations among our team that have transformed how we think about teaching and student learning. The qualitative analysis review sessions in particular have encouraged serious conversations about the nature of student learning and the origins of common misconceptions. We consider the insights derived from this work as the most important motivator of our continued interest in curriculum reform. We have found that student explanations in response to the HPI CI questions hold information that is valuable in revealing how students understand or do not understand HPI concepts. Each

TABLE 1: Protocol for Analysis of HPI CI Student Pre and Post Responses

1. Data from the online CI are downloaded to Excel files.

2. Pre- and post-test means are calculated for each course and tabulated.

3. Student explanations for each question are sorted by distractor choice to facilitate qualitative analysis. Responses for each distractor are sorted alphabetically. Numbers of responses with and without explanations are recorded. For qualitative analysis, responses without explanations are deleted from the working spreadsheet.

4. FLC members meet to review and discuss student performance on the CI. Quantitative data (pre- and post-test means) are reviewed and discussed by the group as a whole. For qualitative analysis, faculty members work in pairs with laptop computers to read and discuss student responses to different subsets of CI questions. Each pair then reports their major findings to the entire group for additional discussion.

5. FLC members summarize findings defining common misconceptions that lead students to select particular distractors.

of us has used insights from HPI CI analysis to inform our teaching in various ways and support our development as informed educators.

Creation of the VT Microbiology Faculty Learning Community

We in the UMD HPI FLC hypothesized that similar deep analysis of student CI responses would motivate the formation and success of new FLCs. To explore this notion, we brainstormed to identify a group of faculty members who might be interested in forming a community motivated by discussion of CI data. We decided to approach colleagues in the Department of Biological Sciences at VT. We chose this route as VT, like UMD, is a research university, we have colleagues in the department with research areas similar to ours, and the department offers a full set of microbiology courses comparable to those at UMD. We found that as at UMD the faculty members at VT had a strong interest in teaching microbiology, however meaningful discussions about student learning and large-scale collaborative projects were not occurring.

We entered into a collaborative agreement with a set of VT faculty members. As a result, in Fall 2010 the VT Microbiology (MICB) FLC was formed with eight members. The VT MICB FLC agreed to use the HPI CI for pre- and post-assessment of student learning in a set of microbiology courses. The group would then meet to discuss the data as we have done (Table 1). To support the VT FLC the UMD group served as mentors in the review and the evaluation of CI data. To this point, the VT MICB FLC has employed the HPI CI as pre- and post-surveys in four courses (two of which are offered every spring and three every fall) since 2011. The discussion of data has supported the development of the learning community as we hypothesized.

THE BENEFIT OF CI DISCUSSION IN MOTIVATING SUCCESS OF A FACULTY LEARNING COMMUNITY

Above, we indicated the value UMD FLC members have placed on the discussion of CI data in motivating participation in the FLC, and in curricular and pedagogical transformation efforts. Similarly, VT faculty members have been engaged by these discussions. On a recent survey of our communities, in response to the question, "What impact has your participation in your FLC had on your teaching?" one VT faculty member reported that participation in the community encouraged him/her to "more formally link learning outcomes with class learning material and assessments." Another VT faculty member wrote, "I have learned more about misconceptions that my students have before they reach my classroom, and the unexpected ways that they think about information I present to them."

The success of the VT MICB FLC is further evidenced by significant curriculum transformation. The community transformed the set of microbiology courses in their department into a full microbiology major program using the HPI CI as the assessment tool. Also, the group is now participating in STEM education research conferences, and several members are involved in national STEM education initiatives. Further, the UMD and VT groups are now working on a collaborative project to define common misconceptions among students entering a general microbiology course. This work is ongoing and we plan to publish it in a microbiology education journal.

LESSONS LEARNED AND APPLICATION OF FINDINGS

The UMD and VT communities have similar and distinct attributes (Table 2). As both communities exist at research universities, we each are composed of significant numbers of tenure/tenure-track faculty members who have both research and teaching responsibilities. Each community meets a few times each semester, with the UMD group meeting more regularly over a longer span of time (10 years). Both communities have one of the community members serving as a facilitator who sets the agenda and prepares meeting materials and reports. Similar drivers motivate both sets of faculty members including a desire for excellence in teaching, concern for student learning of important principles in microbiology, the goal of offering a curriculum where learning in one course builds upon the prior course, and an interest in contributing to the scholarship of teaching and learning. The UMD FLC was formed with a stated main initiative to foster deep and research-oriented learning in HPI, whereas the VT group was motivated by the desire to create a new undergraduate major. For both communities, discussion of HPI CI data served to engage the members in the work of the community. For the UMD group, this began with the development of the tool. Reading student explanations for selection of distractors was necessitated to validate the HPI CI. We found the analysis so interesting and informative that we continued this work beyond the tool development stage, and analysis of CI data became a major part of the community work. VT adopted the UMD HPI CI and found the discussion of the data equally compelling.

The UMD FLC was developed in response to a call for proposals and has had funding from a HHMI grant to the College of Chemical and Life Sciences. The group also successfully competed for NSF funding that provided support for two years. With funding, we benefited from support for a statistician, external evaluator of our work, graduate students, money for travel, and the opportunity to provide lunch at meetings. The VT group has had only limited funding from their Office of Assessment and Evaluation to support a part-time graduate student for one summer. The UMD team has had a science educator as a long-standing member of the team who has introduced science education literature, assisted in curriculum design, pedagogy and assessment implementation, and supported the documentation and dissemination of the work.

Although the communities have distinctions, both have been successful on multiple levels: impacting courses, programs, and their institutions, as well as contributing to the national conversation of STEM reform.

TABLE 2: Attributes of the UMD HPI FLC and the VT MICB FLC

Institution	UMD	VT
Membership	14 members including tenure/tenure track (9) and instructors (5)	8 members including tenure/tenure track (6) and instructors (2)
Meetings	Three times/semester over lunch (1.5 hour) with half day working meetings between semesters	Two times/semester for 2 hours
Duration	2004–2015	2011–2015
Facilitator	One team member serves role as facilitator	One team member serves as facilitator
Motivation for participation	• Desire for excellence in teaching • Concern for student learning of field • Learning progression within program • Interest in producing publications on teaching and learning	• Desire for excellence in teaching • Concern for student learning of field • Learning progression within program • Interest in producing publications on teaching and learning
FLC Main Initiative	Foster deep and research oriented learning in host pathogen interactions	Assessment of new microbiology degree program expected by accreditation
Concept Inventory	Created the HPI CI	Adopted HPI CI
Funding	Funding from NSF and HHMI that allowed support for • Food at meetings—lunch • Science educator • External evaluator • Statistician • Graduate student support • Travel to meetings	Summer support for one part-time graduate student from VT Office of Assessment and Evaluation
Science Education Expertise	Science educator integral part of the team Facilitator participated in science education programs including ASM Biology Scholars	UMD Science Educator provided assistance. Facilitator participated in science education programs including ASM Biology Scholars

There are some crucial elements important for maintaining a vibrant FLC. Each FLC meeting must be planned in an efficient manner to maximize the potential of the teamwork. The role of the community facilitator is essential for pre-meeting preparation, directing the meetings, and documenting progress. Further, there must be a link to the greater science education community, either by members attending conferences and reading the literature, or through the help of a science educator who supports the team in this manner. Limited funding may hamper the success of an FLC if members cannot attend conferences and if there is not sufficient support for data collection and organization.

In conclusion, we believe that FLCs that participate in discussion of assessment data, like that collected from the implementation of a CI, provide the right mix of support and intellectual challenge to engage STEM faculty members and motivate them toward curriculum reform efforts. There is the myth that research faculty members do not value their teaching mission to the same extent as their research. This is evidenced in that it is common for research faculty members to engage in frequent conversations with colleagues about research, while it is rare for these faculty members to discuss their teaching, attend STEM education conferences, or complete a serious analysis of student learning in their courses. Yet we *are* interested in being excellent teachers. The discussion of the CI data with colleagues has provided us an entrée into science education research and terminology, and a community with which to act on our interests in science education.

REFERENCES

American Association for Advancement in Science (AAAS). (2009). *Vision and change: A call to action*. Washington, DC: AAAS.

Arum, R., & Roksa, J. (2011). *Academically adrift: Limited learning on college campuses*. Chicago: University of Chicago Press.

Association of American Universities (AAU). (2011). *Five year initiative for improving undergraduate STEM education*. http://www.aau.edu/WorkArea/DownloadAsset.aspx?id=12590.

Cathcart, L. A., Stieff, M. Marbach-Ad, G., Smith, A. C., & Frauwirth, K. A. (2010). Using knowledge space theory to analyze concept maps. *In 9th International Conference of the Learning Sciences.* Chicago, Illinois

Cole, S. & Barber, E. (2003). *Increasing faculty diversity: The occupational choices of high-achieving minority students*. Boston, MA: The Harvard University.

Cox, M. D. 2004. Introduction to faculty learning communities. *New Directions for Teaching and Learning*, 97, 5–23.

Crouch, C. & E. Mazur. (2001). Peer instruction: Ten years of experience and results. *American Journal of Physics*, 69, 970–977.

D'Avanzo, C. (2008). Biology concept inventories overview, status, and next steps. *Bioscience*, 58, 1079–1085.

Duschl, R., Maeng, S., & Sezen, A. (2011). Learning progressions and teaching sequences: A review and analysis. *Studies in Science Education*, 47(2), 123–182.

Evans, D. L., Gray, G., Krause, S., Martin, C., Midkiff, B., Notaros M. & et al. (2003). Progress on concept inventory assessment tools. In *33rd ASEE/IEEE Frontiers in Education Conference*, T4G1–T4G8. Boulder, CO.

Fisher, K. M. (1983). Amino acids and translation: A misconception in biology. In *International Seminar on Misconceptions in Science and Mathematics*. Cornell University, Ithaca, NY.

Fisher, K. M. (2004). Conceptual assessments in biology: Tools for learning. In *RCN-UBE incubator*. Arlington, VA: National Science Foundation.

Gaff, J. G. & Lambert, L. M. (1966). Socializing future faculty to the values of undergraduate education. *Change*, 28, 38–45.

Green, K. C. (1989). A profile of undergraduates in the sciences. *The American Scientist*, 77, 475–480.

Hake, R. R. 1998. Interactive-engagement vs. traditional methods: A six-thousand-student survey of mechanics test data for introductory physics courses. *American Journal of Physics*, 66, 64–74.

Handelsman, J., Ebert-May, D., Beichner, R., Bruns, P., Chang, A., DeHaan, R., Gentile, J., Lauffer, S., Stewart, J., Tilghman, S. M., & Wood, W. B. (2004). Scientific teaching. *Science*, 304, 521–522.

Henderson, C., Beach, A., & Finkelstein, N. (2011). Facilitating change in undergraduate STEM instructional practices: An analytic review of the literature. *Journal of Research in Science Teaching*, 48, 952–984.

Henderson, C., Beach, A. Finkelstein, N., & Larson, R. S. (2008). Preliminary categorization of literature on promoting change in undergraduate STEM. In *Facilitating change in undergraduate STEM symposium*. Augusta, MI.

Henderson, C., & Dancy, M. H. (2008). Physics faculty and educational researchers: Divergent expectations as barriers to the diffusion of innovations. *Am. J. Phys*, 76, 70–91.

Hestenes, D., Wells, M., & Swackhamer, G. (1992). Force concept inventory. *The Physics Teacher*, 30, 141–158.

Hurtado, S., Newman, C. B., Tran, M. C., Chang, M. J. (2010). Improving the rate of success for underrepresented racial minorities in STEM fields: Insights from a national project. *New Directions for Institutional Research*, 148: 5–15.

Injaian, L., Smith, A. C., German Shipley, J., Marbach-Ad, G. & Fredericksen, B. (2011). Antiviral drug research proposal activity. *Journal of Microbiology & Biology Education*, 12, 18–28.

Johnson, G. M. (1996). Faculty differences in university attrition: A comparison of the characteristics of art, education, and science students who withdrew from undergraduate programs. *Journal of Higher Education Policy and Management*, 18, 75–91.

Libarkin, J. (2008). *Concept inventories in higher education science.* National Academies Press, Washington, DC. Retrieved from http://www7.nationalacademies .org/bose/Libarkin_CommissionedPaper.pdf:

Marbach-Ad, G. & Arviv-Elyashiv, R. (2005). What should life science students acquire in their BSc studies? Faculty and student perspectives. *Bioscene*, 30, 3–13.

Marbach-Ad, G., Briken, V.,Frauwirth, K., Gao, L. Y., Hutcheson, S.W., Joseph, S.W., Mosser, D., Parent, B., Shields, P., Song, W., Stein, D. C., Swanson, K., Thompson, K. V., Yuan, R., Smith. A.C. (2007). A faculty team works to create content linkages among various courses to increase meaningful learning of targeted concepts of microbiology. *CBE—Life Sciences Education*, 6:155–162.

Marbach-Ad, G., Briken, V., El-Sayed, N. M., Frauwirth, K., Fredericksen, B., Hutcheson, S., Gao, L., Joseph, S., Lee, V., McIver, K. S., Mosser, D., Quimby, B. B., Shields, P., Song, W., Stein, D. C., Yuan R. T., & Smith A. C. (2009). Assessing student understanding of host pathogen interactions using a concept inventory. *Journal of Microbiology & Biology Education*, 10, 43–50.

Marbach-Ad, G., McAdams, K. C., Benson, S., Briken, V., Cathcart, L., Chase, M., El-Sayed, N., Frauwirth, K., Fredericksen, B., Joseph, S. W., Lee, V., McIver, K. S., Mosser, D., Quimby, B. B., Shields, P., Song, W., Stein, D. C., Stewart, R., Thompson, K. V., Smith, A. C. (2010). A model for using a concept inventory as a tool for students' assessment and faculty professional development. *CBE—Life Sciences Education*, 10: 408–416.

Marbach-Ad, G., Ribke, M., & Gershoni, J. M. (2006). Using a "primer unit" in an introductory biology course: "A soft landing". *Bioscene*, 32, 13–20.

Merkel, S. 2012. The development of curricular guidelines for introductory microbiology that focus on understanding. *Journal of Microbiology and Biology Education*, 13, 32–38.

Mulford, D. R., & Robinson, W. R. (2002). An inventory for alternate conceptions among first semester general chemistry students. *Journal of Chemical Education*, 79, 739–744.

Osborne, R. & Freyberg, P. (1985). *Learning in science: The implications of children's science.* Hong Kong, China: Heinman.

President's Council of Advisors for Science and Technology (PCAST). (2012). *Engage to excel: Producing one million additional college graduates with degrees in science, technology, engineering, and mathematics.* Available at www.white house.gov/sites/default/files/microsites/ostp/pcast-engage-to-excel-final_2-25 -12.pdf

Quimby, B. B., McIver, K. S., Marbach-Ad G. & Smith, A. C. (2011). Investigating how microbes respond to their environment: Bringing current research into a pathogenic microbiology course. *Journal of Microbiology & Biology Education*, 12, 176–184.

Senkevitch, E., Marbach-Ad, G., Smith, A. C., & Song, S. (2011). Using primary literature to engage student learning in scientific research and writing. *Journal of Microbiology and Biology Education*, 12, 144–151.

Seymour, E. (1995). Guest comment: Why undergraduates leave the sciences. *American Journal of Physics*, 63, 199–202.

Seymour, E., & Hewitt, N. M. (1997). *Talking about leaving: Why undergraduates leave the sciences.* Boulder, CO: Westview Press.

Smith, A. C. & Marbach-Ad, G. (2010). Learning outcomes with linked assessment—an essential part of our regular teaching practice. *Journal of Microbiology and Biology Education*, 11, 123–129.

Smith, J., & Tanner. K. (2010). The problem of revealing how students think: Concept inventories and beyond. *CBE Life Science Education*, 9, 1–5.

Sorensen, K. H. (1999). Factors influencing retention in introductory biology curriculum. In *National Association for Research in Science Teaching*. Boston, MA.

Tanner, K. & Allen, D. (2006). Approaches to biology teaching and learning: On integrating pedagogical training into the graduate experiences of future science faculty. *CBE Life Science Education*, 5, 1-6.94-116.

Thijs, G. D., & van den Berg, E. (1993). Cultural factors in the origin and remediation of alternative conceptions. In *Third International Seminar on Misconceptions and Educational Strategies in Science and Mathematics*. Cornell University, Ithaca, NY.

Wenger, E. (1998). *Communities of practice: Learning, meaning, and identity*. Cambridge: Cambridge University Press.

Wood, W. B. (2009). Innovations in teaching undergraduate biology and why we need them. *Annual Review of Cell and Developmental Biology*, 25, 93–112.

ABOUT THE AUTHORS

Ann C. Smith is the Assistant Dean in the Office of Undergraduate Studies at the University of Maryland in College Park, Maryland.

Gili Marbach-Ad is the Director of the CMNS Teaching and Learning Center of the College of Computer, Mathematical, and Natural Sciences at the University of Maryland in College Park, Maryland.

Ann M. Stevens is a Professor of the Department of Biological Sciences at Virginia Tech in Blacksburg, Virginia.

Sarah A. Balcom is a Lecturer of the Department of Animal and Avian Sciences at the University of Maryland in College Park, Maryland.

John Buchner is a Lecturer of the Department of Cell Biology and Molecular Genetics at the University of Maryland in College Park, Maryland.

Sandra L. Daniel is an Instructor of the Department of Biological Sciences at Virginia Tech in Blacksburg, Virginia.

Jeffrey J. DeStefano is a Professor of the Department of Cell Biology and Molecular Genetics at the University of Maryland in College Park, Maryland.

Najib M. El-Sayed is an Associate Professor of the Department of Cell Biology and Molecular Genetics at the University of Maryland in College Park, Maryland.

Kenneth Frauwirth is a Lecturer of the Department of Cell Biology and Molecular Genetics at the University of Maryland in College Park, Maryland.

Vincent T. Lee is an Associate Professor of the Department of Cell Biology and Molecular Genetics at the University of Maryland in College Park, Maryland.

Kevin S. McIver is a Professor of the Department of Cell Biology and Molecular Genetics at the University of Maryland in College Park, Maryland.

Stephen B. Melville is an Associate Professor of the Department of Biological Sciences at Virginia Tech in Blacksburg, Virginia.

David M. Mosser is a Professor of the Department of Cell Biology and Molecular Genetics at the University of Maryland in College Park, Maryland.

David L. Popham is a Professor of the Department of Biological Sciences at Virginia Tech in Blacksburg, Virginia.

Birgit E. Scharf is an Assistant Professor of the Department of Biological Sciences at Virginia Tech in Blacksburg, Virginia.

Florian D. Schubot is an Associate Professor of the Department of Biological Sciences at Virginia Tech in Blacksburg, Virginia.

Richard W. Selyer, Jr. is an Advanced Instructor of the Department of Biological Sciences at Virginia Tech in Blacksburg, Virginia.

Patricia Ann Shields is a Senior Lecturer of the Department of Cell Biology and Molecular Genetics at the University of Maryland in College Park, Maryland.

Wenxia Song is an Associate Professor of the Department of Cell Biology and Molecular Genetics at the University of Maryland in College Park, Maryland.

Daniel C. Stein is a Professor of the Department of Cell Biology and Molecular Genetics at the University of Maryland in College Park, Maryland.

Richard C. Stewart is an Associate Professor of the Department of Cell Biology and Molecular Genetics at the University of Maryland in College Park, Maryland.

Katerina V. Thompson is the Director of Undergraduate Research and Internship Programs in the College of Computer, Mathematical, and Natural Sciences at the University of Maryland in College Park, Maryland.

Zhaomin Yang is an Associate Professor of the Department of Biological Sciences at Virginia Tech in Blacksburg, Virginia.

Stephanie A. Yarwood is an Assistant Professor of Environmental Science and Technology at the University of Maryland in College Park, Maryland.

SECTION D
Faculty Development

As institutions move to embrace evidence-based, student-centered instructional approaches, the faculty members who are at the center of acting upon that mission need to be supported as they themselves transition to new, unfamiliar, and sometimes daunting approaches. Not only do faculty need to become familiar with new pedagogies and possibly new technologies, they also need to learn how to help their students learn in these environments. The chapters in this section present a variety of perspectives on approaches to and results of different forms of faculty development. While they do not represent the entirety of faculty development literature, they represent the types of efforts that need to be an integral part of institutional transformation in STEM education. The Egan, et al., chapter describes an approach to faculty development that takes into account the specialized needs of the discipline in which faculty are teaching, in contrast to a faculty development approach that serves all faculty on a campus with the same programs. The following chapter on faculty learning communities from Thompson, et al., presents details regarding the benefits and challenges of this approach to supporting faculty change. The next two chapters, Owens and Mack, explore particular techniques that can help faculty develop increased effectiveness and confidence in student-centered approaches to teaching, through the use and interpretation of concept maps or through increasing cultural competence. The last two chapters in this section demonstrate the importance of giving faculty a voice in the process of change and of the faculty development itself. Hill examines the pedagogical content knowledge of faculty, the foundation upon which faculty teaching expertise is built. Carleton, et al., examine the effectiveness of a large-scale transformation project through the perspectives of faculty encountering everyday barriers in their implementation attempts.

1

A Disciplinary Teaching and Learning Center: Applying Pedagogical Content Knowledge to Faculty Development

Gili Marbach-Ad, Laura C. Egan, and Katerina V. Thompson

There is widespread acknowledgement of a looming crisis in science, technology, engineering, and mathematics (STEM) education (Arum & Roksa, 2001; President's Council of Advisors on Science and Technology [PCAST], 2012). This crisis threatens the quantity and quality of future scientists, as well as the educators, entrepreneurs, and policymakers whose work requires scientific literacy. Averting this situation will require large-scale improvements in STEM curriculum and instruction at the undergraduate level. STEM faculty members and graduate students must be the change agents because of their important role in delivering education, and they cannot play this role without support.

Research suggests some common impediments to improving curriculum and instruction: lack of pedagogical training for faculty members, which leads them to rely on the style of teaching in which they were taught; lack of institutional rewards for teaching; lack of time for re-conceptualizing classes and curricula; large class settings, which make it difficult to engage students; and student resistance to unfamiliar teaching styles. Most universities use some combination of two prevalent approaches to help faculty members overcome these impediments. The first approach consists of short but intensive professional development programming (e.g., summer teaching institutes) offered by disciplinary societies. The second approach is a campus-wide teaching and learning center that offers a-la-carte-style workshops and seminars that serve all disciplines. Both approaches can motivate faculty and introduce them to effective teaching approaches, but each has shortcomings. Campus-wide programs may be discounted by faculty members and graduate teaching assistants (GTAs) because these programs rarely offer discipline-specific teaching strategies (Henderson, Beach, & Finkelstein, 2011). Both campus-level programs and short, intensive institutes may lack ongoing support for faculty members trying to develop proficiency with new teaching approaches. Neither of these approaches focuses on institutional change at the level of the department, where scientific professional identity resides (Wieman, Perkins, & Gilbert, 2010), although some campus centers are now placing more emphasis on serving STEM faculty and building relationships with STEM departments.

Building on Wieman et al.'s (2010) call to change departmental cultures at research universities to foster educational innovation in STEM fields, we present the disciplinary teaching and learning center (TLC) as an effective mechanism for supporting faculty members and graduate students to achieve sustained improvements in undergraduate STEM curriculum and instruction. We discuss the structure of the TLC, share evidence of its impact on faculty members and graduate students, and suggest effective practices for adoption of this model at other research universities.

THE DISCIPLINARY TEACHING AND LEARNING CENTER MODEL

The TLC at the University of Maryland was established in 2006 to serve the faculty members and GTAs of the College of Chemical and Life Sciences, which encompassed three biology departments (Biology, Entomology, and Cell Biology and Molecular Genetics) and the department of Chemistry and Biochemistry. These departments are comprised of approximately 200 faculty members (both tenure track and professional track), 400 graduate students, and 2,400 undergraduate students.

Our center provides professional development that reflects the importance of Pedagogical Content Knowledge (PCK) theory. PCK supports the special requirements for teaching within a particular discipline by integrating relevant content and the best pedagogical practices to teach this content (Shulman, 1986). Building on PCK and discipline-specific norms and practices, the TLC familiarizes faculty members and GTAs with evidence-based practices and supports implementation of those practices in the classroom. TLC staff members have combined expertise in science and science education, and regularly collaborate with discipline-based education researchers.

The overarching goals of the TLC are to (1) provide opportunities for science faculty members to collaborate and consult with science education experts, (2) incorporate training in teaching science as part of the standard graduate program, and (3) create a structured environment of teaching and learning communities that support faculty members and graduate students in their efforts to identify appropriate content and adopt effective pedagogies. To accomplish these goals, the TLC provides a wide variety of resources and support to faculty members and GTAs (Figure 1), while continuously evaluating their needs and the effectiveness of resources provided. We elaborate on each of the TLC activities in the following pages.

FIGURE 1: Primary professional development activities of the TLC

ENRICHMENT PROGRAMS

The TLC offers a variety of opportunities for faculty members, postdoctoral fellows, and GTAs to learn about effective teaching practices, and fosters dialogue about teaching and learning. The primary components of these enrichment programs are (1) teaching and learning workshops, (2) Visiting Teacher/Scholars, and (3) travel grants to attend off-campus teaching and learning conferences and workshops. As the TLC has established its value within the college, attendance at TLC workshops and Visiting Teacher/Scholar seminars has grown. Workshops and seminars now average between 20 and 70 attendees representing a variety of departments and positions.

Teaching and Learning Workshops

Each semester, the TLC hosts one or more workshops open to all, as well as several luncheon workshops targeted at faculty members whose responsibilities focus primarily on teaching. Workshops follow varying formats, including individual presentations, panel presentations followed by questions and answers, and small working groups. Some workshops are very specific to a single course or a sequence of courses, while others cover broader topics. Workshop topics are often suggested by faculty members and may highlight a successful practice or reflect a desire to learn more about an emerging approach. Other topics are chosen by the TLC staff members based on salience in science education literature. Examples of recent workshop topics are creating blended learning courses, optimizing the sequence of introductory courses, teaching with models and concept maps, and using student misconceptions to facilitate learning. We describe here two workshops that were designed to impact departmental culture and practices.

One of the first TLC workshops focused on the role of the GTA. In this workshop, faculty members and graduate students collaborated first in homogenous and then in heterogonous small groups to discuss how GTAs contribute to teaching and how their professional development and preparation could be enhanced. Based on participant feedback from end-of-workshop evaluations, both faculty members and GTAs found the workshop valuable and came away with a deepened understanding of the expectations, needs, and experiences of the other. One of the outcomes of this workshop was the implementation of mandatory, team-taught preparatory courses for biology and chemistry GTAs (described below).

A subsequent workshop was organized at the request of the biology departmental chair. This four-hour workshop, held during a departmental retreat, engaged faculty in developing a rubric for peer evaluation of teaching. As a result of this workshop, the biology department implemented a system of peer observation using the rubric they had created to provide formative feedback to faculty on their teaching. Each faculty member now is observed teaching twice each semester, and in turn observes and provides constructive feedback to two other faculty members.

Visiting Teacher/Scholars

The Visiting Teacher/Scholars program brings to campus those rare individuals who can serve as role models for the successful integration of teaching and scientific research. Each semester, the TLC hosts one Visiting Teacher/Scholar who spends two days meeting with faculty members and graduate students and offers a seminar on teaching and learning; some also give a seminar on their scientific research area. Visiting Teacher/Scholars represent a wide range of research areas within the chemical and life sciences, in an effort to appeal to the college's diverse faculty population.

Travel Grants to Attend Off-Campus Teaching and Learning Conferences and Workshops

In addition to offering enrichment activities on our own campus, the TLC offers funding for interested faculty members and GTAs to travel to conferences and workshops on topics related to teaching and learning to enable them to learn about recent research on effective teaching practices and share ideas with the broader science education community. The TLC also encourages faculty members and GTAs to present their teaching approaches and research results at these meetings. The TLC assists presenters with preparing conference proposals, presentations, and proceedings papers. In the first eight years of the TLC (2006–2014), this resulted in 133 conference presentations and 34 publications.

ACCULTURATION OF NEW FACULTY MEMBERS

Incoming faculty members enter into new teaching positions with diverse backgrounds and experiences, but often very little preparation for fulfilling their role as educators. Graduate programs generally prioritize training in research while providing limited or no formal training in teaching (Cox, 1995; Golde & Dore, 2001; Handelsman, Miller, & Pfund, 2007; Luft et al., 2004). As they adjust to their new positions, many faculty members experience difficulties in effectively fulfilling their teaching responsibilities (Boice, 2011). Well-designed institutional support and professional development programs can ease new faculty members' transition and promote their development as effective science educators (Austin, Sorcinelli, & McDaniels, 2007).

In the early years of the TLC, we sought to better understand our new faculty members' backgrounds and needs through a longitudinal study of a cohort of eleven faculty members who came to our college in 2007. We interviewed each new faculty member when they arrived at the university to learn about their teaching philosophies, teaching challenges, and the kinds of support they felt they needed to further develop their teaching skills. We found that most new faculty members enter the college with little previous teaching experience and a wide variety of concerns related to curriculum, pedagogy, and time management (Marbach-Ad et al., 2014a, 2014b; Marbach-Ad, Egan, & Thompson, 2015).

These interviews helped us to develop an array of professional development activities to meet the varied needs of our new instructors.

- Welcome workshops: All new faculty members are invited to participate in a short workshop to open communication with the TLC, to provide an overview of pedagogical approaches in the sciences, and to introduce them to online and institutional resources for teaching and learning.

- Welcome packet: Each new faculty member receives a packet of materials that includes reference books on effective teaching (e.g., Handelsman et al., 2007; McKeachie & Svinicki, 2006), influential reports from national science organizations (e. g., American Association for the Advancement of Science [AAAS], 2010), seminal science education papers, and guides on the effective use of technology in the classroom.
- Workshops and seminars: We encourage new faculty members to participate in the TLC's workshops and seminars on science teaching and learning and connect new faculty members to relevant workshops hosted by the campus-wide Center for Teaching Excellence and the Division of Information Technology.
- Individual consulting: New faculty members who require targeted or supplemental support may make appointments to meet with TLC staff members for individual consulting.
- Faculty learning communities: We encourage new faculty members to join on- and off-campus FLCs that are relevant to their teaching interests.

After three years, we conducted a second round of interviews with these faculty members to gauge their teaching development and acculturation into the college. In these interviews, faculty members discussed how their teaching philosophies had evolved as a result of their participation in TLC activities. The teaching philosophies that were expressed by faculty members generally developed in accord with national recommendations for effective teaching (Handelsman et al., 2007; AAAS, 2010). These philosophies include teaching for understanding instead of memorization, interacting with and engaging students, tailoring instruction to the diversity of students in the class, applying course material to everyday life, teaching with an interdisciplinary approach, and connecting course material to science research (Marbach-Ad et al., 2014a, 2015).

ONE-ON-ONE CONSULTATION WITH FACULTY MEMBERS

In addition to the enrichment programs that are designed to appeal to broad swaths of college faculty members, the TLC offers one-on-one support in response to the specific needs of individuals. Consulting, as a highly personalized and context-specific support, can help faculty members overcome the barriers they face in implementing changes in their teaching (Hativa, 1995). Our consulting relationships generally begin with an individual faculty member coming to the TLC director with a specific question about a course that he or she is teaching. TLC staff members then offer highly individualized support on topics including changing course curriculum, implementing new pedagogies appropriate to the course content and the instructor's strengths, reviewing student course evaluations and responding to issues raised in those evaluations, designing formative and summative assessment tools, and observing classes to provide constructive feedback.

In addition to consulting on curricular change and pedagogical implementation, the TLC also supports faculty members in preparing grant proposals for science education initiatives. In this role, the TLC staff members actively encourage faculty members to apply for grants, share resources and knowledge relevant to the teaching and learning aspects of the grant, and assist with writing components of the grant proposal. Many scientific research grants now require a demonstration of impact beyond the discovery of new knowledge (e.g., National Science Foundation Broader Impacts). TLC staff members assist with measuring this impact, as well as connecting the grant writers to other faculty members with similar education and outreach interests.

CONSULTATIONS FOR FACULTY LEARNING COMMUNITIES

The TLC encourages faculty member involvement in a variety of FLCs that facilitate curricular redesign and support faculty members in their efforts to adopt innovative teaching strategies. These FLCs focus variously on gateway introductory courses, thematically linked sequences of upper-level courses, and the interface between related science disciplines (e.g., biology/mathematics and biology/physics). Our communities foster productive collaborations between lecturers, tenure-track faculty members, graduate students, and science education specialists, and provide opportunities for experienced instructors to mentor novice instructors. The teamwork that develops within communities creates a supportive environment that promotes faculty exploration of innovative pedagogies, and TLC staff members provide resources related to these innovative pedagogies. Such ongoing teamwork also makes it more feasible for faculty members to engage in and obtain grant funding for large-scale initiatives.

One of the longest standing communities in the college is the Host-Pathogen Interaction (HPI) FLC. This community includes approximately 20 faculty members with a shared research interest in HPI who are collectively responsible for nine undergraduate courses that relate to HPI. The HPI FLC meetings are analogous to their regular research meetings, and the group has met monthly since 2004 to discuss topics related to teaching and learning. Through these monthly meetings and periodic full-day retreats, the HPI FLC has mapped curriculum to reduce overlap and fill gaps across courses, created course activities that integrate content across courses, and developed assessment tools for gauging student conceptual understanding (Marbach-Ad et al., 2007, 2009, 2010).

PROGRAMS FOR GRADUATE STUDENTS

Graduate teaching assistants (GTAs) play a pivotal role in undergraduate education by leading laboratory and discussion sessions of high-enrollment introductory courses. Training for GTAs is important, not only to improve their ability to perform this role, but also for their professional development, as many of our GTAs plan to pursue careers that involve teaching in some capacity. Our programming for graduate students in chemistry and biology includes mandatory components for all new GTAs and optional components for graduate students with an interest in gaining further expertise in teaching and learning.

GTA Preparatory Courses

A six-week mandatory preparatory course for all new GTAs is offered every fall in separate sessions for graduate students in biology and chemistry (Marbach-Ad et al., 2010; Marbach-Ad, Schaefer, Kumi, Friedman, Thompson, & Doyle, 2012). These courses were developed around three major goals: (1) to build a community for new GTAs and socialize them into their respective departments; (2) to model good teaching by employing strategies that have a research base in support of their effectiveness; and (3) to help GTAs to understand their roles within the department and the course that they will be teaching. The preparatory courses are team-taught and cover multiple topics, including communication skills, student assessment, and teaching strategies. Every course also integrates veteran GTAs and faculty members who share their experience and answer any questions that the new GTAs may have. Some preparatory courses conclude with individual classroom observations in which one of the preparatory course instructors observes the GTA teaching and provides constructive feedback. A formal evaluation of the preparatory course for chemistry GTAs (Marbach-Ad, Schaefer, Kumi, et al., 2012) showed that those who had completed the preparatory course, on average, received significantly higher scores on student course evaluations than the cohort of new GTAs in the year prior to the establishment of the preparatory course for measures such as effective teaching, respecting students, and the instructor's level of preparation for course sessions.

Prior to the creation of the TLC, the Chemistry and Biochemistry Department did not offer teaching preparatory courses for new GTAs, while in the life sciences each department offered a different version of the course. The TLC staff members not only consulted with departmental representatives in designing these courses, but also helped teach the courses. The TLC also catalyzed the collaboration between the three life sciences departments, which resulted in a team-taught course that leveraged the expertise of the three departments and built community among GTAs from the three departments who were teaching the same course within the introductory curriculum. This collaboration between the TLC and the departments resulted in substantial, long-lasting change to departmental policy and practice, such as making the prep course mandatory for all new GTAs and involving a large number of faculty members and experienced GTAs in delivering instruction.

University Teaching and Learning Program (UTLP)

For graduate students who want additional training and a teaching certificate that is included as a notation in their diploma, the TLC partners with the campus Center for Teaching Excellence to offer an optional, extensive University Teaching and Learning Program (UTLP). The philosophy behind the UTLP is that graduate students' preparation for their future careers would benefit from training in teaching as well as training in research, particularly for students who plan to seek positions involving instruction or, more broadly, communicating science to broad audiences.

The UTLP requires multiple components that enrich graduate students with knowledge, skills, and experience. At a minimum, these components include the following:

- Completing a two-credit science education course
- Participating in seven teaching and learning workshops
- Being mentored in teaching by a science faculty member
- Observing classes taught by experienced faculty members
- Developing a teaching portfolio
- Conducting a teaching project

Students participating in the UTLP found that it strengthened their teaching skills, provided opportunities for publications and for presentations at science education conferences, and enhanced their attractiveness as job candidates (Marbach-Ad et al., in press). As one of the GTAs remarked, "Doing the research project and having more of a background in science education research sets you apart from other applicants in the hard sciences."

MONITORING AND EVALUATION OF THE TLC AND ITS ACTIVITIES THROUGH ONGOING RESEARCH

While the TLC provides a comprehensive package of professional development activities, it also continually monitors the needs of its audience and evaluates the success of its activities. This research employs multiple assessment tools and methods, including pre- and post-surveys, observations, and interviews (see http://cmns-tlc.umd.edu/ for tools and literature). The TLC uses its research findings to inform program activities; to build credibility within the target departments, college, and broader community; and to add to the growing body of literature on evidence-based teaching approaches and effective professional development. These research and evaluation activities require considerable time and effort but are crucial to the success of the TLC.

The TLC's multi-level evaluation plan includes measures of participation, satisfaction, learning, application, and impact (Colbeck, 2003; Guskey, 2000; Kirkpatrick, 1998). Given the growing body of research that suggests that faculty members face significant barriers in implementing desired changes in their teaching (e.g., Dancy & Henderson, 2008; Henderson et al., 2008), we focus a great deal of attention on the application level and how this reconciles with faculty beliefs about how they should be teaching. Faculty beliefs influence their practices; however, sometimes practices do not align with beliefs because of impediments to the adoption of new practices and/or resistance to these new practices at the institutional, instructor, or student level.

The TLC periodically collects data from the three principal populations that are impacted, directly or indirectly, by its services: faculty members, graduate students, and undergraduate majors. Analyzing the three population data sets provides insight into faculty and students' beliefs and use of instructional practices to understand progress in implementing changes in teaching across the college (Marbach-Ad, Schaefer, & Thompson, 2012, 2014b). Additionally, the data that we collect from these three populations afford us a cross-sectional view that can help us understand where we can target our professional development. For example, science education researchers suggest that working in groups and collaborative learning at the undergraduate level is important for course-related learning as well as students' preparation for their future careers (Ebert-May, Brewer, & Allred, 1997; Hake, 1998). Through our surveys, we found that only about half of the faculty (55%) and undergraduates (50%) we surveyed placed importance on working in groups at the undergraduate level. We also found strong correlations between the faculty's rated importance of group work and its use in class (Spearman's r=0.46, p<0.01) and outside of class (Spearman's r=0.31, p<0.05). Our finding that only half of faculty members employ group work inside or outside of class may reflect the logistical difficulties in designing and facilitating productive group-work in large undergraduate classes. Based on this finding, the TLC hosted a Visiting Teacher/Scholar who spoke about and provided examples of effective group activities for large undergraduate courses. TLC staff members also offer ongoing support to individual faculty in the development and implementation of group study activities.

REFLECTIONS AND FUTURE DIRECTIONS

Within a span of eight years, the TLC has developed into a valued resource within the college. It provides a wide variety of support and scaffolding to an expanding population, including faculty members, postdoctoral students, and graduate students in multiple science departments. The TLC conducts regular needs assessments and evaluates the effectiveness and comprehensiveness of the resources provided. This allows us to maintain relevancy and enhance the utility of TLC offerings, so that the TLC adapts to the changing needs and growing capacities of its audience. Conducting research on all activities, as well as periodic surveys of all faculty and graduate students about their teaching beliefs and practices, provides important information for developing new activities, improving existing activities, and reaching out to new audiences.

The Importance of Discipline-Based Professional Development

The disciplinary nature of the TLC is critical to its success. The strength of the disciplinary teaching and learning center model is that the professional development is not only linked to the disciplinary content and PCK, but is also integrated into the departmental community structure in a way that is not possible in campus-wide teaching centers or disciplinary society programs.

Faculty members and graduate students see the professional development provided by the TLC as relevant mainly because the TLC is housed within the college that it serves and is staffed by education specialists with strong credentials and credibility in the target disciplines. Our disciplinary focus also means that all professional development happens within the context of an existing and enduring community. Faculty members who participate in professional development activities do so alongside their departmental colleagues, with whom they collaborate on an ongoing basis in their teaching, research, and service roles. By connecting the professional development with these departmental collaborations, the TLC is able to impact the departmental culture. This model therefore facilitates institutional and cultural change that can be difficult to achieve through summer teaching institutes or campus-level professional development initiatives that generally work with faculty members who do not share an existing and enduring community. Similarly, graduate students benefit from receiving training within their own department. The initial training programs occur within an existing cohort of incoming students, which promotes collaboration within the cohort as well as integration into the broader community of graduate students in the department.

We are aware that there are potential drawbacks to the discipline-based model. One drawback is the potential for redundancy between disciplinary and campus-wide teaching and learning centers. Our TLC was developed as part of an effort of the campus Center for Teaching Excellence to establish disciplinary satellite centers that could extend campus-level activities and provide more specialized support. We work collaboratively with the campus center to promote programs that are of general interest, and we encourage our faculty and graduate students to participate in programs that do not require a disciplinary context. For example, we refer our UTLP graduate students to the teaching portfolio retreat offered by the campus center, instead of duplicating this program.

Another potential drawback is the perpetuation of academic silos and loss of interdisciplinary interactions. To encourage these interactions, we often convene FLCs that are focused on the interface between disciplines and include faculty from other departments. For example, the TLC supports communities that include faculty members from the disciplines of biology, physics, and education. In addition, our workshops and seminars are open to the campus community and typically draw a diverse audience, which sparks interdisciplinary conversation and can lead to more extended collaborations.

We believe that the importance of PCK and embedded, discipline-specific professional development applies not just to the sciences but to all disciplines. While our TLC was developed to serve the specific needs of the chemistry and biology departments within our specific university context, this model is broadly applicable to other STEM disciplines and to other research universities.

REFERENCES

American Association for the Advancement of Science (AAAS). (2010). *Vision and change: A call to action.* Washington, DC: AAAS. www.visionandchange.org /VC_report.pdf.

Arum, R. & Roksa, J. (2001). *Academically adrift: Limited learning on college campuses.* Chicago: The University of Chicago Press.

Austin, A. E., Sorcinelli, M. D., & McDaniels, M. (2007). Understanding new faculty: Background, aspirations, challenges, and growth. In R. Perry & J. Smart (Eds.), *The scholarship of teaching and learning in higher education: An evidence-based perspective,* (pp. 39–89). Dordrecht, Netherlands: Springer.

Boice, R. (2011). *Improving teaching and writing by mastering basic imagination skills.* Paper presented at the Lilly Conference on College & University Teaching, Washington, DC.

Colbeck, C. L. (2003). *Measures of success: An evaluator's perspective.* Paper presented at the CIRTL Forum, Center for the Integration of Research, Teaching and Learning, Madison, WI.

Cox, M. D. (1995). The development of new and junior faculty. In W. A. Wright (Ed.), *Teaching improvement practices: Successful strategies for higher education.* Bolton, MA: Anker.

Dancy, M. & Henderson, C. (2008). *Barriers and promises in STEM reform.* Paper presented at the Commissioned Paper for National Academies of Science Workshop on Linking Evidence and Promising Practices in STEM Undergraduate Education, Washington, DC.

Ebert-May, D., Brewer, C., & Allred, S. (1997). Innovation in large lectures—teaching for active learning. *BioScience,* 47(9), 601–607.

Golde, C. M. & Dore, T. M. (2001). *At cross purposes: What the experiences of doctoral students reveal about doctoral education.* Philadelphia, PA: Pew Charitable Trusts. Retrieved from www.phd-survey.org.

Guskey, T. R. (2000). *Evaluating professional development.* Thousand Oaks, CA: Corwin.

Hake, R. R. (1998). Interactive-engagement vs. traditional methods: A six-thousand-student survey of mechanics test data for introductory physics courses. *American Journal of Physics,* 66, 64–90.

Handelsman, J., Miller, S., & Pfund, C. (2007). *Scientific teaching.* New York: W. H. Freeman and Company.

Hativa, N. (1995). The department-wide approach to improving faculty instruction in higher-education: Qualitative evaluation. *Research in Higher Education,* 36(4), 377–413.

Henderson, C., Beach, A., & Finkelstein, N. (2011). Facilitating change in undergraduate STEM instructional practices: An analytic review of the literature. *Journal of Research in Science Teaching,* 48 (8), 952–984.

Henderson, C., Beach, A., Finkelstein, N., & Larson, R. S. (2008). *Preliminary categorization of literature on promoting change in undergraduate STEM.* Paper presented at the Facilitating Change in Undergraduate STEM symposium, Augusta, MI.

Kirkpatrick, D. L. (1998). *Evaluating training programs: The four levels (2nd ed.).* San Francisco, CA: Berrett-Koehler.

Luft, J. A., Kurdziel, J. P., Roehrig, G. H., & Turner, J. (2004). Growing a garden without water: Graduate teaching assistants in introductory science laboratories at a doctoral/research university. *Journal of Research in Science Teaching,* 41(3), 211–233.

Marbach-Ad, G., Briken, V., El-Sayed, N. M., Frauwirth, K., Fredericksen, B., Hutcheson, S., Gao, L., Joseph, S., Lee, V., McIver, K. S., Mosser, D., Quimby, B. B., Shields, P., Song, W., Stein, D. C., Yuan R. T., & Smith A. C. (2009). Assessing student understanding of host pathogen interactions using a concept inventory. *Journal of Microbiology & Biology Education*, 10, 43–50.

Marbach-Ad, G., Briken, V., Frauwirth, K., Gao, L., Hutcheson, S., Joseph, S., Mosser, D., Parent, B., Shields, P., Song, W., Stein, D., Swanson, K., Thompson, K. V., Yuan, R., & Smith A. C. (2007). A faculty team works to create content linkages among various courses to increase meaningful learning of targeted concepts of microbiology. CBE—*Life Sciences Education*, 6, 155–162.

Marbach-Ad, G., Egan, L. C., & Thompson, K. V. (2015). *A discipline-based teaching and learning center: A model for professional development*. Dordrecht, Netherlands: Springer.

Marbach-Ad, G., Katz, P., & Thompson, K. V. (in press). The value of a disciplinary teaching certificate program for chemistry and biology graduate students. *Journal on Centers for Teaching and Learning*.

Marbach-Ad, G., McAdams, K., Benson, S., Briken, V., Cathcart, L., Chase, M., El-Sayed, N., Frauwirth, K., Fredericksen, B., Joseph , S., Lee, V., McIver, K., Mosser, D., Quimby, B., Shields, P., Song, W., Stein, D., Stewart, R., Thompson, K., & Smith, A. (2010). A model for using a concept inventory as a tool for students' assessment and faculty professional development. CBE—*Life Sciences Education*, 9, 408–436.

Marbach-Ad, G., Schaefer, K. L., Kumi, B. C., Friedman, L. A., Thompson, K. V., & Doyle, M. P. (2012). Prep course for chemistry graduate teaching assistants at a research university. *Journal of Chemical Education*, 89(7), 865–872.

Marbach-Ad, G., Schaefer, K. L., & Thompson, K. V. (2012). Faculty teaching philosophies, reported practices, and concerns inform the design of professional development activities of a disciplinary teaching and learning center. *Journal on Centers for Teaching and Learning*, 4, 119–137.

Marbach-Ad, G., Schaefer-Ziemer, K. L., Orgler, M., & Thompson, K. V. (2014a). New instructors teaching experience in a research intensive university: Implications for professional development. *Journal on Centers for Teaching and Learning*, 5, 49–90.

Marbach-Ad, G., Schaefer-Ziemer, K. L., Orgler, M., & Thompson, K. V. (2014b). Science teaching beliefs and reported approaches within a research university: Perspectives from faculty, graduate students, and undergraduates. *International Journal of Teaching and Learning in Higher Education*, 26(2), 232–250.

Marbach-Ad, G., Shields, P. A., Kent, B. W., Higgins, B., & Thompson, K. V. (2010). Team teaching of a prep course for graduate teaching assistants. *The Journal of Graduate Teaching Assistant Development*, 13, 44–58.

McKeachie, W. & Svinicki, M. (2006). *McKeachie's teaching tips (12th ed)*. New York: Houghton-Mifflin.

President's Council of Advisors on Science and Technology (PCAST). (Feb 2012). Report to the president. *Engage to excel: Producing one million additional college graduates with degrees in science, technology, engineering, and mathematics.* President's Council of Advisors on Science and Technology. http://www.white house.gov/administration/eop/ostp/pcast

Shulman, L. S. (1986). Paradigms and research programs in the study of teaching: A contemporary perspective. In M.C. Wittrock (Ed.), *Handbook of research on teaching* (pp. 3–36). York: Macmillan.

Wieman, C., Perkins, K., & Gilbert, S. (2010). Transforming science education at large research universities: A case study in progress. *Change,* 42(2), 6–14.

ABOUT THE AUTHORS

Gili Marbach-Ad is the Director of the CMNS Teaching and Learning Center of the College of Computer, Mathematical, and Natural Sciences at the University of Maryland in College Park, Maryland.

Laura C. Egan is a Research Associate at Westat in Rockville, Maryland.

Katerina V. Thompson is the Director of Undergraduate Research and Internship Programs in the College of Computer, Mathematical, and Natural Sciences at the University of Maryland in College Park, Maryland.

2

Faculty Learning Communities: A Professional Development Model that Fosters Individual, Departmental, and Institutional Impact

Katerina V. Thompson, Gili Marbach-Ad,

Laura Egan, and Ann C. Smith

Student-centered, active-learning strategies (e.g., engagement with subject matter, problem solving, critical thinking) far surpass traditional, teacher-centered approaches (e.g., listening, reading, rote memorization) in motivating students, supporting deep understanding, and preparing students for future academic and career success (Handelsman et al., 2004; Hurtado et al., 2010; Arum & Roksa, 2011; Derting & Ebert-May, 2011; Freeman et al., 2014; Franklin, Sayre, & Clark, 2014). However, university faculty have been slow to embrace these more effective approaches due to lack of formal training in teaching, minimal interaction with science education experts, time constraints, and weak institutional support (Wieman, Perkins, & Gilbert, 2010).

STEM teaching reform efforts, such as teaching institutes and disseminating evidence of effective practices, have primarily been aimed at the level of the individual and have largely fallen short of expectations (Finelli, Pinder-Grover & Wright, 2014; Ebert-May et al., 2011; Austin, 2011; Fairweather, 2006). This lack of success is not surprising, since teaching reform is not simply a matter of increasing faculty awareness of recommended practices, but of engaging them in an ongoing process of evaluation and reflection that enables them to implement those practices thoughtfully and effectively (Smith & Marbach-Ad, 2010; Marbach-Ad et al., 2010). Faculty members typically work in isolation, rather than approaching teaching as a shared responsibility, which compounds the problem by slowing the propagation of innovative and effective teaching practices. The most promising approaches for catalyzing change are those with an institutional focus that is rooted in disciplinary cultures and that recognizes the essential role of ongoing peer support, such as that provided by faculty learning communities (Fairweather, 2008; Austin, 2011; Henderson, Beach, & Finkelstein, 2011).

Faculty learning communities (FLCs), in which groups of faculty meet regularly around shared interests and goals, have been touted as an efficient, sustainable means of overcoming the isolation and lack of pedagogical expertise that constitute barriers to educational reform (Yoder, 2013). The collaborative approach to improving teaching that is embodied by FLCs has a long history that can be traced back to Dewey (1916). FLCs are a form of community of practice (Wenger, 1998) that provide a supportive structure for faculty to rethink and transform their teaching practices (Cox, 2004; Layne et al., 2002). FLCs are thought to help instructors implement changes to teaching, curriculum, and assessment (Demir & Abell, 2010; Lakshamanan et al., 2011; Vescio, Ross, & Adams, 2008), leading to deeper and more durable student learning (Cox, 2004; Dawkins, 2006; Silverthorn, Thorn, & Svinicki, 2006).

Cox (2004) strongly advocated the use of communities for improving teaching and provided a formal definition for FLCs based on their implementation at his institution, Miami University. He described FLCs as "a cross-disciplinary faculty and staff group of six to fifteen members (eight to twelve members is the recommended size) who engage in an active, collaborative, yearlong program with a curriculum about enhancing teaching and learning and with frequent seminars and activities that provide learning, development, the scholarship of teaching, and community building" (Cox, 2004, p. 8). He further characterized FLCs as being either cohort-based or theme-based. Cohort-based FLCs involved individuals with similar educational ranks, roles, or experiences (e.g., junior faculty or graduate students). Topic-based FLCs involved individuals of differing ranks or roles who gathered around a specific educational issue or goal (e.g., teaching with technology, or enhancing the first-year student experience). In this essay, we take a more expansive view and consider FLCs as any regular gathering of multiple university educators with shared objectives related to enhancing student learning and/or educator teaching expertise. While the explicit goal of a given FLC may relate to one of these objectives, they are so tightly interlinked that, in practice, improving one will almost certainly have a positive impact on the other.

FACULTY LEARNING COMMUNITIES AS AN INSTITUTIONAL CHANGE STRATEGY

The University of Maryland (UMD) has been establishing FLCs for over two decades to enhance faculty teaching expertise and support a range of campus-wide educational initiatives (Benson et al., 2013). The earliest FLCs were a collaborative effort of the UMD Office of Undergraduate Studies and Center for Teaching Excellence, and their membership was drawn from all campus units. The first FLC consisted of a group of about a dozen faculty selected through a competitive application process who met weekly for an academic year for discussion and to carry out a group project related to teaching and learning. This group was responsible for launching several enduring campus-wide initiatives, including the annual Undergraduate Research Day, a campus-wide system for student evaluation of courses, and an annual departmental award for excellence in undergraduate education. Subsequent FLCs have been organized around specific themes that are institutional priorities, for example shifting courses from traditional lecture to a technologically enhanced mode and developing courses that fulfill the requirements of our new General Education framework.

At the level of the College of Computer, Mathematical, and Natural Sciences (CMNS), curricular and professional development initiatives are supported by our discipline-based Teaching and Learning Center (TLC) (Marbach-Ad, Schaefer, & Thompson, 2012; Marbach-Ad, Egan, & Thompson, 2015). The TLC seeks to increase faculty buy-in and participation by focusing specifically on university science teaching and supporting the development of disciplinary instructional expertise (Pedagogical Content Knowledge, Shulman 1986a,b). In addition to hosting workshops and offering individual consultation, the TLC supports a variety of FLCs that promote curriculum redesign, help faculty develop innovative teaching approaches, and gather evidence of the impact of these reforms. Many of these FLCs were established in support of biological sciences curriculum initiatives funded by a succession of grants from the Howard Hughes Medical Institute.

TLC involvement in CMNS-centered FLCs has fostered opportunities for faculty members to have ongoing interactions with discipline-based education researchers and science education specialists. The main roles of TLC staff are to (1) increase faculty awareness of relevant science education literature; (2) assist with the development and adoption of new pedagogies; (3) provide guidance on the selection, development, and validation of instruments to assess student learning and attitudes; and (4) document the science education reform initiatives and disseminate their outcomes via scholarly conferences and journals. FLC members are encouraged to take a leading role in these dissemination efforts, as a way of further integrating them into the national science education community. This has resulted in over 100 presentations at science education conferences and 28 papers in peer-reviewed journals.

FLCs have resulted in tangible changes in the UMD undergraduate curriculum, including the development of new courses (Calculus for the Life Sciences: Thompson et al., 2013b; NEXUS/Physics for the Life Sciences: Redish et al., 2014; Principles of Microbiology: Marbach et al., 2010; Marquee Courses in Science and Technology and I-Series courses, www.gened.umd.edu). In addition, community efforts have garnered over $2.7 million in grant funding to support their efforts, including grants from the National Science Foundation, the University System of Maryland, and the UMD Teaching and Learning Transformation Center.

Our experience has shown that education reform initiatives that originate in communities of faculty have greater impact and sustainability than those that are the province of a single faculty member. Social networks analysis of UMD's 10-year effort to strengthen quantitative reasoning in the biological science curriculum revealed two characteristics that contributed to the success of broad curricular reforms. First, the communities were characterized by ongoing interactions between faculty members from different, but interlinked, disciplines. Second, several key individuals participated in multiple communities over successive years, which facilitated the spread of teaching innovations across the curriculum (Thompson et al., 2013b).

POTENTIAL BENEFITS OF FACULTY LEARNING COMMUNITIES

While individual FLCs may have been established to achieve a specific individual, course, academic program, or institutional goal, it is likely that benefits accrue at multiple levels since these various levels of the institutional organization form a nested series (Fig. 1). For example, FLCs that are convened to advance institutional agendas often simultaneously provide participants with individual benefits comparable to that which they would have acquired in an individually-focused FLC. Below we discuss the types of benefits that might be expected to accrue as a result of faculty participation in FLCs and provide examples from our experience at UMD.

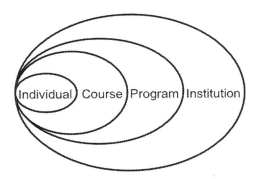

FIGURE 1. Faculty Learning Communities are likely to have benefits at multiple levels of institutional organization, since these form a nested series.

FLCs can provide a variety of benefits to individual faculty members, and many are convened specifically with these goals in mind (Marbach-Ad et al., 2007). Benefits include the opportunity for exchanging of ideas and experiences related to teaching and having near-peer role models for effective teaching. This provides members with moral support for trying out new approaches, as well as the opportunity to troubleshoot and gather feedback on the success of those efforts. Experienced FLC members can serve as mentors for new faculty members, facilitating their transition into teaching. Our research shows that individuals in FLCs place a higher value on and are more likely to use student-centered teaching practices (Marbach et al., 2014), although we could not distinguish whether FLC participation caused changes in beliefs and practices, or whether student-centered faculty were simply more likely to join FLCs. While these FLCs may be established primarily to assist individuals in their development as effective teachers, the interactions that occur within the FLC help establish departmental and institutional expectations related to teaching. Moreover, the very existence of institutionally supported FLCs sends a clear message that teaching is valued.

The most basic form of individually focused FLC at UMD is a loosely structured gathering of individuals with a shared interest in discussing topics related to teaching and learning. This is exemplified by the College of CMNS lecturers' luncheons. The luncheons are organized by the administrative staff of the CMNS dean's office and usually have a specific theme that is salient to the faculty with primarily teaching responsibilities (e.g., teaching approaches for the large lecture class, discouraging plagiarism). These loosely knit gatherings typically attract 15–20 individuals from among the approximately 80 professional-track faculty within the college's 10 departments.

FLCs can also be established to coordinate teaching efforts at the level of a particular course. These course-focused FLCs facilitate effective teaching in large-enrollment courses and maintain consistency of instructional approaches when there are multiple instructors. For example, the UMD General Microbiology course has a well-defined structure that incorporates a variety of collaborative and active learning approaches (e.g., case studies, online discussions, group projects). This curriculum was created by a multi-level teaching team that included instructors, laboratory coordinators, graduate teaching assistants, and undergraduate assistants who engaged in iterative cycles of course assessment and revision (Smith et al., 2005).

Course-focused FLCs can also enable collaborative curriculum development, especially when the FLC is composed of individuals with complementary expertise. This is exemplified by the NEXUS/Physics FLC, which brought together physicists, biophysicists, biologists, and science education specialists to create an innovative, fundamentally multidisciplinary introductory physics for life sciences course (Thompson et al., 2013a, Redish et al., 2014). Since its inception five years ago, this FLC has created new lecture and lab curricula; written a wikitext to replace conventional, disciplinary textbooks; collected assessment data on the effectiveness of the new curriculum; and provided support to new faculty who join the teaching rotation for this course.

FLCs can also provide multiple benefits at the level of the department or academic program, which has been recognized as the critical unit of change within the university (Quardokus & Henderson, 2014; Wieman et al., 2010; Marbach et al., 2015). They can ensure comprehensive content coverage across multiple courses in the curriculum and help minimize redundancy within a degree program. For example, a curriculum mapping project conducted by the UMD's Host-Pathogen Interactions FLC revealed that none of the courses in the microbiology degree program covered a key learning objective set by the group (Marbach-Ad et al., 2009, 2010). FLCs can foster consistency in pedagogical approaches within an academic program. This unified front can help overcome the oft-reported student resistance to instructional approaches that require increased student engagement and effort (Shimazoe & Aldrich, 2010; Henderson & Dancy, 2011; Finelli, Daly, & Richardson, 2014). FLCs allow faculty to build consensus around programmatic learning goals and collect data to evaluate whether these goals have been met, as now required by many higher education accrediting bodies (Beno, 2004; Provezis, 2010). Finally, FLCs built around program-level objectives are particularly well positioned to capitalize on large scale funding opportunities offered by foundations and federal agencies.

One of the longest standing uses of FLCs is to advance institutional educational agendas. These communities often consist of highly accomplished and respected faculty members who can serve as change agents in their respective departments. These FLCs are characterized by a high degree of disciplinary diversity, and as such, are ideal venues for facilitating interdisciplinary collaborations and developing multidisciplinary academic programs. When UMD launched its new General Education framework, FLCs were a critical element of the implementation plan (Benson et al., 2013). The signature element of the new General Education framework is the "I-Series" introductory-level course. Rather than being broad, shallow surveys of a particular discipline, I-Series courses instead focus on how disciplinary experts approach current, complex questions of importance to society. Faculty seeking to develop an I-Series course are expected to participate in an FLC that provides enrichment and opportunities to share their experiences implementing the new courses. This approach has made the creation and continual improvement of these unique courses less onerous for faculty and has ensured that the General Education framework is translated into practice with fidelity to the original vision.

A MODEL FOR FACULTY LEARNING COMMUNITY IMPACT

While there is widespread consensus that FLCs enhance undergraduate education, empirical support for this contention is sparse. In addition, FLCs can vary widely in their goals, format, membership, and longevity, even within a given institution. There is a need for a better understanding of the attributes that contribute to the success of communities and the mechanisms by which these successes are achieved, to provide guidance to academic leaders who seek to use FLCs to facilitate institutional change. We describe a conceptual model for investigating the effect of FLC participation on undergraduate instruction at multiple levels of institutional organization.

We posit that the impact of FLCs on faculty teaching beliefs and practices is mediated by the psychological sense of community, defined by McMillan and Chavis (1986, p. 9) as "a feeling that members have of belonging, a feeling that members matter to one another and to the group, and a shared faith that members' needs will be met through their commitment to be together." Sense of community is a central concept of social psychology and has served as a framework for investigating the dynamics and impacts of a variety of different types of communities, ranging from urban neighborhoods to workplaces to university living-learning programs (Hanley, 2011).

We propose a multi-faceted conceptual model (Fig. 2) for characterizing the effect of FLCs on undergraduate instruction. Briefly, we recognize that there exists considerable variability in **community attributes,** as well as variability in **personal attributes** of the individuals who comprise communities. These individual and community attributes collectively influence the experience of individuals within that community, and are likely to result in affective and attitudinal changes in the participating faculty (i.e., an enhanced **psychological sense of community**).

We posit that individuals with a strong sense of community will show greater involvement in the work of their community (e.g., by regularly attending community meetings and participating in the community over an extended time period). These positive experiences may also motivate them to join additional communities. This essentially creates a positive feedback loop in which faculty members with a growing sense of community increase their level of

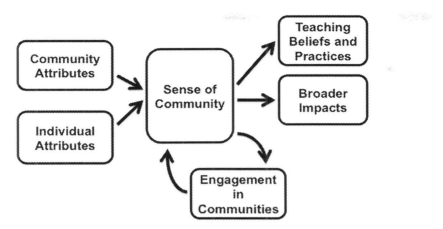

FIGURE 2. A conceptual model for how individual, departmental, and institutional teaching reform might be mediated by the psychological sense of community within faculty learning communities.

engagement and commitment in communities, which, in turn, would further intensify their sense of community.

In addition, we hypothesize that sense of community affects **teaching beliefs and practices**, particularly the degree to which faculty value and use student-centered teaching approaches. We think that it is important to measure beliefs and practices separately, because valuing an approach is a necessary, but not sufficient, step to adopting that approach, and it is possible that FLC participation affects one but not the other. This positive influence of communities on faculty affect and attitudes could ultimately have **broader impacts** beyond individual teaching beliefs and practices. These might include influencing the teaching practices of colleagues who are not themselves FLC members, and catalyzing changes in the culture of teaching at the departmental and institutional levels.

FUTURE DIRECTIONS

Faculty learning communities hold great promise for catalyzing change at multiple levels of institutional organization, but empirical evidence of their impact is scarce. Furthermore, the mechanisms by which these effects are produced are unclear. We propose a conceptual model that can be used to elucidate each of these effects and their mechanisms. The resultant data will provide a roadmap for creating and supporting FLCs as a means of transforming institutions.

TABLE1. Examples of Dimensions that Could be Used to Characterize Faculty Learning Communities and the Individuals who Participate in Them

Individual attributes	Community attributes
• Gender • Ethnicity • Scholarly discipline • Position within the university (e.g., tenured faculty member, non-tenure-track faculty, graduate assistant) • Level of teaching expertise (e.g., expert, novice) • Length of time at the university	• Number of participants • Frequency of meetings (e.g., weekly, biweekly, or monthly) • Duration (e.g., one semester, one year, or ongoing) • Type of leadership (e.g., one leader or distributed leadership) • Directionality of communication within the group (e.g., unidirectional or multidirectional) • Degree to which participation is voluntary • Source of community mission (e.g., communally decided or externally imposed) • Diversity of membership in terms of position within the university (e.g., similar or diverse faculty ranks) • Diversity of membership in terms of scholarly discipline (e.g., disciplinary or multidisciplinary) • Whether food is used as an incentive for participation • Whether monetary compensation (e.g., stipend) is used as an incentive for participation

REFERENCES

Arum, R., & Roksa, J. (2001). *Academically adrift: Limited learning on college campuses.* Chicago: The University of Chicago Press.

Austin, A. E. (2011). *Promoting evidence-based change in undergraduate science education.* A Paper Commissioned by the National Academies National Research Council Board on Science Education.

Beno, B. A. (2004). The role of student learning outcomes in accreditation quality review. *New Directions for Community Colleges, 126,* 65–72.

Benson, S.A., Smith, A., & Eubanks, E. B., (2013). Developing scholarly teaching at a research university: Using learning communities to build capacity for change. In D. J. Salter (Ed.), *Cases on quality teaching practices in higher education* (pp. 212–227). Hershey, PA: IGI Global.

Cox, M. L. (2004). Introduction to faculty learning communities. *New Directions for Teaching and Learning, 97,* 5–23.

Dawkins, P. W. (2006). Faculty development opportunities and learning communities. In N. Simpson & J. Layne (Eds.), *Student learning communities, faculty learning communities, & faculty development* (pp. 63–80). Stillwater, OK: New Forum.

Demir, A., & Abell, S. (2010). Views of inquiry: Mismatches between views of science education faculty and students of an alternative certification program. *Journal of Research in Science Teaching, 47,* 716–741.

Derting, T. L., & Ebert-May, D. (2010). Learner-centered inquiry in undergraduate biology: Positive relationships with long-term student achievement. *CBE Life Science Education*, 9, 462–472.

Dewey, J. (1916). *Democracy and education: An introduction to the philosophy of education*. New York: Macmillan.

Ebert-May, D., Derting, T. L., Hodder, J., Momsen, J. L., Long, T. M., & Jardeleza, S. E. (2011). What we say is not what we do: Effective evaluation of faculty development programs. *BioScience*, 6(17), 550–558.

Fairweather, J. (2008). *Linking evidence and promising practices in science, technology, engineering, and mathematics (STEM) undergraduate education: A status report for the National Academies National Research Council Board on Science Education*. Commissioned Paper for the National Academies Workshop: Evidence on Promising Practices in Undergraduate Science, Technology, Engineering, and Mathematics (STEM) Education. Accessed on 01/10/15 from https://www.nsf.gov/attachments/117803/public/Xc--Linking_Evidence--Fairweather.pdf

Finelli, C. J., Daly, S. R., & Richardson, K. M. (2014). Bridging the research-to-practice gap: Designing an institutional change plan using local evidence. *Journal of Engineering Education—Special Issue on the Complexities of Transforming Engineering Higher Education*, 103(2), 331–361.

Finelli, C. J., Pinder-Grover, T., & Wright, M. C. (2011). Consultations on teaching. Using student feedback for instructional improvement. In C. Cook & M. Kaplan (Eds.), *Advancing the culture of teaching at a research university: How a teaching center can make a difference* (pp. 65–79). Herndon, VA: Stylus.

Franklin, S. V., Sayre, E. C., & Clark, J. W. (2014). Traditionally taught students learn; actively engaged students remember. *American Journal of Physics*, 82: 798–801.

Freeman, S., Eddy, S. L., McDonough, M., Smith, M. K., Okoroafor, N., Jordt, H., & Wenderoth, M. P. (2014). Active learning increases student performance in science, engineering, and mathematics. *Proc Natl Acad Sci U S A*. doi: 10.1073/pnas.1319030111

Handelsman, J., Ebert-May, D., Beichner, R., Bruns, P., Chang, A., DeHaan, R., Gentile, J., Lauffer, A., Stewart, J., Tilghman, S. S., & Wood, W. B. (2004). Scientific teaching. *Science*, 304, 521–522.

Hanley, J. R. (2011). *A study of psychological sense of community within living-learning environments*. Master's Thesis. Paper 151. Retrieved 07/21/14 from http://thekeep.eiu.edu/theses/151

Henderson, C., Beach, A., & Finkelstein, N. (2011). Facilitating change in undergraduate STEM instructional practices: An analytic review of the literature. *Journal of Research in Science Teaching*, 48, 952–984.

Henderson, C., & Dancy, M. H. (2011, February). Increasing the impact and diffusion of STEM education innovations. Invited paper for the *National Academy of Engineering, Center for the Advancement of Engineering Education Forum*,

Impact and Diffusion of Transformative Engineering Education Innovations, available at: http://www.nae.edu/File.aspx.

Hurtado, S., Newman, C. B., Tran, M. C., & Chang, M. J. (2010). Improving the rate of success for underrepresented racial minorities in STEM fields: Insights from a national project. *New Directions for Institutional Research,* 148, 5–15.

Lakshamanan, A., Heath, B. P., Perlmutter, A., & Elder, M. (2011). The impact of science content and professional learning communities on science teaching efficacy and standards-based instruction. *Journal of Research in Science Teaching,* 48, 534–551.

Layne, J., Froyd, J., Morgan, J., & Kenimer, A. (2002). Faculty learning communities. *Frontiers in Education,* 2002. FIE 2002. 32nd Annual (Vol. 2, pp. F1A–13– F1A–18 vol.2). doi:10.1109/FIE.2002.1158114

Marbach-Ad, G., Briken, V., El-Sayed, N. M., Frauwirth, K., Fredericksen, B., Hutcheson, S., Gao, L., Joseph, S., Lee, V., McIver, K. S., Mosser, D., Quimby, B. B., Shields, P., Song, W., Stein, D. C., Yuan R. T., & Smith, A. C. (2009). Assessing student understanding of host pathogen interactions using a concept inventory. *Journal of Microbiology & Biology Education,* 10, 43–50.

Marbach-Ad, G., Briken, V., Frauwirth, K., Gao, L. Y., Hutcheson, S.W., Joseph, S.W., Mosser, D., Parent, B., Shields, P., Song, W., Stein, D. C., Swanson, K., Thompson, K. V., Yuan, R., & Smith. A. C. (2007). A faculty team works to create content linkages among various courses to increase meaningful learning of targeted concepts of microbiology. *CBE—Life Sciences Education,* 6, 155–162.

Marbach-Ad, G., Egan, L. C., & Thompson, K. V. (2015). *A discipline-based teaching and learning center: A model for professional development.* Dordrecht, Netherlands: Springer.

Marbach-Ad, G., McAdams, K. C., Benson, S., Briken, V., Cathcart, L., Chase, M., El-Sayed, N., Frauwirth, K., Fredericksen, B., Joseph, S. W., Lee, V., McIver, K. S., Mosser, D., Quimby, B. B., Shields, P., Song, W., Stein, D. C., Stewart, R., Thompson, K. V., & Smith, A. C. (2010). A model for using a concept inventory as a tool for students' assessment and faculty professional development. *CBE—Life Sciences Education,* 10, 408–416.

Marbach-Ad, G., Schaefer, K., & Thompson, K. (2012). Faculty teaching philosophies, reported practices, and concerns inform the design of professional development activities of a disciplinary teaching and learning center. *Journal on Centers for Teaching and Learning,* 4, 119–137.

Marbach-Ad, G., Ziemer, K., Orgler, M., & Thompson, K. (2014). Science teaching beliefs and reported approaches within a research university: Perspectives from faculty, graduate students, and undergraduates. *International Journal of Teaching and Learning in Higher Education,* 24(2), 232–250.

McMillan, D. W., & Chavis, D. M. (1986). Sense of community: A definition and theory. *Journal of Community Psychology,* 14(1), 6–23.

Provezis, S. J. (2010). Regional accreditation and learning outcomes assessment: Mapping the territory. Unpublished PhD thesis, University of Illinois at Urbana-Champaign. Retrieved on 01/09/2015 from http://hdl.handle.net/2142/16260

Quardokus, K., & Henderson, C. (2014). Using department-level social networks to inform instructional change initiatives. *Proceedings of the NARST 2014 Annual Meeting*, April 1, 2014, Pittsburgh, PA.

Redish, E. F., Bauer, C., Carleton, K. L., Cooke, T. J., Cooper, M., Crouch, C. H., Dreyfus, B. W., Geller, B., Giannini, J., Svoboda Gouvea, J., Klymkowsky, M. W., Losert, W., Moore, K., Presson, J., Sawtelle, V., Thompson, K., Turpen, C., & Zia, R. K. P. (2014). NEXUS/Physics: An interdisciplinary repurposing of physics for biologists. *American Journal of Physics*, 82(5), 368–377.

Shimazoe, J., & Howard, A. (2010). Group work can be gratifying: Understanding & overcoming resistance to cooperative learning. *College Teaching*, 58(2), 52–57.

Shulman, L. S. (1986a). Those who understand: Knowledge growth in teaching. *Educational Researchers,* 15(2), 4–14.

Shulman, L. S. (1986b). Paradigms and research programs in the study of teaching: A contemporary perspective. In M.C. Wittrock (Ed.), *Handbook of research on teaching* (pp. 3–36). York: Macmillan.

Silverthorn, D. U., Thorn, P. M., & Svinicki, M. D. (2006). It's difficult to change the way we teach: Lessons from the integrative themes in physiology curriculum module project. *Advances in Physiology Education, 30*, 204–214.

Smith, A. C., & Marbach-Ad, G. (2010). Learning outcomes with linked assessment— An essential part of our regular teaching practice. *Journal of Microbiology and Biology Education,* 11, 123–129.

Smith, A. C., Stewart, R., Shields, P., Hayes-Klosteridis, J., Robinson, P., & Yuan, R. (2005). Introductory biology courses: A framework to support active learning in large enrollment introductory science courses. *J Cell Biology Education*, 4, 143–156.

Thompson, K. V., Chmielewski, J. A., Gaines, M. S., Hrycyna, C. A., & LaCourse, W. R. (2013a). Competency-based reforms of the undergraduate biology curriculum: Integrating the physical and biological sciences. *CBE—Life Sciences Education*, 12, 162–167.

Thompson K. V., Cooke, T. J., Fagan, W. F., Gulick, D., Levy, D., Nelson, K. C., Redish, E. F., Smith, R. F., & Presson, J. (2013b). Infusing quantitative approaches throughout the biological sciences curriculum. *International Journal of Mathematical Education in Science and Technology*. DOI: 10.1080/0020739X.2013.812754.

Vescio, V., Ross, D., & Adams, A. (2008). A review of research on the impact of professional learning communities on teaching practice and student learning. *Teaching and Teacher Education*, 24(1), 80–91. doi:10.1016/j.tate.2007.01.004

Wenger, E. (1998). *Communities of practice: Learning, meaning, and identity.* Cambridge: Cambridge University Press. ISBN 978-0-521-66363-2.

Wieman, C. E., Perkins, K., & Gilbert, S. (2010). Transforming science education at large research universities: A case study in progress. *Change* 42(2), 7–14.

Yoder, B. (2013). Faculty development using virtual communities of practice. Paper presented at the 120th *ASEE Annual Conference & Exposition*, June 23–26, 2013, Atlanta, GA. Retrieved from http://www.asee.org/file_server/papers/attachment/file/0003/4446/ASEE_VCP_Final_Paper.pdf

ABOUT THE AUTHORS

Katerina V. Thompson is the Director of Undergraduate Research and Internship Programs in the College of Computer, Mathematical, and Natural Sciences at the University of Maryland in College Park, Maryland.

Gili Marbach-Ad is the Director of the CMNS Teaching and Learning Center of the College of Computer, Mathematical, and Natural Sciences at the University of Maryland in College Park, Maryland.

Laura C. Egan is a Research Associate at Westat in Rockville, Maryland.

Ann C. Smith is the Assistant Dean in the Office of Undergraduate Studies at the University of Maryland in College Park, Maryland.

3

STEM Faculty Perceptions of Concept Map Assessments

Lindsay Owens, Chad Huelsman and Helen Meyer

In their 2011 literature review of undergraduate STEM faculty instructional practices and changes, Henderson, Beach, and Finkelstein developed an analytical framework for delving into changes in STEM instruction in institutions of higher education. This desire for improving faculty instructional practices was seen in the reports of different national commissions and increased funding for research in undergraduate STEM teaching (Labov, Singer, George, Schweingruber & Hilton, 2009). These efforts mirror attempts to enhance STEM teaching and learning in K–12 education, which we now see codified into A Framework for K–12 Science Education: Practices, Crosscutting Concepts, and Core ideas (National Research Council, 2012). With these dual points of pressure, national reports of teaching and learning in higher education and new standards in K–12, there is hope that college STEM faculty are increasing their awareness and use of instructional practices that result in greater student learning. The importance of improved instruction by college faculty is critical when we consider that college STEM faculty are often seen as instructional leaders, deemed crucial in the development of K–12 STEM teachers, by leading professional development (PD) activities based on the new standards.

However, Coppola and Krajcik (2013) argue that college STEM faculty often lack the foundational pedagogical knowledge and appropriate instructional skillsets to support K–16 learners. They suggest that, even though college faculty have earned advanced degrees, they do not necessarily have the interest, awareness, or ability to convey the complexities of their field to the more novice learner. Part of this inability could be related to instructional strategies used by STEM college faculty in their delivery. Seymour and Hewett in 1997 reported on the poor teaching practices pervasive in most college STEM courses, which focused on a teacher-centered information delivery system, as opposed to a learner-centered knowledge construction system. Padilla and Cooper (2012) further shed light on the issue with a focus on STEM faculty's use of traditional assessment techniques promoting rote recall of factual information. Even though traditional instructional and assessment techniques have a place within the classroom, Dauer, et al., (2013) have asserted it is time for STEM faculty to change traditional behaviors and diversify their "teaching toolbox" to include techniques promoting conceptual understanding.

In 1991, Novak built a case for the use of concept maps: a graphic metacognitive tool used to organize and assimilate existing knowledge with the meaningful learning of new concepts. As end-point (summative) assessments, or pre-instruction and formative assessments, concept maps have been shown to assist higher education faculty in reflecting about their teaching (Irons, 2008). In addition, concept maps make learning, as opposed to grading, the centerpiece of assessment. Based on the arising tension between STEM classes as an avenue for learning versus simply instruction, we argue if college STEM faculty can be enticed to diversify their assessment practices through the use of concept maps we can assist in the improvement of STEM instruction.

The goal of this research was to understand how a small group of STEM faculty interpreted student concept maps as a reflection of learning. The faculty were presented with 1) a quantitative gain score resulting from the pre- and post-course concept maps; and 2) qualitative structural changes represented in the pre- and post-course concept maps. The following research questions were addressed in this research:

- How do science and engineering faculty understand student concept maps as reflections of students' content knowledge?
- How do science and engineering faculty interpret the changes in the students' pre- and post-course concept maps as a reflection of students' learning?
- Do science and engineering faculty reflections regarding changes in student pre- and post-course concept maps, impact faculty reflections regarding changes in their teaching?

THEORETICAL FOUNDATIONS

Since 1972, concept maps (Figure 1) have been used to support learners with the organization and assimilation of concepts (Novak, 1998) in a wide range of science disciplines, including: chemistry, psychology, nursing, pediatrics, and engineering (Jacobs-Lawson & Hershey, 2002; Markow & Lonning, 1998; Srinivasan, McElvany, Shay, Shavelson, & West, 2008; West, Pomeroy, Park, Gerstenberger, & Sandoval, 2000).

The visualization process that occurs while students arrange their conceptual structures on maps can be useful to both the learners and teachers. Concept maps are useful as a formative instructional tool to understand students'

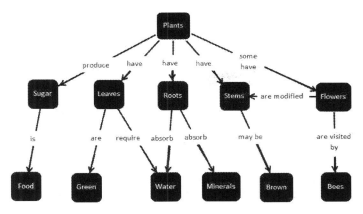

FIGURE 1. Simple plant concept map. The concept map demonstrates a two-dimensional representation of structural and functional relationships existing in plants. Adapted from *Online Concept Maps: Enhancing Collaborative Learning by Using Technology with Concept Maps,* by Canas, A. J., Ford, K. M., Novak, J. D., Hayes, P., Reichherzer, T. R., & Suri, N. 2001, Science Teacher, 68(4), p.50.

content misconceptions (Novak & Gowin, 1984; Novak, 1998). Students' science misconceptions often go unaddressed or unchallenged by instructors, which can leave students with an incomplete or erroneous understanding of the conceptual connections within the content. Concept maps are also useful for educators as a form of assessment that accurately displays students' conceptual understanding of a topic. Concept maps are metacognitive tools that allow learners to visualize their "conceptual structures" (Chinn & Samarapungavan, 2009). Specifically, a student's misconceptions are not part of his/her knowledge, however, they are still part of the student's conceptual structure of the topic or concept. Finally, Chinn and Samarapungavan (2007) stress the idea of conceptual structure, or organizational structure that includes hierarchies and the relationships between concept components.

In summary, concept maps are useful formative and metacognitive tools for the learners, as well as assessment instruments for instructors. They represent students' organizational structures and assist in the organization of core teaching concepts for educators. Despite concept mapping's empirically documented benefits, extant research that addresses how concept maps act as a resource to impact teacher's pedagogy and instructional practices is limited. By allowing university STEM faculty to reflect on their students' learning with concept maps, we anticipate changes in the faculty's beliefs about using techniques promoting conceptual learning.

METHODS

Study Overview and Participants

The research project integrated pre- and post-course concept mapping assessments into four summer classes instructed by five University of Cincinnati STEM faculty members. Two instructors co-taught the mathematics summer course. The university faculty taught courses designed for secondary science and mathematics teachers as part of an NSF funded Math and Science Partnership (MSP). The intent of the courses was to deepen the teachers' content knowledge within the specific discipline, as well as model for the teachers how to integrate engineering into traditional content instruction. Table 1 provides basic background information about the faculty involved in the study.

TABLE 1. Participants

Pseudonym	Course Taught	Pre-Knowledge of Concept Maps
Christine	Chemistry	One workshop; did not use in instruction
Phillip	Physics	No exposure
Mark	Engineering Math	One workshop; did not use in instruction
Mary	Engineering Math	No exposure
George	Geology	Had used concept-maps in instruction

Only George maintained an active disciplinary research agenda. The other faculty members' responsibilities were primarily teaching in freshman sequences within their disciplines. Prior to the start of the summer courses, the third author worked with the faculty to develop concept maps specific to their course objectives. The pre-course concept maps were obtained from the teacher participants during the summer program orientation. The post-course concept maps were completed on the last full instructional day. In addition to concept maps, the faculty used assessments of their own design in order to produce a grade for the teacher participants.

Designing and Conducting the Interview

All concept maps were quantitatively scored using a scoring system modified from West et al. (2000) and qualitatively analyzed by the authors as part of the overall program evaluation. Using the pre- and post-course concept map quantitative scores, an overall "gain score" was calculated for each student in one of the summer courses and an average course gain score was calculated and reported to the STEM faculty. The qualitative analysis was used to create paired sets of pre- and post-course concept maps for each course. Features such as concept organization were noted. The qualitative analysis allowed the authors to track if misconceptions from a pre-course concept map persisted on the post concept map, (Figure 2). Eventually, the paired map sets represented different levels of knowledge change, in the quantitative gain score or the qualitative structure or both, from the summer course participants.

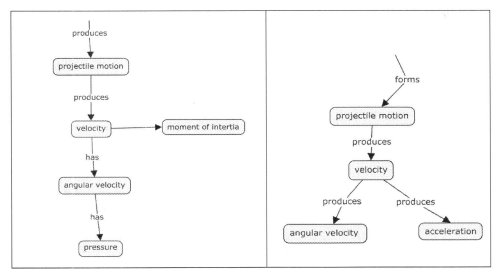

FIGURE 2. Misconception in Physics Concept Map. The left shows a portion of a physics pre-course concept map in which the student states that angular velocity has pressure. This misconception does not appear in the same student's post-course concept map (on the right) in which angular velocity is no longer linked to pressure.

At the conclusion of the summer program, each participating faculty member was interviewed about the use of, and results from, the concept maps from his or her course. During the interviews, the researchers shared the representative paired pre- and post-course maps with the faculty member. The focus of the interviews was to have the faculty members: 1) explore the paired maps' quantitative gain scores and qualitative organizational structures, 2) discuss how the concept maps reflected their instructional goals, and 3) reflect on how the concept map results influenced their thoughts about student learning.

RESULTS AND DISCUSSION

At the start of the interviews, most of the STEM faculty were interested in the quantitative data; specifically, in the class average gain scores. Phillip, like many of the faculty members, was surprised at the number of students who received a score of zero on either a pre- or post-course concept map. This prompted the interviewer to show an example of a student's concept map which received a score of zero, as shown in Figure 3 below.

When Phillip observed the concept map, he noted that while many of the physics concepts were present, the student did not use many prepositional phrases to connect the concepts. Since a prepositional phrase was required for a valid concept link, the concept map contained no valid links, and thus received a score of zero. Phillip stated,

> Looks like it would have gotten lots of points. [. . .] This person's had a physics class before, I think. They didn't put a verb there. If they'd have put a verb there they'd have gotten more points. Yeah, this person had a physics class before. (Phillip, lines 900–904).

How do faculty understand student concept maps as reflections of students' content knowledge?

Faculty discussed the background knowledge of their students at the start after noticing several patterns in the pre-course concept maps. As Phillip mentioned, he was able to discern the student had likely had physics prior to the summer

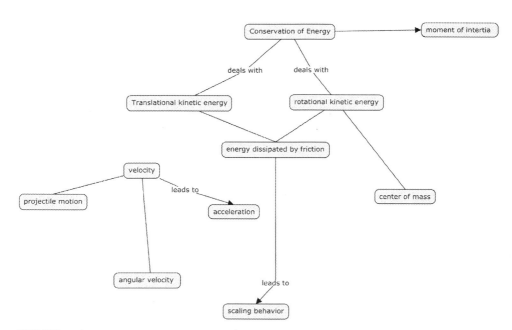

FIGURE 3. Pre-course concept map in physics.

course. In addition, other faculty discovered that many students were not familiar with certain terms used in the concept map word bank. For example, in Mark's class, students created a mathematical model to represent an electric circuit containing a battery, capacitor, and resistor. Capacitor and resistor were two of the words in the word bank. After observing several pre-concept maps, Mark stated, "People didn't know what capacitors and resistors were coming in and if you looked through to the entire stack of pre-maps they're kind of thrown in every which place" (Mark, lines 1289–1291).

George, who had used concept maps in his courses previously, looked closely at the organization and hierarchy of each concept map. When discussing the concept map used in Figure 4, George claimed,

> Yeah, again I would say the hierarchy is missing; everything else seems to be alright, and these cross relationships are okay. But at the same level, seismologists, earthquakes and ground shakings are put at the same level. I would put, for example, earthquakes are the ground shakings, and then studied by seismologists and they use two scales, intensity and magnitude. (George, lines 682–687).

In the end, the faculty felt the concept maps could provide a good representation of knowledge. They illustrated students' thoughts about *how* concepts related, instead of simply knowing that concepts have some connection. In a discussion of a pre-course concept map for chemistry, a student had made an

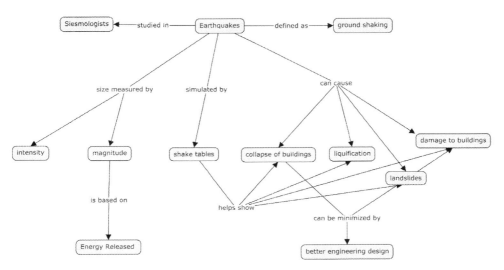

FIGURE 4. Geology Pre-Course Concept Map. Semiologist, earthquakes, and ground shaking are all listed on the same level (top) of hierarchy.

effort to link related concepts, though it was obvious the student did not under-stand the relationship. Christine stated:

> [The student] just didn't know what was getting the linkages in the right places. That's very interesting. Oh, I don't know whether it shows that they know that they should have been connected, but they just didn't quite know what to put between. Oh, that's very interesting, isn't it? (Christine, lines 980–987).

How did faculty interpret the changes in the pre- and post-course concept maps as a reflection of learning?

In addition to seeing learning reflected in students' post-course concept maps, some faculty members also commented that their students had a deeper level of understanding of content, even content they were familiar with entering the course. Christine compared a student's pre- and post-course chemistry maps,

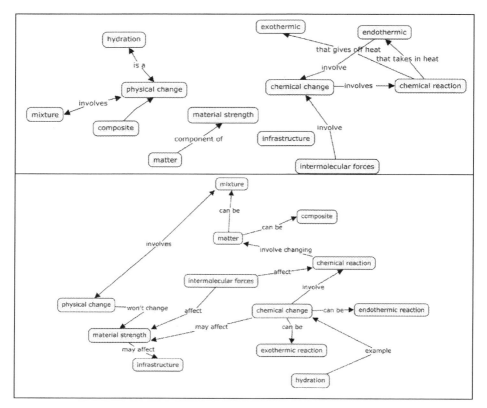

FIGURE 5. Chemistry pre-course concept map (top) and post-course concept maps (bottom), from the same student.

stating "This definitely, the post-map, definitely looks more sophisticated [. . .] Yeah, I think they have certainly, um, understood that things are connected differently from what they thought before they started"(Christine, 690–695).

George was able to see his students understood the importance of hierarchy in their post-course concept maps. "Notice that in terms of hierarchy they put energy release and seismologists at the same level; here earthquakes as the highest priority and then everything else is coming down, which is what we want," (George, lines 639–641).

Do faculty reflections on the pre- and post-course concept maps lead them to reflect on changes in their teaching?

After discussion of the quantitative and qualitative results, the faculty were asked to reflect on how concept maps could be used in their own courses in the

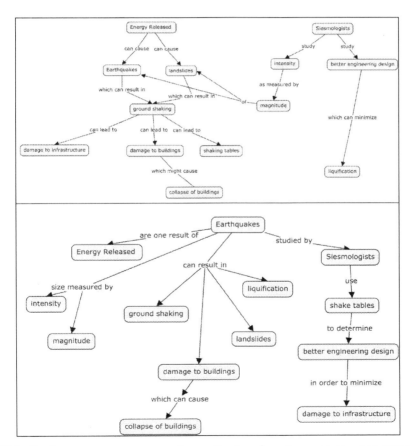

FIGURE 6. Geology pre-course concept map (top) and post-course concept maps (bottom) from the same student.

future, even courses not associated with the project. Christine and Mark both believed pre-course concept maps were inherently useful for the identification of misconceptions at entry to a course. For example, Christine was surprised to find a number of students not understanding that matter was the central theme of chemistry. The centrality of matter was a concept Christine never mentioned in her course because she assumed all the teachers would know this. She stated, "It's also interesting how they don't really appear to, um, they don't really appear to have any concept that matter is the central connecting theme" (Christine, lines 672–674).Christine indicated that a pre-course concept map evaluation would have illuminated the gap between what the teachers knew and what she thought they knew.

Both Christine and Mark indicated the use of a pre-concept map would be useful to guide the curriculum of their courses. Mark stated he felt he used math topic words, such as exponential equations, that were too domain specific to allow students to make the broad connections between mathematics and engineering he was hoping to see. Mark indicated in the future, he would revise his word bank to include broader phrases, though he did not specify any potential phrases in the interview.

In summary, four of the faculty were able to relate pre-course concept maps to student misconceptions. They saw student knowledge gains in the comparison of the pre- and post-course concept maps. Lastly, they saw subtleties that existed in a student's concept map, independent of the quantitative score. At the completion of the interview, Christine, Mark, and Phillip expressed that they were excited to use concept maps again in their courses. George had been using concept maps in his courses prior to this summer program and indicated that he would continue to utilize them.

Mary, on the other hand, indicated she felt concept maps in her college mathematics courses would not provide useful information because qualitative/conceptual assessments do not demonstrate whether students can solve mathematics problems. We probed Mary further, asking what she would like to see on a concept map. She stated, "If they can put an algebraic formula there or something [like an] algebraic statement [in the concept map], that would sure help" (Mary, lines 733–734). Throughout Mary's interview, it was evident that she felt assessments of learning must be quantitative in nature.

> I was trained as a statistician so concept maps are like 'Oh my, I have no idea how I would [interpret this].' I'm used to seeing numbers and seeing 'Does this number change?' This is so subjective to me. [. . .] I mean even more than I ever want to quantify it. (Mary, lines 643–650).

Finally, both Mark and Christine made suggestions as to the need to refine their word banks to better align with the key course concepts and learning goals in their course. Mark indicated a goal for students was to see broad connections between mathematics and engineering. Therefore, he should modify his word bank to include broader mathematics terms which focus on making connections, rather than categorizing mathematics terms into mathematical domains. This suggested change is not a subtle revision, such as be more clear about "X," rather Mark was reconsidering what was of value to know and be able to do with the mathematics in engineering.

IMPLICATIONS AND LIMITATIONS

Although this research was limited to a small number of STEM faculty teaching in a specialized program, it does provide important insights for the transition of STEM instruction in higher education from a teacher-centered system to a learner-centered system. The use of concept maps in large, freshman level STEM courses is likely to be prohibitive. Effective implementation and map analysis is time consuming, making it unlikely to be embraced in courses with hundreds of students. However, we have seen in this limited setting, the use of pre- and post-course concept maps, and guided reflections on them can result in faculty gaining an understanding of what learning looks like beyond a simple numeric score.

Our research provides a frame for other STEM faculty interested in exploring conceptual learning. It allows them to revisit the use of concept maps as a learning and assessment strategy by learning from firsthand experiences. Further, it provides faculty development teams insight into STEM faculties' attitudes and beliefs about conceptual assessments. Assessments such as concept maps can be used formatively to design new PD with the aim of improving pedagogy, curriculum, and assessment practices.

CONCLUSIONS

The use of a conceptual assessment tool provided an opportunity for faculty to engage with students' learning. We observed faculty identifying misconceptions, missing concepts, and misplaced hierarchies central to the disciplines. This reflection-on-practice model (Schon, 1983) is an area where many STEM faculty need support to develop assessments that measure actual learning rather than discrete knowledge bits.

Our study also revealed resistance to change deeply embedded in many STEM disciplines, such as Mary, who only saw value in quantitative assessments. These attitudes could hinder the use of conceptual assessments and new pedagogies in STEM classrooms. Counter to much of the rhetoric, typical traditionalist STEM cultures often do not reward faculty who focus on improving pedagogy. The reliance on simple quantitative measures of learning is efficient in our large freshman-level classes. Conceptual assessments lack efficiency and therefore conflict with customary teaching styles and grading systems. A remedy to this situation begins with discussions with STEM faculty about learning based on student products of learning in an open and un-repercussive environment that supports reflection. Small steps through advocacy can lead to positive changes in curriculum and instructional, peer collaboration, and student learning.

REFERENCES

Canas, A. J., Ford, K. M., Novak, J. D., Hayes, P., Reichherzer, T. R., & Suri, N. (2001). Online concept maps: Enhancing collaborative learning by using technology with concept maps. *Science Teacher, 68*(4), 49–51.

Chinn, C. A., & Samarapungavan, A. (2009). Conceptual change—multiple routes, multiple mechanisms: A commentary on Ohlsson (2009). *Educational Psychologist, 44*(1), 48–57. doi:10.1080/00461520802616291

Coppola, B. P., & Krajcik, J. S. (2013). Discipline-centered post-secondary science education research: Understanding university level science learning. *Journal of Research in Science Teaching, 50*(6), 627–638.

Dauer, J. T., Momsen, J. L., Speth, E. B., Makohon-Moore, S. C., & Long, T. M. (2013). Analyzing change in students' gene-to-evolution models in college-level introductory biology. *Journal of Research in Science Teaching, 50*(6), 639–659. doi:10.1002/tea.21094

Henderson, C., Beach, A., & Finkelstein, N. (2011). Facilitating change in undergraduate STEM instructional practices: An analytical review of the literature. *Journal of Research in Science Teaching, 48,* 952–984.

Irons, A. (2008). *Enhancing learning through formative assessment.* New York: Routledge.

Jacobs-Lawson, J., & Hershey, D. A. (2002). Concept maps as an assessment tool in psychology courses. *Teaching of Psychology, 29*(1), 25–29.

Labov, J., Singer, S., George, M., Schweingruber, H., & Hilton, M. (2009). Effective practices in undergraduate STEM education part 1: Examining the evidence. *CBE—Life Sciences Education, 8,* 157–161.

Markow, P. G., & Lonning, R. A. (1998). Usefulness of concept maps in college chemistry laboratories: Students' perceptions and effects on achievement. *Journal of Research in Science Teaching, 35*(9), 1015–1029.

National Research Council. (2012). *A framework for K–12 science education: Practices, crosscutting concepts, and core ideas.* Committee on a Conceptual Framework for New K–12 Science Education Standards. Board on Science Education, Division of Behavioral and Social Sciences and Education. Washington, DC: The National Academies Press.

Novak, J. (1998). *Learning, creating, and using knowledge: Concept maps as facilitative tools in schools and corporations.* Mahwah, N.J: L. Erlbaum Associates.

Novak, J. (1991). Clarify with concept maps. *Science Teacher, 58*(7), 44–49.

Novak, J. D., & Gowin, D. B. (1984). *Learning how to learn.* Cambridge, UK: Cambridge University Press.

Padilla, M., & Cooper, M. (2012). From the framework to the next generation science standards: What will it mean for STEM faculty? *Journal of College Science Teaching, 41*(3), 6–7. Retrieved from http://search.proquest.com/docview/1321 110963?accountid=2909

Schön, D.A. (1983). *The reflective practitioner: How professionals think in action.* New York: Basic Books.

Seymour, E., and Hewett, N. (1997). *Talking about leaving: Why undergraduates leave the sciences.* Boulder, CO: Westview.

Srinivasan, M., McElvany, M., Shay,J., Shavelson, R., & West, D. (2008). Measuring knowledge structure: Reliability of concept mapping assessment in medical education. *Academic Medicine, 83*(12), 1196–1203.

West, D. C., Park, J. K., Pomeroy, J. R., & Sandoval, J. (2000). Concept mapping assessment in medical education: A comparison of two scoring systems. *Medical Education, 36*(9), 820–826. doi: 10.1046/j.1365-2923.2002.01292.x

ABOUT THE AUTHORS

Lindsay Owens is a Graduate Research Assistant in the College of Education, Criminal Justice, and Human Services at the University of Cincinnati in Cincinnati, Ohio.

Chad Huelsman is a Graduate Research Assistant in the College of Education, Criminal Justice, and Human Services at the University of Cincinnati in Cincinnati, Ohio.

Helen Meyer is an Associate Professor in the College of Education, Criminal Justice, and Human Services at the University of Cincinnati in Cincinnati, Ohio.

4

Teaching to Increase Diversity and Equity in STEM (TIDES): STEM Faculty Professional Development for Self-Efficacy

Kelly M. Mack and Kate Winter

CHALLENGE

By the end of this decade, the U.S. economy will annually create over 120,000 new jobs requiring a bachelor's degree in computer science for emerging fields like cloud architecture, forensic investigation, and geospatial technology (Evans, Mckenna, & Schulte, 2013). However, currently, only approximately 41,000 computer science baccalaureates are produced each year (NSF, 2013). Further complicating this supply-demand mismatch are improved international economies that no longer allow the U.S. to rely on foreign-born talent to meet its STEM workforce demands; and the persistent underrepresentation of women, minorities and persons with disabilities who now comprise the fastest growing undergraduate population (NSF, 2013). Already, women of all racial and ethnic backgrounds account for nearly 60% of all U.S. college undergraduates (NSF, 2013); and the Western Interstate Commission for Higher Education (2008), in *Knocking at the College Door,* projects that, by 2022, the number of public and non-public high school graduates who are from minority populations will significantly increase, while that of white, non-Hispanic high school graduates will decrease. Werf and Sabatier (2009) predict that this trend will result in minority students outnumbering whites on U.S. college campuses by as soon as 2020.

The projected shifts in undergraduate student composition not only make it increasingly likely that all institutions of higher education will experience significant growth in their overall underrepresented student enrollments, but also emphasize the need and urgency for immediate implementation of pedagogical reform (PCAST, 2012; Tsui, 2007) that is evidence-based and culturally sensitive to the lived experiences of these populations (Froyd, 2008). However, mastery of new pedagogy commonly poses a substantial challenge for STEM faculty who oftentimes lack the substantive knowledge of and proficiency in teaching strategies that would enable students to master STEM content while becoming skilled learners (Froyd, Srinivasa, Maxwell, Conkey, & Shryock, 2005). Also, many current professional development interventions aimed at exposing faculty to enhanced STEM teaching strategies continue to overlook the role of cultural competence in teaching, fail to inextricably link culturally sensitive STEM pedagogies with advanced research, are non-reflective, and are devoid of the elements necessary for achieving sustained behavioral change in STEM teaching practices and patterns. As a result, modern teaching strategies continue to be wrongly directed toward "fixing" the student, and implemented at varying levels of precision with only modest gains in STEM student success.

Quintessential to overcoming these challenges is providing professional development opportunities for STEM faculty that not only depart from the traditional workshop model, but also effectively pair cultural consciousness with advanced pedagogy, and integrate the elements of self-efficacy needed for long-term implementation. Many scholars have noted that self-efficacy is a strong determinant of behavioral change (Bandura, 1977; DeChenne, Enochs, & Needham, 2012; Mohamadi & Asadzadeh, 2012). Indeed, infusion of its core elements—performance accomplishment, vicarious experiences, verbal persuasions and psychological states—into the professional development activities of STEM faculty not only results in the kind of sustained behavioral changes required for relevant and modernized STEM teaching, but also mitigates deficient coping behavior that can arise in the face of institutional barriers and positively impacts STEM student learning (DeChenne et al., 2012; Mohamadi & Asadzadeh, 2012).

To that end, the Association of American Colleges and Universities (AAC&U) and its Project Kaleidoscope (PKAL) have initiated Teaching to Increase Diversity and Equity in STEM (TIDES) as a three-year program, generously funded by the Helmsley Charitable Trust, to increase the self-efficacy of STEM faculty in implementing culturally competent pedagogies.

INNOVATION

It is widely accepted that active-learning strategies increase undergraduate student performance in STEM disciplines (Freeman et al., 2014), particularly for underrepresented students (Tsui, 2007). However, in a recent survey of department chairs at the nation's top 200 research universities, over 50% of respondents noted that they did not see a need to significantly change introductory course instructional methods in order to retain more underrepresented students (Bayer Corporation, 2012). Instead, it was noted that more co-curricular support, principally in the form of mentoring and tutoring, was necessary; and it was believed that faculty should assume primary responsibility for such.

Despite this expressed need for increased faculty engagement, few efforts to recruit or retain underrepresented STEM students include faculty professional development as a viable strategy. Of those professional development opportunities that do exist for faculty, most offer only cursory interventions that are primarily aimed at "fixing the student," and fail to either recognize the whole student, appreciate the value of the student's diverse perspective, or address the ways in which the underlying implicit associations of faculty contribute to their underrepresentation.

To address these inherent weaknesses, Freeman et al. (2014) recommend a second generation of STEM education research that relies upon the advances of educational and social psychology and explores the aspects of faculty behavior that most significantly impact student performance. By focusing on STEM faculty self-efficacy as a means of achieving pervasive institutional change, the AAC&U TIDES program promotes a novel approach that more completely addresses the structural barriers that threaten underrepresented STEM student learning, interest, competencies, and retention in the computer sciences and related STEM disciplines.

IMPLEMENTATION

Overview

Guided by its core values of inclusive excellence, collaboration and accountability, TIDES is committed to leading STEM higher education reform through the following goals:

1. (Re)Designing computer science courses to:
   a. **maximize the likelihood of success** (higher test scores, increased pass rates, increased retention rates, increased graduation rates) for diverse students; and
   b. attract and engage traditionally underrepresented groups in STEM courses to **increase self-efficacy and the likelihood for identification with STEM.**
2. **Increasing utilization of culturally competent pedagogy** by:
   a. **raising STEM faculty awareness** and **consciousness** of why and how to be culturally responsive;
   b. **increasing STEM faculty confidence** in implementing culturally competent pedagogy; and
   c. **creating a Networked Improvement Community** (i.e., Community of Practice) to support, promote and deploy project resources for immediate adoption, adaptation and widespread change in

STEM content delivery; and to **ensure support for institutional project success**.

The specific strategies that are employed include: 1) a national call for proposals, which serves to identify cross-disciplinary courses that can be meaningfully integrated with the concepts, methods, technologies, and tools foundational to the computer/information technology sciences; 2) a STEM Institute, which provides STEM faculty with the opportunity for introspection, reflection, exposure to cutting-edge research related to cultural competence in STEM pedagogies, and real time practice with implementation of new pedagogies; and 3) development and administration of new assessment instruments for self-efficacy in culturally responsive STEM pedagogy. The overview of the project's activities and goals (logic model) is provided in Figure 1.

Currently, this project supports 19 institution teams (comprised of three faculty/administrators per institution) over a three-year period to develop, refine, and implement cross-disciplinary courses related to enhancing computational

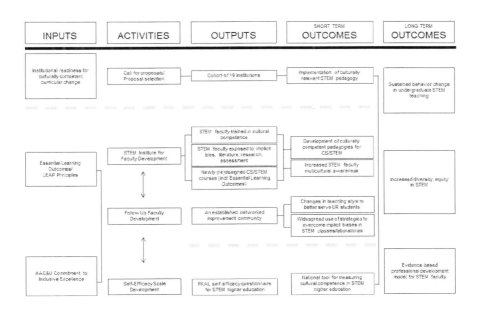

FIGURE 1. TIDES Logic Model.
Description: The Logic Model was developed to outline specific project outcomes and the various strategies for achieving them.

thinking skills.[1] Newly developed and/or re-designed STEM courses not only focus on developing the skills required for students to persist in the computer sciences, but also:

1. promote promising pedagogical practices that are culturally sensitive;
2. foster the inclusion of historically underrepresented groups in the STEM disciplines at unprecedented levels; and
3. lend themselves to sustained impact on STEM teaching strategies at multiple institution sites across the country.

Assuming a typical teaching load for faculty (up to 12 credit hours/semester) and an average lower-level STEM class size of approximately 100 students per class, the TIDES project has the potential to positively impact the computational skills of nearly 100,000 STEM majors, particularly those from underrepresented populations. Additionally, through a robust project identity and dissemination plan, the STEM faculty professional development materials will be deployed throughout the PKAL regional networks and AAC&U national meetings to STEM faculty beyond those participating in TIDES.

The overarching philosophy of the faculty professional development is grounded in evidence that training and support facilitates one's ability to thrive and advance in one's professional setting (Webster, 1998). In keeping with this philosophy, a cohort of awardee STEM faculty and administrators participate in an annual week-long STEM Institute that provides opportunities to: 1) reflect upon prior pedagogical performance, 2) practice culturally competent teaching strategies in real time, 3) contribute to a national community of practitioners, and 4) develop and maintain a networked improvement community through synchronous and asynchronous web-based and virtual platforms.

National Call for Proposals

Over 400 faculty from more than 300 institutions participated in an informational webinar, which highlighted the vision and goals of the TIDES program. Prior to actual proposal submission, letters of intent were received from 341 institutions, with 112 (or 33%) representing HBCUs, HSIs, women's colleges, and tribal colleges; 168 full proposals were received and reviewed, including 30% from diverse institution types.

---

1. The TIDES Program has provided funds for five institution teams to participate in key project activities, although their campus-based projects were not selected to receive funding through the competition. These institutions are referred to as "Honorable Mention" and are considered full members of the TIDES Networked Improvement Community.

Based upon careful review of proposals and reviewer critiques and comments, 14 projects were selected to receive funding, with another five institutions receiving Honorable Mention. More than half of the funded institutions are HBCUs, HSIs, women's colleges, or tribal colleges, where the likelihood of impacting diverse STEM baccalaureates is highest. Details of the selected institutions are provided in Table 1.

TABLE 1. TIDES Projects.
Description: Nineteen institutions of higher education participate in the AAC&U TIDES initiative.

Institution	Classification	Control	Other
Bryn Mawr College, PA	Baccalaureate	Private, not-for-profit	Women's
California State University Northridge, CA	Master's Large	Public	HSI
Fairleigh Dickinson University, NJ	Master's Large	Private, not-for-profit	
Fayetteville State University, NC	Master's Medium	Public	HBCU
Howard University, DC	Research Universities (high)	Private, not-for-profit	HBCU
Lawrence Tech University, MI	Master's Large	Private, not-for-profit	
Montgomery College, MD	Associate	Public	
Morgan State University, MD	Doctoral/Research Universities	Public	HBCU
Salish Kootenai College, MT	Associate/Baccalaureate	Private, not-for-profit	Tribal College
Smith College, MA	Baccalaureate	Private, not-for-profit	Women's
University of Dayton, OH	Research Universities (high)	Private, not-for-profit	
University of Puerto Rico–Humacao, PR	Baccalaureate	Public	HSI
Westminster College, UT	Master's Medium	Private, not-for-profit	
Wright State University, OH	Research Universities (high)	Public	
Connecticut College, CT*	Baccalaureate	Private, not-for-profit	
Knox College, IL*	Baccalaureate	Private, not-for-profit	
Ohio Northern University, OH*	Baccalaureate	Private, not-for-profit	
Pitzer College, CA*	Baccalaureate	Private, not-for-profit	
Queens College CUNY, NY*	Master's Large	Public	

*Honorable Mention

STEM Institute and Networked Improvement Community

As noted, mastery of pedagogy oftentimes poses a challenge for STEM faculty. AAC&U has learned, through its century of experience in empowering institutions of higher education, that these challenges can be permanently overcome through a prescribed sequence of events consisting of: 1) reflection, 2) professional development/coaching, and 3) development of a networked improvement community. The TIDES program integrates all of these components and provides STEM faculty with theoretical and practical exposure to cultural competence as a lifelong practice in STEM content delivery.

Reflection: At the outset of the STEM Institute, participants complete a reflective analysis of STEM course(s) that is(are) proposed for (re)design. This phase also introduces participants to the underlying theories and practices of cultural competence in STEM pedagogies that are grounded in social science and educational psychology research, including critical race theory, implicit bias and social cognitive career theory; as well as the role that they can play in facilitating underrepresented STEM student success.

Professional Development/Coaching: During afternoon sessions of the STEM Institute, participants have carefully designed opportunities to engage in difficult dialogues with an institution coach who is a recognized leader in institutional change, gender equity, race-gender intersectionality and/or minority STEM student retention. Institution coaches are relied upon to provide the kind of verbal persuasion that is known to contribute to the development of self-efficacy (Bandura, 1977).

Networked Improvement Community: At varying points throughout the academic year, awardee institution participants convene via teleconference and/or webinars for post-professional-development meetings that serve as a forum for discussing major issues, challenges and/or successes in implementing culturally sensitive STEM pedagogies. Participants also engage in follow-up capstone experiences at the annual AAC&U STEM Conference. These experiences serve to provide: 1) an open discussion of individual STEM courses within particular institutional contexts; 2) opportunity for cross-institutional synthesis that yields both generalized institutional practices that are easily transferred to other similarly situated institutions; and 3) the vicarious experiences that have been noted by Bandura (1977) as an essential source for promoting self-efficacy. This phase of the TIDES program is also critical to preventing faculty from reverting back to more traditional modes of undergraduate STEM teaching.

Additionally, during the first TIDES Institute, faculty completed surveys about their level of confidence in their own culturally competent STEM pedagogy, their confidence in their department to support and engage diverse students in STEM, the climate and culture of the university, their perceptions of various potential outcomes related to culturally competent pedagogy in STEM, and aspects of their teaching style.

Self-Efficacy Instruments

The TIDES Program is supported by the development of appropriate instruments that provide both an understanding of baseline self-efficacy regarding culturally responsive STEM pedagogy and assessment of changes during the initiative. Self-efficacy instruments must be domain specific to provide useful results (Bandura, 2006)1997. As such, the team reviewed instruments specific to both culturally responsive pedagogy, or multi-cultural teaching, and STEM undergraduate education. Because no single instrument exists to suit the needs of TIDES, a self-efficacy instrument was created by using or adapting relevant items from various instruments. The resulting TIDES self-efficacy instrument captures: 1) faculty responses about their level of confidence in their own culturally competent STEM pedagogy; 2) their confidence in their department to support and engage diverse students in STEM; 3) the climate and culture of the university with regards to diversity; 4) their perceptions of various potential outcomes related to culturally competent pedagogy in STEM; and 5) aspects of their teaching style.

Faculty completed a beta version of the survey during the first TIDES Institute. Administrators completed an abbreviated version of the instrument that omitted the teaching-focused sections. The data were then used to establish baselines and explore the validity of measures of group efficacy, self-efficacy in teaching, and expected outcomes, which are key indicators of the likelihood of sustained behavior change (Bandura, 2000, 2001, 2006; Siwatu, 2007).

RESULTS AND CONCLUSION

Participant feedback on the first TIDES Institute indicated that it was thought provoking, informative, and valuable to refining next steps for the institutions' projects. Insights on future topics and content include the need to incorporate additional coverage of the theoretical bases supporting active learning, the obstacles faced by underrepresented students in STEM, and the need for cultural change in STEM.

Exploratory factor analysis and confirmatory factor analysis of the faculty and administrator survey instruments supports the validity and internal reliability of a group efficacy measure (GE-all) comprising 9 items ($\alpha = 0.910$) in two sub-scales (GE1: 6 items, $\alpha = 0.903$; GE2: 3 items, $\alpha = 0.890$), a faculty

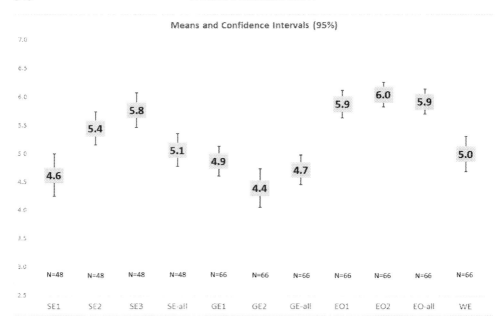

FIGURE 2. Means and Confidence Intervals for Measures.
Description: This table provides the mean score and confidence interval (95%) for each of the developed measures discussed in this chapter, along with the number of responses used in each analysis.

self-efficacy measure (SE-all) comprising 15 items (α = 0.940) in three sub-scales (SE1: 8 items, α = 0.947; SE2: 4 items, α = 0.867; SE3: 3 items, α = 0.899), an expected outcomes measure (EO-all) comprising 12 items (α = 0.946) in two sub-scales (EO1: 9 items, α = 0.946; EO2: 3 items, α = 0.802), and a three-item measure of how welcoming the department is for diverse faculty, staff, and students (WE, α = 0.936). Items asked about confidence level in either the department or oneself to engage in various activities, or level of agreement with expected outcomes and were scored on a 7-point scale (1= To an extremely small extent, 2=To a very small extent, 3=To a small extent, 4=To a moderate extent, 5=To a large extent, 6=To a very large extent, 7=To an extremely large extent). At the .05 level, there were no statistically significant differences between faculty and administrators on the shared measures. Means and confidence intervals (95%) for the measures are provided in Figure 2.

Scores for the expected outcomes of supporting students through culturally competent pedagogy (EO1, EO2, EO-all) were the highest, while the scores for the group efficacy items (GE1, GE2, and GE-all) were among the lowest, particularly for items that explicitly dealt with addressing mismatches between students' native cultures and those present in higher education. Among the faculty-only self-efficacy items, the lowest score was for the measure of culturally competent pedagogy. SE1 focuses on aspects that explicitly address the needs of underrepresented students in STEM classrooms, while the other two faculty SE measures are more broadly applicable to good teaching practices, generally (i.e., providing hands-on support to students, reflecting on the efficacy of one's teaching, and learning and basing instruction on students' interests, etc.). The measure of confidence in the department's ability to create a welcoming environment fell toward the middle of the other measures, demonstrating that there is perceived room for improvement, but that perceptions are relatively positive. An additional faculty-only section includes aspects related to background and preparation for teaching, but preliminary analysis is ongoing.

Additional baselines include single-item measures of perception of institutional climate and culture regarding diversity and diverse students. Overall, statistically significant differences between faculty and administrator responses existed for only two of the survey items (Table 2). Perceptions of institutional support for developing culturally competent pedagogy were relatively lower than other items. Faculty indicated higher levels of confidence in preparation to teach diverse students than administrators indicated in their preparation to develop/approve curriculum appropriate for diverse students. Interestingly, faculty indicated feeling less responsible for their students' academic success in STEM than did the administrators. Both administrators and faculty indicated relatively highly that they perceive their institution as being welcoming of diverse students and that it is an institutional goal to recruit and retain a diverse student body. However, their confidence in their departments to recruit and retain diverse students was almost a full point lower than their institutional scores.

In response to an open-ended question at the conclusion of the survey section on institutional culture and climate,[2] faculty responses indicated a wide range of institutional commitment to supporting diverse learners. These two examples best demonstrate the extremes:

- "My institution realizes that part of our mission is to shepherd students from dependence to independence. There are many challenges for all students in this endeavor. We help them learn who they are and challenge them to be more than they ever thought they could be."
- "My institution is not interested in changing its teaching philosophies in order to create more active learners."

---

2. "Please feel free to provide any comments you have related to your responses to items in this section."

TABLE 2. Descriptive Statistics for Additional Baseline Measures.
Description: This table provides descriptive statistics for single-item measures, with breakdowns by faculty and administrator responses only for items with a statistically significant difference.

	Mean	CI (95%)	SD	Min	Max
To what extent do you feel satisfied as (an administrator) (a faculty member)?	5.1	4.73–5.41	1.30	2	7
To what extent do you feel that you are supported by your institution in the development of culturally competent pedagogy?	4.9	4.48–5.28	1.52	2	7
To what extent do you feel that your (involvement with curriculum design/approval) [teaching] influences your students' learning in STEM?	5.6	5.28–5.94	1.26	1	7
To what extent do you feel that your institution is welcoming of traditionally underrepresented students?	5.9	5.61–6.19	1.11	3	7
To what extent do you perceive that it is a goal of your institution to recruit and retain diverse students?	5.9	5.47–6.22	1.45	1	7
To what extent are you confident in the ability of your department to recruit and retain diverse students?	5.0	4.69–5.38	1.33	1	7
To what extent do you agree that exposure to a variety of teaching methods will help students to be successful?	6.0	5.69–6.25	1.15	2	7
To what extent do you feel prepared (to develop or approve curricula appropriate for) [to teach your subject to] diverse groups of students?*					
Administrators	5.0	4.34–5.66	1.28	2	7
Faculty	5.7	5.35–6.05	1.20	3	7
To what extent do you feel responsible for your students' academic success in STEM?*					
Administrators	6.3	5.94–6.65	0.69	5	7
Faculty	5.5	5.08–5.94	1.47	1	7

* Difference is significant at the 0.05 level.

LESSONS LEARNED AND TRANSFERABILITY

Quality of teaching is one of the strongest and most consistent predictors of underrepresented student interest and retention in science, both as a major and as a career (Tsui, 2007). To that end, professional development of STEM faculty emerges as integral to increasing the number of underrepresented STEM students who persist to degree; and is best facilitated by a dual-pronged approach that: 1) highlights those effective teaching practices that depart from traditional

STEM teaching strategies and differentially impact underrepresented STEM student success; and 2) incorporates the elements of STEM faculty self-efficacy needed to support sustained behavior change in modes of STEM content delivery. Because this approach lends itself to providing STEM faculty with tools necessary for creating the kind of inclusive classroom experiences that support learning and success for *all* students, it is highly adaptable and appropriate for broadening the participation of any historically underrepresented group, including students with disabilities and those from low socioeconomic backgrounds.

Lesson #1: Culturally Competent STEM Teaching Differentially Improves Underrepresented Student Success, But Is Not Always Readily Transferable.

Since 2005, AAC&U's Liberal Education and America's Promise (LEAP) initiative has sought to improve undergraduate teaching and educational outcomes by providing faculty with frameworks that address multiple dimensions of undergraduate learning and lead to essential learning outcomes, authentic assessment, and implementation of various high-impact practices, including collaborative problem-solving, undergraduate research, learning communities, and experiential learning.

While successful implementation of these pedagogies advantages all students, researchers suggest that utilizing these approaches creates the kind of supportive environments that are essential and differentially beneficial to URM student success (Kuh, 2008). Indeed, AAC&U has demonstrated that such educational practices not only enhance student gains in critical thinking and problem solving, but also disproportionately improve the four-year graduation rates and educational gains for underrepresented students (Finley & McNair, 2013). However, the specificity of some culturally relevant practices is not often clearly elucidated in mainstream STEM literature, is disproportionately focused on "fixing" the student as opposed to the institutional barriers to success, and does not always lend itself well to ease of transferability across institution types (Dhunpath, 2000).

Thus, the TIDES Initiative, because of its unique design, captures both the theories that support *equally* evidence-based and culturally relevant practices, interventions, and the successful pedagogies that are grounded in institutional contexts and relevant cultural themes. By bringing together and translating theories from various disciplines into STEM practices, the TIDES program is uniquely poised to: 1) distill these effective practices into professional development materials and resources for STEM faculty, 2) diffuse them throughout higher education for universal adoption and/or adaptation to any institutional setting at the level of detail needed for precise application and sustained impact, and 3) develop a conceptual framework for conducting qualitative research on broadening the participation of underrepresented groups in STEM fields.

Lesson #2: STEM Faculty Communities of Practice Can Lead to Higher Levels of Student Engagement.

Addressing the need for better STEM teaching methods relies heavily upon the combined scientific acumen, cultural competence, and pedagogical prowess of STEM faculty. However, few of these reports offer specific details on the kind of holistic approach to professional development that will fully support STEM faculty in implementing *equally* evidence-based and culturally relevant STEM teaching strategies.

PKAL has historically used its regional STEM faculty networks to promote modern changes to traditional modes of undergraduate STEM teaching. Recent preliminary data have revealed that levels of student engagement among first-year STEM students—a common predictor of underrepresented STEM student success—are significantly higher than those for the general student population in institutions where STEM faculty have participated extensively in PKAL professional development activities.

In keeping with this trend, the TIDES program also utilizes didactic presentations along with scaffolded introspection and highly interactive sessions where faculty are presented with opportunities to reflect upon and demonstrate proficiency with culturally competent STEM pedagogical techniques. Added to this professional development curriculum is the influence of peer support through capstone experiences, webinars and institution coaches.

REFERENCES

Bandura, A. (1977). Self-efficacy: Toward a unifying theory of behavioral change. *Psychological Review, 84*(2), 191–215. doi:10.1037/0033-295X.84.2.191

Bandura, A. (2000). Exercise of human agency through collective efficacy. *Current Directions in Psychological Science, 9(3), 75–78.*

Bandura, A. (2001). Social cognitive theory: An agentic perspective. *Annual Review of Psychology, 52*(1), 1–26.

Bandura, A. (2006). Guide for constructing self-efficacy scales. In T. Urdan & F. Pajares (Eds.), *Self-efficacy beliefs of adolescents* (pp. 307–337). Greenwich, Connecticut: Information Age Publishing.

Bayer Corporation. (2012). Bayer facts of science education XV: A view from the gatekeepers-STEM department chairs at America's top 200 research universities on female and underrepresented minority undergraduate STEM students. *Journal of Science Education & Technology, 21*(3), 317–324.

DeChenne, S., Enochs, L., & Needham, M. (2012). Science, technology, engineering, and mathematics graduate teaching assistants teaching self-efficacy. *Journal of the Scholarship of Teaching and Learning, 12*(4), 102–123.

Dhunpath, R. (2000). Life history methodology: "Narradigm" regained. *Qualitative Studies in Education, 13*(5), 543–551.

Evans, C., Mckenna, M., & Schulte, B. (2013). Closing the gap: Addressing STEM workforce challenges. *EDUCAUSE Review, 48*(3).

Finley, A., & McNair, T. (2013). *Assessing underserved students' engagement in high-impact practices.* Retrieved from Association of American Colleges and Universities website: http://www.aacu.org/sites/default/files/files/assessinghips/AssessingHIPS_TGGrantReport.pdf

Freeman, S., Eddy, S. L., McDonough, M., Smith, M. K., Okoroafor, N., Jordt, H., & Wenderoth, M. P. (2014). Active learning increases student performance in science, engineering, and mathematics. *Proceedings of the National Academy of Sciences of the United States of America*, 1–6. doi:10.1073/pnas.1319030111

Froyd, J. E. (2008). White paper on *Promising practices in undergraduate STEM education decision-making framework for course/curriculum development.* College Station, Texas. Retrieved online at: http://nsf.iupui.edu/media/b706729f-c5c0-438b-a5e4-7c46cc79dfdf/HhFG1Q/CTLContent/FundedProjects/NSF/PDF/2008-Jul-31_Promising_Practices_in_Undergraduate_STEM_Education.pdf

Froyd, J. E., Srinivasa, A., Maxwell, D., Conkey, A., & Shryock, K. (2005). A Project-based approach to first-year engineering curriculum development. *Frontiers in Education, 2005. FIE '05. Proceedings 35th Annual Conference.* Indianapolis, Indiana. doi:10.1109/FIE.2005.1611955

Kuh, G. D. (2008). *High-impact educational practices: What they are, who has access to them, and why they matter.* Washington, DC: Association of American Colleges & Universities.

Mohamadi, F. S., & Asadzadeh, H. (2012). Testing the mediating role of teachers' self-efficacy beliefs in the relationship between sources of efficacy information and students achievement. *Asia Pacific Education Review, 13*(3), 427–433. doi:10.1007/s12564-011-9203-8

National Science Foundation; National Center for Science and Engineering Statistics. (2013). Women, minorities, and persons with disabilities in science and engineering: 2013. *Special Report NSF 13-304.* Arlington, VA: Special Report NSF 13-304. Retrieved from http://www.nsf.gov/statistics/wmpd/

Siwatu, K. O. (2007). Preservice teachers' culturally responsive teaching self-efficacy and outcome expectancy beliefs. *Teaching and Teacher Education, 23*(7), 1086–1101. doi:10.1016/j.tate.2006.07.011

Tsui, A. S. (2007). From homogenization to pluralism: International management research in the academy and beyond. *Academy of Management Journal, 50*(6), 1353–1364.

Webster, W.J. and R. Chadbourn. (1989). *The longitudinal effects of SEED instruction on mathematics achievement and attitudes: Final report.* Dallas, TX: Dallas Independent School District, TX Department of Research, Evaluation and Information Systems.

ABOUT THE AUTHORS

Kelly M. Mack is the Vice President for Undergraduate STEM Education in the Association of American Colleges and Universities in Washington, D.C.

Kate Winter is the President of Kate Winter Evaluation in Melbourne, Florida.

5

A Social Constructivist Perspective of Teacher Knowledge: The PCK of Biology Faculty at Large Research Institutions

Kathleen M. Hill

U.S. occupational employment projections report that of the 30 fastest growing occupations, with growth rates at 27% or greater, many are science- and technology-related (Bureau of Labor Statistics, 2007). Policy makers reason that by improving undergraduate education, more science and engineering students will persist in these fields and fill the future science and engineering positions in the United States (National Research Council [NRC], 2006; NRC, 2012). Additionally, improving undergraduate education in the sciences will support all students in higher education to become scientifically literate (NRC, 2006; Brewer & Smith, 2011). That is, they will be able to make better health-related decisions and reason through scientific claims that are shared in the media (e.g., personalized medicine, genetics).

One particular pathway for improving undergraduate education in the science fields is to reform undergraduate teaching. Publications from the National Academy Press conclude that undergraduate education should embrace (a) active-learning environments, (b) fewer key concepts, and (c) cooperative learning groups (NRC, 1998, 2000, 2003). These methods give undergraduates an opportunity to build deep knowledge in certain areas, and it allows them to learn how to effectively operate within learning communities. While it is important that these strategies are implemented in science classrooms of postsecondary institutions of all sizes, colleges and universities with large student enrollment have the potential impact the education of significant numbers of science and non-science majors.

Efforts have been made to reform undergraduate science teaching, and numerous studies (e.g., Armbruster, Patel, Johnson, & Weiss, 2009; Brownell, Kloser, Fukami, & Shavelson, 2012; Hoskinson, Caballero, & Knight, 2013) have been performed to evaluate the success of these reform efforts. Through these efforts, various barriers to reform have been identified, such as time constraints, lack of support from administration, reward and tenure practices, and learning challenges of students (Ebert-May, 2011; Wright & Sunal, 2004). Other research in higher education has emphasized the role of teachers' beliefs in influencing their teaching practices (Trigwell & Prosser, 1996; Trigwell, Prosser, Martin, & Ramsden, 2005; Trigwell, Prosser, & Taylor, 1994). In the context of a professional development workshop, one important study revealed that the biology faculty's self-reported use of innovative teaching strategies was not consistent with observed classroom practices (Ebert-May, 2011). Although these teachers perceived themselves as making use of more learner-centered strategies, they primarily engaged in the transmission model of teaching. To support the actual implementation of reformed teaching practices, it is important to acquire an understanding of instructors' knowledge about teaching as it can serve to inform the design of faculty development programs.

In contrast with secondary-level educational research, very few studies have investigated faculty knowledge for teaching. In 1986, Shulman put forth a framework to explore the types of knowledge that teachers use and integrate to form a unique knowledge base for teaching that he called pedagogical content knowledge (PCK). This framework has been used extensively in research performed at the kindergarten through 12th grade (K–12) levels of schooling (e.g., Geddis, A. N., 1993; Gess-Newsome, J., 1999; Grossman, P.L., 1990; Lee, E., Brown, M. N., Luft, J. A., & Roehrig, G. H., 2007; Lee, E. & Luft, J.A., 2008; Magnusson, S., Krajcik, J., & Borko, H., 1999). However, only a limited number of studies of teachers' PCK have been performed in higher education, and these studies investigated teachers' general PCK across disciplines or PCK for teaching specific content topics (Davidowitz, B., & Rollnick, M, 2011; Fernández-Baboa, J.M., & Stiehl, J., 1995; Padilla, K., Ponce-de-León, A. M., Rembado, F. M., & Garritz, A., 2008; Padilla, K., & Van Driel, J., 2011). This study investigated the nature of the PCK of biology faculty who are teaching at large doctoral/research institutions by (1) identifying the types of knowledge used by biology faculty in teaching large introductory-level courses, and (2) exploring the experiences, interactions and teaching perspectives influencing the PCK of biology faculty.

Results of this study serve to inform faculty, researchers, and administrators about how postsecondary teachers develop knowledge to teach science. Research in this area will provide a better understanding of the knowledge base influencing teaching practices, which ultimately impacts student science learning at the postsecondary level. As teachers of both science-majors and non-science majors, biology faculty at large, doctoral/research institutions impact the science learning of significant numbers of undergraduate students. Findings serve to inform instructional reform efforts to target identified knowledge domains that may require strengthening in postsecondary science teaching, as well as aid administrators in developing policies and/or incentive structures for faculty to promote improved undergraduate science teaching and learning.

RESEARCH IN PEDAGOGICAL CONTENT KNOWLEDGE

What is Pedagogical Content Knowledge?

The concept of pedagogical content knowledge was first introduced by Lee Shulman (1986, 1987) through a series of professional publications. Motivated to redirect educational research, he put forth a new framework emphasizing the types of knowledge required for teaching subject matter to others. He contended that an elaborate knowledge base exists for teaching that extends beyond the isolated domains of content and pedagogy.

Shulman (1987) described these types of knowledge as that which distinguish a teacher from a specialist within a particular discipline. He asserted that studies of teachers had revealed that content knowledge and pedagogical knowledge were important for teaching, yet the sum of the two knowledge domains produced an inadequate depiction of the total knowledge for teaching. In 1986, he introduced the term, pedagogical content knowledge (PCK). Given the complexity and specificity of PCK, Shulman (1986) compared the knowledge development of teachers to that of doctors and lawyers, who become proficient in skills, cases, and procedures demonstrated in practice. He urged educational researchers and policy-makers to view teachers as "professionals" rather than "skilled workers" and provided a conceptual framework for examining teachers' knowledge. In 1987, Shulman presented a comprehensive list of the types of knowledge that teachers use and develop in their profession as follows:

- "content knowledge;
- general pedagogical knowledge, with special reference to those broad principles and strategies of classroom management and organization that appear to transcend subject matter;
- curriculum knowledge, with particular grasp of the materials and programs that serve as 'tools of the trade' for teachers;
- pedagogical content knowledge, that special amalgam of content and pedagogy that is uniquely the province of teachers, their own special form of professional understanding;
- knowledge of learners and their characteristics;
- knowledge of educational contexts, ranging from the workings of the group or classroom, the governance and financing of school districts, to the character of communities and cultures; and
- knowledge of educational ends, purposes, and values, and their philosophical and historical grounds." (p. 8)

Shulman theorized PCK to be a unique and discrete knowledge domain that developed as teachers engaged in practice. PCK "represents a blending of content and pedagogy into an understanding of how particular topics, problems, or issues are organized, represented, and adapted to the diverse interests and abilities of learners, and presented for instruction" (Shulman, 1987, p. 8). Other researchers have expanded upon Shulman's definition of PCK. Based upon the multitude of definitions found in the literature, Park and Oliver (2008) developed a comprehensive definition:

> PCK is teachers' understanding and enactment of how to help a group of students understand specific subject matter using multiple instructional strategies, representations, and assessments while working within the contextual, cultural, and social limitations in the learning environment. (Park et al., 2008, p. 264)

Although many contend that PCK remains an elusive construct, Shulman was successful in redirecting educational research as his conceptual framework of teachers' knowledge has been utilized in studies of educators at all levels and across disciplines. These studies have resulted in the development of various models of PCK.

Studies using Models of PCK

Many educational researchers embraced Shulman's conception of teachers' knowledge domains including PCK. Over the past two decades, alternative models of PCK have been introduced and used in studying the knowledge of teachers. A limited number of studies have explored the categories of PCK through examination of teachers' ideas and practices using a grounded approach. One of the studies investigated the "generic" PCK of university faculty across multiple disciplines (Fernandez-Balboa, 1995). Another examined the conceptualization of PCK by secondary science teachers (Lee & Luft, 2008).

In 1995, Fernandez-Balboa and Stiehl conducted a study of 10 experienced university professors from multiple disciplines to explore the "generic" components of PCK across subjects. The participants were a purposeful sample of faculty that were identified by five college deans as exceptional teachers based upon student evaluations, peer reviews, and teaching awards. Data were collected through in-depth semi-structured interviews during which the participants reflected on their teaching experiences. The authors presented the findings as the faculty's collective PCK, rather than the PCK of individual teachers. The results of the study revealed five components of PCK that emerged from the data analysis; these include knowledge of: (1) subject matter, (2) students, (3) instructional strategies, (4) the teaching context, and (5) one's teaching purposes. As one of the first studies of faculty PCK, these findings provided a window into the types of knowledge used by teachers at the tertiary level.

Lee and Luft (2008) conducted research designed to elicit the components of PCK from four experienced secondary teachers. The sample included teachers with more than 10 years of teaching experience and more than three years of mentoring beginning science teachers. Given that the teachers actively taught more than one science discipline, the study explored the general or subject-specific PCK of science teachers. The secondary science teachers were remarkably consistent in the identification of knowledge domains included in PCK. Results of the analysis revealed seven components of PCK, which were identified by all four teachers as important for teaching science: (1) knowledge of science, (2) knowledge of goal, (3) knowledge of students, (4) knowledge of curriculum organization, (5) knowledge of teaching, (6) knowledge of assessment, and (7) knowledge of resources. However, the teachers' conceptions of PCK varied in the interactions between knowledge domains.

METHODS

The epistemological perspective of constructionism views a knowledge base as being dynamic as humans continually engage in and interpret the world around them. However, this knowledge construction does not take place simply as individuals encounter phenomena and make sense of them; humans enter a world in which meaning already exists. It is influenced by the socially agreed upon meanings of objects and phenomena of the world (Crotty, 1998). Using the interpretivist theoretical framework, the complex process of PCK development is considered to involve changes to an interrelated set of knowledge domains and experiences specific to the profession (Lee, et al., 2007; Magnusson, et al., 1999). Meaning making is the mechanism through which teachers develop PCK as they engage in the processes of planning, reflection, and teaching of specific subject matter (Magnusson et al., 1999). Using a qualitative approach, this study identified and explored emergent themes regarding (1) the types of knowledge used by biology faculty in teaching large introductory-level courses, and (2) the experiences and social interactions influencing the PCK of biology faculty.

Setting and Participants

This study included a purposeful sample of six biology faculty members. The study participants came from three large doctoral/ research institutions located in various regions of the United States. Relevant demographics for these universities are included in Table 1. Identification of the instructors occurred by selecting teachers who were engaged in teaching an introductory-level biology course during the study. An "introductory-level" course is defined as a class designed for first-year college students, which may include courses designed

TABLE 1. Relevant Demographics of the Three Institutions Included in the Study

University	12-month undergraduate enrollment for 2010	Number of bachelor's degrees in biology* compared to total degrees awarded in 2009
Red University	28,082	525 / 6,490
Blue University	37,989	834 / 8,223
Green University	60,204	647 / 11,810

Note. Bachelor's degrees include fields of biological and biomedical sciences.

for either biology majors or non-majors. Participants indicated their willingness to participate via e-mail in response to correspondence explaining the proposed study and inviting their participation. Study participants were assigned gender-neutral pseudonyms to protect the anonymity of the faculty and the universities.

Data Collection

Data were collected from multiple sources during the study and used to characterize participants' knowledge for teaching large introductory biology courses, as well as document the experiences and social interactions that influence their PCK. Study participants provided their curriculum vitae, which provided information about the educational background and professional experiences of the biology faculty, as well as the current context in which they work. Additional data were collected through a series of interviews and classroom observations. A pre-observation semi-structured interview was conducted immediately prior to the classroom observations. The participants were interviewed about their experiences and beliefs around teaching large introductory-level biology courses. Each study participant was observed teaching about concepts in an introductory biology course. The researcher recorded field notes of classroom observations, which focused on the interactions between the teacher and their environment. Any available artifacts that were generated as part of the observed lesson were obtained from the biology faculty. Following the classroom observations, post-observation interviews were conducted. The participants were interviewed about their thinking and experiences while teaching the lesson.

Data Analysis and Interpretation

The processes of data collection and data analysis occur simultaneously in qualitative research (Creswell, 2007). As collected, data (documents, classroom observation data, artifacts, and transcriptions of interviews) were converted into electronic format and uploaded into NVivo, a qualitative analysis software program. Data were "coded" through a process of assigning words or phrases to segments of text or images. Initial coding was conducted to identify the participant's use of specific domains of knowledge. This analysis made use of coding categories that were selected a priori to the data analysis using Shulman's (1987) seven knowledge domains or that emerged during the reading of the data. A second process of coding was performed to identify interactions that influence the participant's teaching practices. This analysis made use of coding categories that emerged from the data. Final coding was performed to look for confirming and disconfirming evidence to support prior coding. At the completion of the data collection process, the data were analyzed by comparing, contrasting, aggregating, and ordering of the collected data (Creswell, 2007). Themes were generated through inductive analysis of the data. These themes were grouped and organized to develop six individual faculty profiles.

Validity and Reliability

Strategies were employed to address validity issues as recommended by Creswell (2007), including triangulation of data and peer debriefing. Data were collected through multiple data sources including documentation, survey responses, artifacts, interviews, and classroom observations. Additional science education researchers served as peer examiners. These individuals reviewed codes and themes in order to provide critical feedback from multiple perspectives. Reliability of the data was ensured through use of a codebook to maintain consistency in the meaning of codes (Creswell, 2007).

Limitations of the Study

In this qualitative study, the sample size was limited to a small number of individuals (n = 6) who will likely not represent the general population of biology faculty. The participants in the study were purposefully selected based upon their employment at a large doctoral/research university and active engagement in teaching an introductory-level biology courses. As such, the ability of researchers to relate the findings to a broader population was extremely limited, however, these findings will help to begin to characterize the nature of the PCK of biology faculty at large doctoral/research institutions.

RESULTS

Following the development of the six individual faculty profiles, a cross-profile analysis was conducted. Based upon the analysis, eight distinct knowledge domains were identified as making up the PCK of the six biology instructors. In addition, four social interactions and experiences were found to influence the PCK of the biology faculty.

Knowledge Domains

In the process of analyzing the six faculty profiles, eight domains emerged which made up the PCK of the biology faculty. Seven domains were identified in prior PCK research: (1) knowledge of content, (2) knowledge of context, (3) knowledge of learners and learning, (4) knowledge of curriculum, (5) knowledge of instructional strategies, (6) knowledge of representing biology, and (7) knowledge of assessment. Of these, the faculty differed primarily in two areas: knowledge of learners and learning, and knowledge of instructional strategies. In addition, the faculty PCK included a domain not previously described in the literature: (8) knowledge of building rapport with students. The following results include a description of the these three knowledge domains—knowledge of learners and learning, knowledge of instructional strategies, and knowledge of building rapport with students—as they pertain to these six biology faculty members.

Knowledge of learners and learning

All of the instructors indicated that students arrive to the course with some level of prior knowledge about biology, which can vary from a well-developed understanding to naïve and/or poorly developed conceptions. However, those instructors that actively reviewed students' written responses in assignments and assessments had a more sophisticated knowledge of individual learners' incoming conceptions, as well as their developing ideas and difficulties in learning biology, over the course of a semester.

Each of the biology faculty members had knowledge of student learning, however, the level of knowledge in the area of student learning varied among participants, with some emphasizing independent actions and behaviors and others emphasizing collaboration. All of the participants indicated that learning requires more than listening to the delivery of explanations by experts during lecture. In general, they agreed that students need to practice working with new information and to make use of feedback to assess their own level of understanding. However, the faculty differed in their understanding of learning theory and their ability to describe specific actions or behaviors, beyond reading the textbook, that support student learning.

Knowledge of instructional strategies

The study participants reported having knowledge of instructional strategies, which included lecture, questioning techniques, small group work, and whole group discussion. However, the faculty differed in their knowledge and practice of implementing these strategies within the context of a large auditorium-style classroom. The repertoires of strategies varied between the instructors. The instructional strategies described and utilized by the study participants are included in Table 2.

TABLE 2. Repertoires of Faculty Instructional Strategies

Instructional Strategy	Alex	Chris	Pat	Sam	Terry	Morgan
Lecturing	X	X	X	X	X	X
Questioning Techniques	X	X	X	X	X	X
Small Group Work			X	X	X	X
Whole Group Discussion					X	X

Knowledge of building rapport with students

The newly identified knowledge domain was the faculty's knowledge of building rapport with students. Different from having knowledge of learners and learning, building rapport with students, either collectively or on an individual level, involved relationship-building. Five of the six instructors described various ways of developing a connection with students. The study participants reported having knowledge of strategies for building rapport with students in the large lecture setting by (1) providing structure, and (2) incorporating elements of fun. The faculty indicated that providing structure was a means of reducing student anxiety by setting clear expectations and maintaining regular communication (e.g., conversations before and after class, meetings during office hours, electronic mail, discussion boards, and online course management systems). Nearly all of the participants indicated incorporating elements of fun, such as humor and music, into the classroom to develop a positive rapport with students. Only one of these five instructors directly connected building rapport with supporting student learning.

Social interactions and experiences

The analysis of the data yielded four social interactions and experiences: teaching experience, models and mentors, collaborations and interactions with faculty, and science education research and literature. In designing faculty development and support efforts that will help sustain new approaches to teaching, it is important to consider the social context that influences instructor PCK, particularly because not all of these influences will encourage the most effective approaches to teaching.

Teaching experience

Many of the faculty reported gaining knowledge of teaching from their own experiences in teaching. Although they varied with regard to teaching experience, four of the participants stated that they learned to teach through "trial and error" in the classroom. Two of the participants did not refer to teaching experience as influencing their development as postsecondary teachers, however, this may be due to their being early career instructors with relatively lower levels of classroom experience.

Models and mentors

All of the study participants reported gaining knowledge of teaching from observed models of instruction. While one instructor reported observing primarily "bad" models of teaching, the other five instructors stated that they had observed "good" models of instruction at the secondary and/or postsecondary level. For some, mentors also played a role in their learning to teach. Three of the participants indicated that they had "good" mentors who played an important role in their development as a teacher during the early part of their careers as college instructors.

Collaborations and interactions with other faculty

Five of the six study participants reported having interactions with other faculty that influenced their teaching practices. The instructors' perceptions of their professional obligations and the teaching/research load of their position played a role in shaping the amount and types of interactions. Those participants who were in non-tenure-track positions perceived teaching to be the primary role of the position. However, their interactions with other faculty were often limited to administrative meetings. Others who were in tenure-track positions perceived teaching and research to be the responsibilities of the position. These participants described attending regularly scheduled meetings and having informal interactions with other faculty who were engaged in similar teaching and research.

The individuals with whom the participants interacted also differed with respect to their level of expertise in education. Four of the participants reported having little to no interaction with faculty who specialized in science education in their career. However, two of the instructors stated that they had regular interactions with faculty who are scholars in the field of science education.

Science education research and literature

For this study, faculty engaging in science education research and reading and/or discussing science education literature are included as social interactions and experiences. The study participants varied with respect to their level of engaging in and/or having experience with scholarly work in science education. Two participants reported having little to no knowledge of the science education literature or experience in the area of performing educational research. Two other instructors reported being familiar with science education literature, as they both engaged in teaching science education courses. Both indicated that reading the primary literature and discussing it with students increased their knowledge about teaching and learning. The two remaining participants reported engaging in multiple science education research projects that focused on teaching and learning in the context of large introductory biology courses. Both indicated that this research served to directly inform their teaching practices.

DISCUSSION

What are the types of knowledge that biology faculty use for teaching?

Results of this study revealed that the six participants had knowledge in eight distinct domains. Seven of these domains were previously identified in the literature pertaining to either secondary and/or postsecondary level studies (Geddis, 1993; Gess-Newsome, 1999; Grossman, 1990; Lee, Brown, Luft, & Roehrig, 2007; Lee & Luft, 208; Magnusson et al., 1999): knowledge of content, knowledge of context, knowledge of curriculum, knowledge of learners and learning, knowledge of representing biology, knowledge of instructional strategies, and knowledge of assessment. Of these domains, the participants varied in their knowledge of learners and learning, and knowledge of instructional strategies. In addition, the faculty reported having knowledge in a new domain: building rapport with students.

The biology faculty differed in their knowledge of learners and learning and their knowledge of instructional strategies. Interestingly, the faculty knowledge was similar between those participants at the same institutions. At Green University, the two faculty had limited knowledge of students' prior conceptions and naïve conceptions in biology, as well as the variation in students' ability levels. In addition, the faculty differed in their knowledge of the ways in which students learn science. The two instructors from Green University emphasized learning as primarily an independent endeavor in which students engage in making sense of concepts on their own. The remaining four instructors focused on the importance of student collaboration in the learning process based on their knowledge of the social constructivist theory of learning.

Faculty knowledge of instructional strategies spanned a continuum with one end of the spectrum including lecture with minimal social interaction and the other end involving an amalgam of approaches. The instructional strategies employed by the faculty at the same institution were similar. At Green University, the two instructors primarily made use of lecturing with minimal questioning. The remaining four instructors had knowledge of strategies for small group discussions and activities; however, the two faculty from Red University instructors also had knowledge of facilitating whole group discussions. The four participants from Blue University and Red University were those that described student learning as being consistent with the social constructivist theory of learning.

Based upon the variations of knowledge within these domains, three categories of faculty PCK emerged: (1) PCK as an expert explainer, (2) PCK as an instructional architect, and (3) a transitional PCK, which fell between the two prior categories. The PCK of the faculty at the same institution was similar regardless of their levels of teaching experience and amount of collaboration. The two faculty at Green University included those who developed PCK as an expert explainer. They emphasized their ability to explain complex biological ideas to students and to develop presentation materials for lecture that logically communicated the concepts. In addition, the expert explainers were those who viewed student learning as an independent endeavor with lecture sessions being a venue for delivering course content. The two faculty at Red University were those who developed PCK as an instructional architect. They focused on designing collaborative learning experiences for students to engage in during class meetings. While these instructors made use of direct instruction, they primarily employed other student-driven instructional strategies, which were based on their social constructivist view of learning. The two faculty at Blue University were those who developed a transitional PCK.

One new domain was identified from the study: knowledge of building rapport with students. This new domain is distinct from the knowledge of learners and learning, as well as the knowledge of instructional strategies. While teachers' knowledge of learners and learning refers to their knowledge of the ways in which students learn science along with students' prior conceptions and difficulties in learning science (Lee & Luft, 2008; Magnusson, et. al, 1999), this new domain refers to the ways in which an instructor establishes a relationship with students either as a collective body or as individual learners. Fernandez-Balboa and Stiehl (1995) included "creating a learning environment" and "motivational strategies" as part of the knowledge domain of instructional strategies. These included strategies such as "[be] fun and exciting", "bug them to learn on their own", "give a pep talk", "use humor", and "take a break". For this study of biology faculty, the aforementioned strategies would not be included in the knowledge domain of instructional strategies, which is reserved for those strategies tied to the literature on learning. The new domain of knowledge of building rapport with students encompasses strategies, such as having regular communication with students and incorporating elements of fun, that are intended to positively influences students' affective domain. Although they are considered to be distinct domains, affect and cognition both play important roles in learning science (Ainley and Ainley, 2011). In most cases, the faculty participants reported making use of strategies designed to reduce student anxiety, which can be a barrier to learning.

What social interactions and experiences influence the PCK of biology faculty?

Studies of secondary science teachers identified classroom experience as an essential component of PCK development (Geddis et al., 1993; Gess-Newsome, 1999; Grossman, 1990). These studies indicate that effective PCK develops as teachers spend more time with students in the classroom. However, among the biology faculty, the variation in amount of teaching experience was not aligned with their PCK. The instructors with more sophisticated knowledge in the areas of instructional strategies and learners and learning had varying amounts of teaching experience. Similarly, those with less knowledge in these domains varied in terms of their teaching experience.

The literature lacks information about the social interactions and experiences that influence the PCK of instructors at the postsecondary level. Studies have reported on faculty-identified barriers to implementing reformed teaching practices (Fernandez-Balboa & Stiehl, 1995; Ebert-May, 2011; Wright & Sunal, 2004). In contrast, this study sought to identify the institutional constructs that influence social interactions and experiences impacting instructors' PCK development. In addition to teaching experience, models and mentors played an important role in the development of the biology faculty as teachers. Engaging in collaborations and interactions with other faculty, which focus on teaching and learning, were also influential. Performing educational research and/or reading and discussing science education literature was also significant in developing effective PCK and served to directly inform the teaching practices of some faculty.

IMPLICATIONS AND FUTURE DIRECTIONS

This study was intended to inform faculty, researchers, and administrators about the PCK of biology faculty teaching large introductory-level courses. Although the six biology faculty had similar knowledge across multiple domains, their knowledge varied in the areas of learners, learning theory, and instructional strategies. Although these differences were evident, the faculty knowledge in these domains was similar between the instructors at the same institution. This suggests that constructs within the institutional setting can influence the PCK of faculty. As such, the knowledge of the teachers at an institution as well as the context in which they operate must be clearly understood in designing effective professional development programs (Luft & Hewson, 2014). Based upon a review of relevant science education literature, van Driel, Berry, and Meirink (2014), suggested that professional development target teacher knowledge that focuses on students learning science content and the practice of science teaching. Faculty could benefit from the study of student work products to gain a sense of novice or developing conceptions. In addition, Wright and Sunal (2004) recommended that faculty form collaborative teams across disciplines, as well as connect with various groups and networks involved in developing exemplary curricular materials. As a means of further expanding their repertoire of instructional approaches, they also suggested that faculty make site visits to other institutions to observe the use of innovative strategies within similar learning environments.

From this and other studies, faculty considered instructional strategies and strategies of building rapport with students as a collective set of teaching practices. It is important that faculty be able to distinguish between these types of strategies. While the approaches for building rapport with students can influence their affective domain and support them to persist in a course, they are not directly tied to strategies for designing and implementing instruction. Further, the faculty's knowledge of strategies of building rapport with students was limited as it emphasized "providing structure" and incorporating "elements of fun". With an emphasis on building the knowledge of faculty in the area of instructional strategies, faculty development programs ought to incorporate research-based connections between instructional practices and student motivation, such as goal setting, student choice, relevancy, individual and group responsibilities, and frequent instructor feedback (Schunk, Pintrich, & Meece, 2008).

The six faculty provided insight into the social factors that affected their PCK. More senior faculty who served as models and/or mentors were deemed important by all six participants. As early career instructors, they reported looking to these colleagues for guidance on teaching undergraduate biology courses. Departments ought to develop a selection process of identifying particular faculty to serve as active mentors in an effort to move instruction toward more reformed-based teaching practices. Furthermore, instructors ought to be given opportunities to plan and teach in more innovative ways with support and structured feedback from mentors and peers (van Driel, Berry, and Meirink, 2014). The participants also indicated that they gained knowledge about teaching through collaborations and interactions with other faculty, as well as through performing educational research and/or discussing science education literature. Faculty development programs ought to support the formation of collaborative groups for instructors to delve into the science education literature for relevant ideas and course innovations. This group should serve to support faculty in designing and conducting active research projects to assess the effectiveness of various instructional approaches (Wright and Sunal, 2004).

From this study, it is apparent that further research is needed with regard to biology faculty PCK development over time. It would be important to study how PCK transforms over years of teaching experience. For this, conducting a longitudinal study focusing on the PCK development of biology faculty would be useful. Future research could also examine various institutional settings and constructs and seek to reveal those that support the development of effective PCK for biology faculty.

REFERENCES

Ainley, M., & Ainley, J. (2011). Student engagement with science in early adolescence: The contribution of enjoyment to students' continuing interest in learning about science. *Contemporary Educational Psychology, 36*(1), 4–12.

Armbruster, P., Patel, M., Johnson, E., & Weiss, M. (2009). Active learning and student-centered pedagogy improve student attitudes and performance in introductory biology. *CBE-Life Sciences Education, 8*(3), 203–213.

Brewer, C. & Smith, D. (2011). *Vision and change in undergraduate biology education: A call to action.* Washington, DC.

Brownell, S. E., Kloser, M. J., Fukami, T., & Shavelson, R. (2012). Undergraduate biology lab courses: Comparing the impact of traditionally based "cookbook" and authentic research-based courses on student lab experiences. *J Coll Sci Teach, 41*(4), 36–45.

Bureau of Labor Statistics, Office of Occupational Statistics and Employment Projections. (November 2007). *Employment outlook: 2006–16: Occupational employment projections to 2016.* Retrieved from http://www.bls.gov/opub/mlr /2007/11/ art5full.pdf.

Creswell, J.W. (2007). *Research design: Qualitative, quantitative, and mixed methods approaches* (3rd ed.). Thousand Oaks, CA: Sage.

Crotty, M. (1998). *The foundations of social research: Meaning and perspective in the research process.* London: Sage.

Davidowitz, B., & Rollnick, M. (2011). What lies at the heart of good undergraduate teaching? A case study in organic chemistry. *Chemistry Education Research and Practice, 12*(3), 355–366.

Ebert-May, D., Derting, T. L., Hodder, J., Momsen, J. L., Long, T. M., & Jardeleza, S. E. (2011). What we say is not what we do: Effective evaluation of faculty professional development programs. *Bioscience, 61*(7), 550–558.

Fernández-Baboa, J.-M., & Stiehl, J. (1995). The generic nature of pedagogical content knowledge among college professors. *Teaching and Teacher Education, 11*(3), 293–306.

Geddis, A. N. (1993). Transforming subject-matter knowledge: The role of pedagogical content knowledge in learning to reflect on teaching. *International Journal of Science Education, 15,* 673–683.

Gess-Newsome, J. (1999). PCK: An introduction and orientation. In J. Gess-Newsome and N. Lederman (Eds.) *Examining PCK: The construct and its implications for science education,* (pp. 3–20). Norwell, MA: Kluwer Academic Publishers.

Grossman, P. L. (1990). *The making of a teacher: Teacher knowledge and teacher education.* New York: Teachers College Press.

Hoskinson, A. M., Caballero, M. D., & Knight, J. K. (2013). How can we improve problem solving in undergraduate biology? Applying lessons from 30 years of physics education research. *CBE-Life Sciences Education, 12*(2), 153–161.

Kane, R., Sandretto, S., & Heath, C. (2002). Telling half the story: A critical review of research on the teaching beliefs and practices of university professors. *Review of Educational Research, 72*(2), 177–228.

Lee, E., Brown, M. N., Luft, J. A., & Roehrig, G. H. (2007). Assessing beginning secondary science teachers' PCK: Pilot year results. *School Science and Mathematics, 107,* 52–68.

Lee, E., & Luft, J. A. (2008). Experienced secondary science teachers' representation of pedagogical content knowledge. *International Journal of Science Education, 30*(10), 1343–1363.

Luft, J. A., & Hewson, P. W. (2014). Research on teacher professional development programs in science. In N. Lederman and S. Abell (Eds.), *Handbook of research in science education,* (pp. 889–909). New York, NY: Routledge.

Magnusson, S., Krajcik, J., & Borko, H. (1999). Nature, sources and development of pedagogical content knowledge for science teaching. In J. Gess-Newsome & N. G. Lederman (Eds.), *Examining Pedagogical Content Knowledge, (pp.* 95–132). Norwell, MA: Kluwer Academic Publishers.

National Research Council (NRC). (1998). *Transforming undergraduate education in science, mathematics, engineering, and technology.* Washington, DC: National Academies Press. www.nap.edu/catalog.php?record_id6453 (accessed April 4, 2012).

NRC. (2000). *How people learn: Brain, mind, experience, and school: Expanded edition.* Washington, DC: National Academies Press. www.nap.edu/catalog.php ?record_id9853 (accessed April 4, 2012).

NRC. (2003). *Evaluating and improving undergraduate education in science, technology, engineering, and mathematics.* Washington, DC: National Academies Press. www.nap.edu/catalog.php?record_id10024 (accessed April 4, 2012).

NRC. (2006). *Rising above the gathering storm: Energizing and employing America for a brighter future.* Washington, DC: The National Academies Press.

NRC. (2012). *Discipline-based education research: Understanding and improving learning in undergraduate science and engineering.* Washington, DC: The National Academies Press.

Padilla, K., Ponce-de-León, A. M., Rembado, F. M., & Garritz, A. (2008). Undergraduate professors' pedagogical content knowledge: The case of "amount of substance". *International Journal of Science Education, 30*(10), 1389–1404.

Padilla, K., & Van Driel, J. (2011). The relationships between PCK components: The case of quantum chemistry professors. *Chemistry Education Research and Practice, 12,* 367–378.

Park, S. & Oliver, S. J. (2008). Revisiting the conceptualization of pedagogical content knowledge (PCK): PCK as a conceptual tool to understand teachers as professionals. *Research in Science Education, 38,* 261–284.

Schunk, D. H., Pintrich, P. R., & Meece, J. L. (2008). *Motivation in education: Theory, research, and applications.* Upper Saddle River, NJ: Pearson.

Shulman, L. S. (1986). Those who understand: Knowledge growth in teaching. *Educational Researcher, 15*(2), 4–14.

Shulman, L. S. (1987). Knowledge and teaching: Foundations of the new reform. *Harvard Educational Review, 57,* 1–22.

Trigwell, K., & Prosser, M. (1996). Congruence between intention and strategy in university science teachers' approaches to teaching. *Higher Education, 32*(1), 77–87.

Trigwell, K., Prosser, M., Martin, E., & Ramsden, P. (2005). University teachers' experiences of change in their understanding of the subject matter they have taught. *Teaching in Higher Education, 10*(2), 251–264.

Trigwell, K., Prosser, M., & Taylor, P. (1994). Qualitative differences in approaches to teaching first year university science. *Higher Education, 27*(1), 75–84.

van Driel, J.H., Berry, A., and Meirink, J. (2014). Research on science teacher knowledge. In N. Lederman and S. Abell (Eds.) *Handbook of research in science education* (pp. 848–870). New York, NY: Routledge.

Wright, E. L., & Sunal, D. W. (2004). Reform in undergraduate science classrooms. *Reform in undergraduate science teaching for the 21st century* (pp. 33–52). Greenwich, CT: IAP.

ABOUT THE AUTHOR

Kathleen M. Hill is a STEM Education Outreach Specialist and Assistant Professor of Education at Pennsylvania State University in State College, Pennsylvania.

6

Culture, Policy and Resources: Barriers Reported by Faculty Implementing Course Reform

Loran Carleton Parker, Omolola Adedokun, Gabriela C. Weaver

Course reform is a growing practice in higher education across the disciplines. Reforms focus on evidence-based practice, course design, and student-centered pedagogies. Existing reform initiatives have championed various active-learning strategies aimed at increasing student engagement and improving student conceptual learning (Prince, 2004). Examination of the success of these initiatives tends to focus on success at improving student-level indicators.

Research examining faculty perception of barriers to their reform's success has been limited. Brownell and Tanner explored the barriers to faculty pedagogical change and identified barriers dealing with the professional identity of the professoriate (2012). Walczyk, Ramsey and Zha reported on large-scale survey results that identified obstacles to instructional innovation for faculty (2007). Their survey was based on Wright and Sunal's (2004) previously identified institutional barriers and was primarily aimed at gauging the presence of these barriers at universities in Louisiana. Henderson and Dancy (2007) interviewed small numbers of STEM faculty who were not part of course reform efforts and identified individual and situational barriers that may inhibit these faculty from effecting change. These studies, however, did not pay close attention to the particular views and perceptions of faculty who were "on the ground" implementing change. As course reform initiatives move from individual scholarship to institutional-scale reform, it becomes increasingly important to consider the ways in which faculty experience and enact these changes. As course reform initiatives become institutionalized (rather than grassroots), research must consider faculty perception of these efforts in order to help design systems/organizations that can best support and sustain faculty teaching innovations

Purdue University, a large research university, has undertaken an unprecedented, university-wide course reform project. This project serves as the context for the current study's examination of faculty-perceived barriers to course reform. Specifically, this study seeks to answer the question, "What barriers and challenges do faculty at a research institution encounter during an institutional reform initiative?" The purpose of this study is to provide a preliminary snapshot of the barriers perceived by faculty as they are "in the trenches" of course reform and to identify areas of special concern for leaders of institutional reform efforts, as well as future areas of research.

STUDY CONTEXT

The impetus for this study is the evaluation of a university-wide course reform project on the main campus of Purdue University. Purdue University is a large land grant research university in the Midwestern United States, with approximately 29,255 undergraduate students, 9,500 graduate and professional students, 1,800 tenured/tenure-track faculty, and offering nearly 200 undergraduate majors and graduate degrees in just over 70 academic programs on their West Lafayette, Indiana, campus. In 2011, Purdue began a campus-wide reform initiative. The goals of the Instruction Matters: Purdue Academic Course Transformation (IMPACT) project are to improve student learning and success by making the learning process more active and engaging for students. IMPACT achieves these goals by assisting instructors in redesigning their courses to be more learner-centered. IMPACT intentionally targets large, foundational courses from all disciplines, supplementing several pre-existing discipline-specific course redesign efforts. The broad scope of the IMPACT initiative requires a program fostering best practices in course redesign as well as flexibility for faculty and departments to enact reforms that meet program specific needs and contexts. IMPACT initially targeted ten courses and has since expanded to reach up to 60 courses each year. (Morris, Parker, Nelson, Pistilli, Hagen, Levesque-Bristol & Weaver, 2014).

IMPACT's focus on learner-centered classrooms is situated in the research base on learner-centered psychological principles as described in Morris and colleagues, (2014):

> These principles synthesize the bodies of knowledge about learning and instruction, and the social and individual factors that influence the learning process. Learner-centered instructional practices are characterized by: (i) the inclusion of "learners in decisions about how and why they learn and how that learning is assessed;" (ii) valuing of "each learner's unique perspectives;" (iii) respecting and accommodating "individual differences in learners' backgrounds, interests, abilities, and experiences;" and (iv) treating "learners as co-creators and partners in the teaching and learning process" (McCombs, 2001, p. 186). These characteristics extend from a theory of learning that posits learning is an active, constructive process building upon learner prior knowledge

and experience and is mediated by social interactions in the learning environment. (Morris, et al., 2014, p 305)

The redesign process begins for a faculty member when they apply to join a cohort of IMPACT faculty fellows. Once accepted, instructors participate in a professional development program that places instructors in a supportive learning community with other instructors and education specialists. During the first semester, instructors attend weekly workshop sessions. The workshops are topical and focus on active learning pedagogies, theories of motivation and learning, instructional technology resources, the scholarship of teaching and learning, and the development of learning objectives and outcomes for course design and assessment. During this workshop semester, fellows create their deliverables: redesign research question, course learning outcomes and course assessment map. These deliverables are entirely focused on the needs of the fellow and their course. Faculty are not required to follow any specific template for their redesign. Beginning with their entrance to the program and continuing as long as desired by the instructor, each cohort member is supported by a small group of fellow faculty and staff. The Faculty Learning Community prioritizes connection among peers. To realize these connections, fellows from previous cohorts serve as invited guests for several of the sessions, providing expertise and direct experience from their redesign. Fellows and their small support groups work together to create a timeline for, implement, and study the outcomes of the redesigned course.

STUDY METHODS

IMPACT faculty fellows from the first four cohorts of the program were interviewed in the semester after they implemented their redesigned course for the first time. Fellows in the first cohort were interviewed individually; representatives from the later cohorts were interviewed in small groups. All fellows were invited to participate in the interviews, however, in all but the first cohort, some fellows did not participate in the interviews. In total, 27 instructors participated. Interviews lasted approximately one hour and consisted of questions regarding the specific changes that were made to the course and to the fellow's teaching practice, how decisions were made regarding the course redesign, discussion of deviations from the planned implementation, expectations versus reality when implementing a redesigned course, successes and challenges during the implementation process, benefits of the redesigned course for instructor and students, future plans for the course, processes and plans for disseminating information about the course redesign and its outcomes.

Data from the interview sessions were audio recorded and transcribed verbatim. Data analysis proceeded through several phases. The first phase consisted of reading through transcripts three times and identifying sections in which fellows discussed barriers or challenges that they perceived or directly experienced during both the development and implementation of their redesigned course. The second phase included open coding of the textual data into categories that represented described barriers. The third phase of analysis was organizing the codes into broader categories and overarching themes that represented the type or domain of barrier described and confirming that these broader themes and category labels continued to be in line with the textual data.

FINDINGS

Overall, the analysis revealed two types of barriers: (1) barriers associated with the cultural norms of teaching and learning in research universities (e.g., differences between student and instructor expectations, and negative perceptions/attitudes of other faculty and administrators); and (2) barriers related to existing institutional teaching structures (e.g., promotion policies and limited classroom resources). These barriers, and the categories (and sub-categories) that emerged under them are discussed in detail in the following sections.

Cultural Norms

A primary category of barrier identified through analysis involved the cultural expectations associated with teaching and learning. This is represented as a theme that is present in multiple contexts identified by the faculty fellows as obstacles they had to navigate during the implementation of their redesigned course. Figure 1 displays this theme and the categories and subcategories of associated barriers described in the data. The theme of cultural expectations operated in two domains: students and colleagues of the faculty fellows. Each of these contextual domains was associated with specific barriers as illustrated in Figure 1.

Barriers associated with student views

Regarding student expectations of teaching and learning, faculty fellows described barriers due to differing expectations between themselves and their students. A typical representation of this barrier would include descriptions of students who believe that learning involves passively listening, reading and absorbing information and demonstrating that memorization on an exam. IMPACT faculty fellows, both through their interest and the IMPACT professional

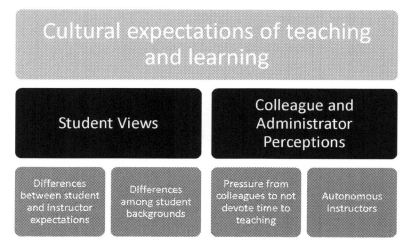

FIGURE 1. Barriers Related to Cultural Expectations of Teaching and Learning

development experience, have very different definitions of teaching and learning. As described previously, IMPACT seeks to create active learning environments for students in which they are expected to direct at least some of their own learning. Faculty fellows described resistance from students due to differing expectations of the roles and responsibilities of instructor and student during the course.

> "They think of this as like a train station. You put money in. You get a ticket out. You get to go on the train. You put your money in. You spend your time here. You get this piece of paper out; that's your ticket to a career."

> "Specifically students told me that this is too much examples and work. 'We'd rather just memorize.'"

> "What didn't work well was that too many of the students just weren't listening and participating. And I think the reason for that is that a lot of freshmen tend to be oriented towards 'What piece of information do I have to repeat for the exam?' And that's the whole thing I was trying to turn around and get them out of that mode . . . it's very hard to get them out of that."

Additionally, faculty fellows identified differences in expectations for teaching and learning among students as a barrier to the success of their redesigned course. They described difficulties getting students to collaborate effectively in group settings because students with different backgrounds held very different understandings and definitions of the learning process and varying communication skills. This difference was marked in domestic and international students, but also present between domestic students who had experienced different K–12 learning environments.

> "We've got kids that won't participate, because it's a cultural difference and they just don't have the skill and they kind of power down."

> "It causes resentment among the [domestic] students, the American students who are like, 'Why am I having to deal with this?'"

Barriers associated with colleague and administrator perceptions

The perception of colleagues and administrators was the second domain in which faculty fellows reported barriers created by cultural expectations. Faculty fellows stated that spending too much time on teaching could be viewed negatively by their colleagues and administrators. Specifically, faculty fellows felt that, while it was acceptable to be an average instructor if their research was stellar, caring too much about their teaching put them at risk of being perceived as a less than stellar researcher. Although these sentiments were expressed by fellows of both genders, women fellows expressed concern that their gender increased the risk of colleagues minimizing their research accomplishments if they put too much effort into teaching.

> "Well, I mean the thing is, at least in our discipline, the teaching end of it, even though there's lip service [to the effect that] it's gonna be viewed beneficially, frankly, [it isn't]."

> "So having the IMPACT thing kind of helped a little bit with [building prestige for quality instruction], but I fear that I'm getting, you know, sort of stereotyped as the teaching professor and actually it's something that I have to be concerned about, particularly being female."

> "I wish I knew how to tell that story [about why to use active learning techniques] better, I think when I tell that story some of my colleagues are feeling threatened and it shouldn't be threatening, it should be joyful, right?"

> "I guess I'll have to make it work, and I'll have to convince the department that this is a good idea."

> "... convincing [departmental colleagues] that there's a different way to [teach large, foundational courses], is gonna be a monumental sell job."

"I wish I knew that my department valued this. Yeah, I wish I knew how to measure and convince my department of the value of this. . . ."

Additionally, faculty fellows had a strong sense of autonomy regarding their teaching and understood that college instructors were expected to have a large amount of control over the "what" and "how" of their instruction. However, several of the fellows felt that this expectation fostered an absence of coordination among colleagues and a vacuum of leadership by administrators that was a barrier to the sustainability of their course redesign.

"I do not agree that an individual faculty member should be coming up with what those [learning] outcomes are, I think that has to be the whole department saying, here's what we value and here's the role that this course plays and we've decided that because we know that these are the professional outcomes we expect of our students, here are the types of careers that they go into, here are the courses that they're going to, that this is foundational too in terms of their future experience, I mean that really needs to be the whole department deciding what the measures of success are gonna be . . ."

Institutional Structures

The second barrier theme identified through analysis was existing institutional structures. This theme was present in multiple contexts relating to policies and resources. Figure 2 displays the theme and the categories and subcategories of associated barriers described in the data.

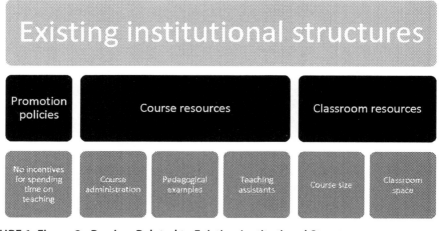

FIGURE 1. Figure 2. Barriers Related to Existing Institutional Structures

A very prominent domain in which existing institutional structures became a barrier to the implementation of the redesigned course was in the institution's tenure and promotion policies. Faculty fellows who were tenure-track felt that there was no incentive related to their promotion process for them to spend much time on their course redesign, particularly if it meant sacrificing time that could be spent on research.

> ". . . For an assistant professor to be responsible for flipping an entire course, I wouldn't think that would be a good thing. When you're only gonna be mostly evaluated on your research . . . unless you really are going to be given benefit in the tenure process. And maybe many disciplines will. But I know in our department I doubt it."

> "And if the time commitment that you put into this is not allowing you to finish your grant, this is in a completely different topic area and stuff, then that would be a distraction, a detriment."

> "There's no rewards for being a good teacher at Purdue other than winning an award or if you get approved for a program like IMPACT or something like that. But there's no rewards. We're all judged on creative endeavor or research."

Regarding resources, fellows identified a lack of resources for course and classroom support that were creating barriers to the implementation and sustainability of their redesigned course. They indicated that the administration of courses was not aligned to support large, foundational courses that differed from traditional lecture, recitation, and laboratory formats. For example, the course registration system was not well equipped to distinguish and appropriately label redesigned courses from their traditional counterparts. This was especially problematic when not all sections of a large enrollment course had been redesigned.

> "When the students signed up for this class they had to sign up for two different sections, because the university has to be able to track them and so there's actually two [course numbers] for this hybrid course, the one for the 75 minutes they're in class, and another one for the 75 minutes they're watching, and to me that's just frustrating in that we can't come up with a new classification. . . ."

Additionally, faculty fellows noted that a large barrier to implementing and sustaining their redesign was the availability of trained teaching assistants; they noted that they were unaware of any university resources that had been devoted to developing more of the skilled teaching assistants needed to facilitate large enrollment courses. Faculty fellows also felt that few concrete examples were available to serve as inspiration and motivation when creating and documenting their redesigned course.

> "TA training is gonna be a huge problem, you can't just be people like me who are continued lecturers who are here full time and have vested interest in the long-term success of this course; it's gotta be able to be a transferable skill, so that TAs who walk in, these kids are 24 years old, 23 years old, right out of college, and are they gonna be in charge of 60 kids in a classroom, and are they gonna be interactive with them? They have enough trouble standing in front of 30, and (inaudible) lecture sometimes. God love them but it's a skill that you have to learn over time: how to interact with students."

> "I go through the educational literature also in vain . . . Now give me an example about [discipline specific concept] that shows how I can do this. Give me a problem that I can present to my student that has these kinds of aspects to it that you are recommending. Because otherwise what happens is, like, oh, here's the abstract description of it. And there's me over here trying to figure out how on Earth am I gonna do that, and again that takes a huge amount of time to make that very easy to describe in the abstract."

Regarding classroom resources, faculty primarily described course size and classroom space as barriers. They indicated that it was not likely that their course would be getting any smaller, but that very few classroom spaces designed for interactive teaching were large enough to accommodate their course.

> "The competition for that room is gonna start getting a little, it's gonna get a little high, and so . . . We need more of those rooms . . . You know, you can't show us the promised land and then shut the door on us. So if we're gonna do this, we gotta do it, and my concern is that all those momentums are gonna get lost because we're gonna have a space problem and then, suddenly, it's just gonna die and then that's it."

> "I teach [science] but there's no way that I can do [science] demonstrations 'cause there's no, you know, laboratory bench [in the classrooms designed for active learning]."

DISCUSSION

This preliminary study has identified two main types of faculty-reported barriers to course redesign and implementation: cultural norms for teaching and learning, and institutional structures. These two types of barriers weave through multiple domains at the university, including interactions with students and colleagues, decisions about how best to plan for successful promotion and tenure, and how to navigate and secure university resources, such as space, for redesigned courses.

Many of the barriers identified here have been described previously, including student resistance, promotion and tenure policies and professional identity as an instructor at a research university (Henderson & Dancy, 2007; Brownell & Tanner, 2012; Walczyk et al., 2007). However, several issues raised by faculty at this large, research university are new or have been less than robustly described in the current literature. Professional identity is rarely described as a barrier to innovative teaching; however, it weighed very heavily on the minds of the tenure-track faculty who participated in this study. Although many disciplines consider teaching to be a lower-status and lower-paid task (Brownell & Tanner, 2012), the impact of status as a barrier to teaching innovation is rarely described as a gendered phenomenon. However, some of the women faculty participating in this study felt that the potential for loss of status as a result of being too closely identified with innovative teaching was magnified because of their gender.

Student resistance is often described in the literature as an obstacle to pedagogical innovation. However, much of this discussion focuses on domestic students and American K–12 education. This study highlights a related issue that is important to consider due to the increasingly international student pool at research universities—the large difference in students' understandings of the roles, responsibilities and even purposes of post-secondary teaching and learning and the importance of communication fluency in active-learning classrooms.

Finally, while resources such as space and technology are often discussed as barriers to reform, longstanding and entrenched systems, such as those that govern course registration, scheduling and designation, are often left out of the conversation. Frustrations with these barriers can quickly sap the energy and momentum of instructors or departments trying to engage in course reform.

This study offers insight into faculty perceptions with regard to barriers to educational reform at a large research university. While these findings may be applicable to other similar universities, the mission, structure, and cultural norms at other institutions may substantially impact perceived faculty barriers in these contexts. Nonetheless, an understanding of the barriers faculty perceive with regard to educational transformation is a key element in designing successful intervention strategies.

REFERENCES

Brownell, S. E., & Tanner, K. D. (2012). Barriers to faculty pedagogical change: Lack of training, time, incentives, and . . . tensions with professional identity? *CBE-Life Sciences Education, 11*(4), 339–346. doi: 10.1187/cbe.12-09-0163

Henderson, C., & Dancy, M. H. (2007). Barriers to the use of research-based instructional strategies: The influence of both individual and situational characteristics. *Physical Review Special Topics—Physics Education Research, 3*(2), 020102.

Morris, R. C., Parker, L. C., Nelson, D., Pistilli, M. D., Hagen, A., Levesque-Bristol, C., & Weaver, G. (2014). Development of a student self-reported instrument to assess course reform. *Educational Assessment, 19*(4), 302–320. doi: 10.1080/10627197.2014.964119

Prince, M. (2004). Does active learning work? A review of the research. *Journal of Engineering Education, 93*(3), 223–231.

Walczyk, J. J., Ramsey, L. L., & Zha, P. (2007). Obstacles to instructional innovation according to college science and mathematics faculty. *Journal of Research in Science Teaching, 44*(1), 85–106.

Wright, E. L., & Sunal, D. W. (2004). Reform in undergraduate science classrooms. In D.W. Sunal, E.L. Wright, & J. Bland (Eds.), *Reform in undergraduate science teaching for the 21st century* (pp. 33–51). Greenwich, CT: Information Age Publishing Inc.

ABOUT THE AUTHORS

Loran C. Parker is an Assessment Specialist in the Discover Learning Center at Purdue University in West Lafayette, Indiana.

Omolola Adedokun is the Academic Support Services Director for the Center for Health Services Research Department at the University of Kentucky in Lexington, Kentucky.

Gabriela C. Weaver is the Vice Provost for Faculty Development and Director of the UMass Institute for Teaching Excellence and Faculty Development and Professor in the Department of Chemistry at the University of Massachusetts, Amherst in Amherst, Massachusetts.

SECTION E
Metrics and Assessment

In order for researchers and administrators involved in educational transformation to make informed decisions about their next step, they need to have data they can rely on. For this goal, they need to engage in serious efforts to probe the progress of their efforts, relative to their goals, and to have the tools that will allow them to measure the efficacy of their efforts. The chapters in this section describe a selection of approaches to developing metrics, selecting evaluation tools, and carrying out assessment. Reimer and her colleagues begin this section by addressing what happens when an innovation moves beyond the context of its original design. They look at the widespread use of clickers to gain a better understanding of their effectiveness outside the initial carefully prepared intervention. Lawrie and her colleagues follow with an examination of first-year chemistry programs in five Australian institutions, where faculty collaborated on the development of diagnostic tools, formative feedback, and multiple face-to-face and self-regulated online study modules. The faculty aim to transform instructional and assessment practices for diverse STEM cohorts at the beginning of their undergraduate education. The third chapter follows with Walter and her colleagues presenting the development and large-scale, interdisciplinary use of two instruments designed to evaluate climate and institutional practices: the Survey of Climate for Instructional Improvement (SCII) and the Postsecondary Instructional Practices Survey (PIPS). In the final chapter of this section, Fairweather, Trapani, and Paulson explore how to determine what data serve the goal of measuring institutional level change, and how to engage a disparate set of institutions in gathering data jointly so as to make progress in understanding widespread institutional change. They present an approach to gathering data across a set of AAU institutions about faculty attitudes and practices, and about the perceptions of institutional leaders on both faculty views and the institutional climate for evidence-based teaching.

1

Clickers in the Wild:
A Campus-Wide Study of Student Response Systems

**Lynn C. Reimer, Amanda Nili, Tutrang Nguyen,
Mark Warschauer, and Thurston Domina**

Global labor markets increasingly demand professionals with sophisticated skills in science, technology, engineering, and mathematics (STEM; Lansiquot et al., 2011; Vergara et al., 2009). However, too few U.S. college graduates possess these skills (Goldin & Katz, 2009; Levy & Murnane, 2012). Instructional methods in undergraduate STEM courses may be partly to blame. These courses typically take place in large lecture halls in which expert teachers transmit knowledge with minimal student interaction. Several observers argue that this course design contributes to the high level of attrition seen in STEM majors during the early undergraduate years (Baillie & Fitzgerald, 2010; Kyle, 1997; McGinn & Roth, 1999; Mervis, 2010; NAE, 2005). Student response systems (SRS), commonly known as *clickers,* may improve student outcomes in STEM by facilitating real-time classroom interaction, providing immediate feedback to both students and teachers, and creating opportunities for students to practice solving real-world problems. For this mixed-methods study, we observed 43 courses (enrolling nearly 15,000 students) over two academic terms, conducted 41 instructor interviews, and analyzed institutional data from the University of California, Irvine (UCI). Special attention was given to student outcome measures and instructor implementation strategies. We found that students earn slightly higher grades in courses that use clickers, with heterogeneous effects for females.

BACKGROUND

A key challenge for undergraduate STEM education involves student achievement and persistence. Clickers are being implemented across STEM undergraduate courses as a means to improve the learning experience for students. The literature on undergraduate clicker implementation addresses four domains: (1) student engagement and perception of learning, (2) student achievement, (3) gender inequality in STEM, and (4) innovative instructional practices. We briefly review each domain as background for discussion of clickers, instructional pedagogies, and research questions.

Student Engagement and Perceived Learning

Clickers have been implemented in many large lecture courses as a means to increase student engagement. Multiple studies suggest that students feel more positive about the learning experience and more engaged when clickers are implemented (Blasco-Arcas, Buil, Hernandez-Ortega, & Sese, 2013; DeGagne, 2011; Wolter, Lundeberg, Kang, & Herreld, 2011; Cain, Black, & Rohr, 2009). Specifically, Han and Finkelstein (2013) considered students' own perceptions of engagement and learning over four semesters in multiple lower and upper division STEM courses that incorporated clickers into instruction. The project aimed to enhance learning and engagement using inquiry-based practices with peer interaction and real-time feedback. Results were gathered via a questionnaire and suggested that students perceived clicker assessment and feedback as increasing both engagement and learning.

Student Achievement

Few studies have addressed how clicker use may be associated with student achievement. However, one study examined the academic achievement across three lecture courses using different clicker strategies: (1) clicker technology for group questioning, (2) group questioning without clickers, and (3) no form of group questioning or clickers (Mayer et al., 2009). The study compared the standardized exam scores (midterms and finals) of the three groups. Overall, the clicker group scored one-third grade point higher in academic achievement compared with the other groups. These results suggest that the immediate feedback from clickers has an advantage over the costs of using paper-based questioning, which provides feedback to students after a significant delay. However, the study focused on upper division students—who have had an opportunity to gain the skills necessary for academic success in these types of lecture courses—and the average enrollment of the course was moderate (119 students). To determine if clickers work as an intervention for students who need help the most—those who are new to the university setting—additional study is needed to replicate and validate findings with large, lower division gateway STEM courses.

In another study of clicker use in a larger (enrollment of 175) second year physiology course, student outcomes were analyzed by dividing the students into low-, middle-, and high-achieving cohorts based on performance in the prerequisite course (Gauci, Dantes, Williams, & Kemm, 2009). The authors found that students from the low achieving cohort earned significantly higher exam scores when compared with low achieving cohorts in the same course previously taught without clickers. Exam score differences were minimal for middle- and high-achieving cohorts. Put another way, those students least likely to succeed in this STEM course showed significant achievement gains when clickers were implemented as compared with previous years without the intervention.

Both studies focused on a single class taught by the same instructor, who was presumably motivated to improve instruction, as clicker implementation was compared to previous years. Thus, other factors in the aforementioned studies may interfere with addressing whether clickers, as an instructional tool, are strongly associated with student achievement. Further study that includes many courses and instructors is necessary. However, both studies may offer some insight into one possible instructional tool that may help close the achievement gap for women in large, undergraduate STEM courses (Xie & Shauman, 2003; Hill, Corbett, & St. Rose, 2010).

Women in STEM

A recent study on the use of clickers in introductory biology courses at nine different institutions, including 12 faculty and 1,457 lower division undergraduates, found women and non-science majors significantly preferred courses taught with clickers, compared with the courses taught without clickers (Wolter, Lundeberg, Kang, & Herreld, 2011). Faculty taught six to eight topics alternating between lecture and case studies with clickers. Although student preference might be linked to increased student academic performance, further study is needed to validate the link between clicker usage and student achievement, especially for women in STEM. Clickers may offer an instructional reform for large, gateway STEM courses that particularly benefits women in STEM without disadvantaging other groups.

While evidence is growing regarding how clicker use affects engagement, perceived learning, student achievement, and benefits for women, we need a better understanding of the related instructional pedagogies and innovations that may be associated with clicker use and positive student outcomes. Further study is needed across disciplines and instructors, in large gateway courses, and during the same time period to eliminate instructor bias.

Clicker Implementation

Because increased class sizes, typical in undergraduate STEM courses, dilute the opportunities for students to actively participate in class discussions, thoughtful implementation of clickers creates the potential for a more active, engaged learning environment where instructor queries become dynamic and dialectical. All students can respond simultaneously and anonymously, providing equal access for students who may be reluctant to raise their hands or speak out in class. All students receive real-time feedback, as does the instructor—if a significant number of students fail to answer the question correctly, she can immediately reteach the concept. This creates an opportunity for instructors to offer deeper explanation, conceptual development, and problem-solving during class (Patterson et al., 2010). Additionally, students can interact with one another for collaborative brainstorming (Mayer et al., 2009).

Our study fills a significant gap in the research on clickers by analyzing how student achievement in large, gateway STEM courses may be enhanced when clickers are implemented, including instructional methods of implementation of clickers. To date, studies have either focused on one course taught by one instructor, or on a single discipline (e.g., introductory biology). Because the use of clickers is becoming more common in large lecture courses, a naturalistic, population-based approach to the study of clicker implementation and associated student outcomes across a range of STEM disciplines is essential. Additionally, for those instructors interested in implementing clickers, research that highlights the ways in which clickers are used is particularly important. Thus, two research questions framed this work: First, does clicker use in STEM gateway courses positively impact student achievement, with special attention to underrepresented groups (i.e., women and URMS)? Second, what instructional pedagogies appear to coincide with clicker implementation?

METHOD

Sample

These questions were addressed through the *WIDER—EAGER: Documenting Instructional Practices in STEM Lecture Courses* research program at the University of California, Irvine. During fall and winter terms, we observed and videotaped gateway STEM courses (N = 43) across seven departments: Biological Sciences, Computer Science, Engineering, Chemistry, Mathematics, and Physics. Observations were made within the first two weeks of instruction and within the last two weeks of instruction for each course. Average course enrollment was 322 (*SD* = 103). In the case of multiple course sections, the section with the largest enrollment was observed. The instructors we observed and interviewed included 18 full professors, eight associate professors, four assistant professors, nine lecturers, and three graduate students; four instructors declined to participate and one declined to be interviewed. The undergraduate sample included over 7,000 students (49% Asian, 20% Hispanic, 13% White, 1.5% African American, 16.5% other, 55% first generation, 41% low income status, and 52% female); many were enrolled in more than one STEM gateway course ($N = 14,722$). Students were informed of the study through the campus learning management system consistent with the requirements of the UCI Institutional Review Board.

Measures

Observation protocol

The research team developed a class observation protocol entitled *Simple PRotocol for Observing Undergraduate Teaching* (SPROUT). We adapted content from three well-known observation protocols: *U-Teach Observation Protocol* (UTOP; Walkington et al., 2012); the *Reformed Teaching Observation Protocol* (RTOP; Sawada et al., 2002); and *Teaching Dimensions Observation Protocol* (TDOP; Hora & Ferrare, 2014). Overall inter-rater agreement for SPROUT was robust (Cohen's κ = .80). SPROUT includes both quantitative and qualitative pieces. Dimensions used in this study include (a) whether clickers were used; (b) number of questions; and (c) the relationship of clicker use to instructional pedagogies, including: the instructor solving problems in front of the class, peer interaction, and feedback from the instructor (the instructor modifies the lecture after seeing responses to a clicker question). Table 1 offers descriptive data on clicker usage and associated instructional practices.

Instructor interviews

In addition to SPROUT, we conducted semi-structured interviews with the instructors at the conclusion of each course. Interview prompts covered the same topics noted in SPROUT. For co-taught classes, instructors were given

TABLE 1. Descriptive Data on Clicker Usage in Observed Courses

	Fall 2013	Winter 2014
Courses using clickers	6/23 (30%)	9/20 (45%)
Biology	3/3	4/4
General Chemistry	1/4	3/6
Physics	2/2	2/2
Methods of Implementation		
Courses using problem-solving with clickers	5/6	8/9
Average # of questions	3.73	2.78
Courses where instructors modified the lecture after a clicker question	5/6	6/9
Courses using peer interaction with clickers	4/6	4/9

the option to be interviewed together or separately. Interview prompts included (a) strategies to promote student engagement, (b) formative assessment of student understanding in class, (c) instructor interaction styles with students during and after class, (d) student interaction styles with each other during and after class, (e) technology use in class, and (f) problem-solving strategies.

Institutional data

Student achievement outcomes (course grade and course progression) for observed classes were evaluated using data obtained from the university's Office of Institutional Research (OIR). Baseline achievement data included high school grade point average, SAT scores, and average score on STEM advanced placement exams. Current achievement data included grade in the observed courses, enrollment in subsequent course (operationalizing persistence), and subsequent course grade.

Analysis

Quantitative analysis

Logistic and ordinary least squares regression examined clicker use in observed course as the independent variable and current course grade, progression to next course, and grade in subsequent course as the dependent variables. We used interaction terms for gender, ethnicity, first generation, and low-income status for all three regressions to identify heterogeneous effects. We also used interaction terms for instructional methods associated with clicker implementation. We controlled for prior academic achievement (i.e., high school grade point average, SAT verbal and math scores, and average score of STEM advanced placement exams), along with gender, ethnicity, first generation, and low-income status. We also analyzed a subset of introductory chemistry courses to offer a quasi-experimental approach of comparing the same course with and without clickers. To preserve data on all cases for variables of interest, missing data were controlled for on the following variables: math and verbal SAT scores, high school GPA, and AP exam scores, as international and transfer students do not have to report any of these. In some instances, gender and/or ethnicity was not provided through the institutional data.

Qualitative analysis

The descriptive measures on SPROUT and instructor interviews provided rich qualitative data on the use of clickers and instructional pedagogy. Interviews were transcribed and reviewed for emerging themes. Three themes emerged that were linked to clicker implementation: (1) problem-solving, (2) peer interactions, and (3) feedback in the form of correct answer and subsequent explanations (Ryan & Bernard, 2003; Siedman, 2006). These themes are supported in the literature as potentially leading to positive student outcomes, including grades and persistence in STEM (NAE, 2005; Nielsen, 2011; Singer, Nielsen, & Schweingruber, 2012). These strategies were then cross-referenced with descriptions of in-class instructor practices from the observation protocol.

RESULTS

Student Achievement and Women in STEM

The first research question considered whether clicker use in large, introductory STEM courses positively impacted student achievement (i.e., grade in course, course progression, grade in subsequent course) and if there were any heterogeneous effects. Table 2 shows the impact of clicker use on grade in the observed course. Clicker use is associated with a 0.15 ($p < 0.001$) positive effect for all students on the grade in current course. An additional positive effect of 0.05 ($p < 0.01$) for women ameliorates the disadvantage that women experience in these courses. However, clicker use did not improve overall course progression or grade in the next course, compared with courses not using clickers. Our additional analysis of introductory chemistry courses afforded the comparison of clicker use (N = 4) and nonuse (N = 6) across different sections of the same course as a robustness check. This course is a pre-requisite course for nearly all STEM majors, the first introductory chemistry course in a three-course series. The results were consistent with the overall sample.

Clicker Implementation

The second research question sought to identify those instructional methods associated with clicker implementation. Again, Table 1 offers descriptive data on clicker usage and associated instructional practices. We observed clickers in all introductory biology and physics courses and a few general chemistry courses. Clickers were used in the context of problem solving by 13 of the 15 instructors, such as case studies in biology or multi-step, calculation-based questions about real-world problems in chemistry or physics. The average number of clicker questions was approximately three per class, with most frequent use observed in biology courses. Eleven of the 15 instructors who used clickers modified their lecture based on immediate student responses. In cases where students' responses were evenly divided, instructors modified their lecture by spending additional time addressing misconceptions or having the students talk to one another and reconsider their answers. In some instances, students were given a second chance to answer the same question, particularly when clicker results were initially split 50-50. Additionally, eight of the 15 instructors

TABLE 2. The Effect of Clicker Use on Student Achievement

	Grade in Observed Course
Clickers	0.150***
	(0.019)
Female	-0.037*
	(0.018)
Clickers x Female	0.045**
	(0.016)
Blacks	0.072
	(0.078)
Clickers x Blacks	-0.061
	(0.071)
Hispanics	-0.148***
	(0.029)
Clickers x Hispanics	-0.081**
	(0.026)
Asians	0.020
	(0.022)
Clickers x Asians	-0.046*
	(0.020)
First Generation	0.043*
	(0.020)
Clickers x First Generation	-0.032
	(0.018)
Low Income	-0.028
	(0.020)
Clickers x Low Income	-0.006
	(0.018)
N	14722
R^2	0.206

allowed or encouraged peer interaction when figuring out a clicker question. Table 3 examines differential effects on achievement when clickers are used in conjunction with the following practices: (1) whether the instructor was solving problems in front of the class connected to the clicker questions; (2) the

TABLE 3. The Impact of Clickers on Student Achievement

Methods of Implementation					
Clickers	0.15***	0.19***	0.15***	0.18***	0.16***
	(0.02)	(0.02)	(0.02)	(0.02)	(0.02)
Problem Solving		-0.03**			
		(0.01)			
Clickers x Problem Solving		-0.04***			
		(0.01)			
Total # Questions			0.02		
			(0.01)		
Clickers x Total # of Questions			-0.01		
			(0.00)		
Modifying Lecture				-0.04*	
				(0.02)	
Clickers x Modifying Lecture				-0.02	
				(0.01)	
Peer Interaction					0.10*
					(0.05)
Clickers x Peer Interaction					-0.06*
					(0.02)
N	14722	14722	14722	14722	14722
R^2	0.206	0.209	0.206	0.207	0.207

Note: Standard errors in parentheses. Interaction terms are designated by a product term. Student achievement is measured as grade in observed course.

*$p < 0.05$. **$p < 0.01$. ***$p < 0.001$.

number of clicker questions; (3) whether instructors modified the lecture following a clicker question; and (4) peer interaction during clicker questions. We found that the positive effects of clickers were consistent across each of these methods of implementation. In many of the classes that we observe clickers, the additional instructional practices, such as peer interaction, are happening in the context of clickers. Thus the additional benefit is negligible. In the case of problem solving (small negative effects), those are offset by the larger effect of clickers. Problem solving does occur in some other classes, whereas peer interaction occurs only in conjunction with clickers.

In addition to descriptive data from the SPROUT, instructors provided detailed responses in semi-structured interviews regarding instructional pedagogies and potential insight into why they used clickers. A majority linked clicker implementation with current topics or review of previous lecture topics. At times, this was extended to include content review from the prior term. Most instructors cited student engagement as a motivation for clicker incorporation. In the words of one biology instructor:

> I found it [clickers] to be a fun tool to engage the students because it was entertaining. I could, for example, talk about how one section got tripped up on this question. All of a sudden I have their attention. They want to outperform the other [class] section. Other times I go through a section of material and I will ask them a pretty hard question and 80% will get it wrong. Then I'll spend a couple of slides explaining why they got it wrong, but that [clicker] attention grabber is what I was after. They stop, they pause, they take it in, and they understand the explanation. Then I'll ask another question and it'll flip. Now 80% get it right.

This example associates engagement with formative assessment in clicker feedback; students are afforded the opportunity to answer questions a second time. Related to this discussion, instructors noted a fruitful association between student interactions and problem solving where clickers were implemented. A general chemistry instructor observed:

> Usually everyone gets the right answer but sometimes there's a discrepancy. It might be 50-50 or 75-25. "Okay! So there are two possible answers, find someone near you who gave the other answer and convince them they are wrong." Then we do it again. I give them two and three minutes to talk about it and then generally it goes from 50-50 to 75-25. So not everyone gets it, but it's the idea for the student that coming to class matters and thinking, "I might have to do something in class today," which I feel helps.

This instructor taught general chemistry off-sequence, meaning these students either failed the same course in the previous term or could not enroll any earlier because their SAT math score did not meet a threshold of 600. Two instructors shared plans to use clickers in the future as a means of increasing student interaction and feedback. No instructors described clickers as difficult to implement; some simply preferred alternative methods to promote feedback and engagement. In fact, ease of implementation is one reason cited for their inclusion in all first year biology courses at UCI.

DISCUSSION

Student Achievement

Increasing persistence in STEM undergraduate programs continues to be a major research focus. The present study investigated the use of clickers as a tool to promote student success and accessibility in undergraduate STEM programs. We found a positive association between clicker use and student achievement, with added benefit for women, in large gateway lecture courses. This builds on a wealth of literature of small single course studies that have found the same results. The lack of any association to course progression merely affirms that students likely to persist will do so with or without innovative instruction. Our concern is for students who are less likely, namely women and other underrepresented groups, to succeed in STEM majors. It was disappointing that clicker use was not able to ameliorate the disadvantage that Hispanic students seem to have in these introductory STEM courses. Further study on other instructional innovations may reveal differential benefits for these underrepresented groups.

Clickers are simple to implement, making it increasingly difficult to justify their absence from lecture halls, especially given their proven ability to improve student achievement, particularly for women, in STEM classes. To obtain an expanded understanding of clickers and underrepresented groups, we intend to collect more nuanced data on the kinds of clicker questions (true/false or multiple choice), the amount of time required to arrive at an answer (e.g., is the question simple or more complex), and to what extent student collaboration is encouraged in the context of clicker questions. With additional course observations over time, we will be able to study persistence beyond lower division STEM courses by tracking students through their major program to graduation.

Women in STEM

We found that clicker use in gateway STEM courses was positively associated with student grades, especially for women, closing the gender gap. This validates smaller studies that found women and non-science majors preferred STEM classes that incorporated clickers, compared with those that do not, and that clickers had the greatest impact on low-achieving students (Wolter et al., 2011; Gauci et al., 2009). Clickers may offer a simple instructional reform that increases success for underrepresented groups, including women, in STEM. The present study suggests that student collaboration during lecture provides opportunities for students to "talk out" the content, which may be a preferred learning strategy for students with a weaker skill set or lower self-efficacy. Because clickers enable anonymous, formative assessment, and an opportunity to practice and improve critical thinking skills, this may explain the positive association between increased grades for all students and closing the gap for women in courses that incorporate clicker usage. Furthermore, the greatest benefit may exist for underrepresented students—those *over*represented in off-sequence STEM gateway courses—by giving them a non-threatening chance to address their misconceptions. Further research is needed to assess what other instructional practices may be beneficial for women and other underrepresented groups in STEM courses.

Clicker Implementation

Feedback

Clickers enable increased formative (i.e., real-time feedback) rather than summative (i.e., linked to grade) assessment (Mayer et al., 2009; Cain, Black, & Rohr, 2009; Guthrie & Carlin, 2004), a pedagogical strength of clicker use. Instructors who modify their lectures in response to student answers presumably help students stay on track and identify gaps in topical understanding with anonymity—students can safely check their answers for accuracy against the class and/ or discover that incorrect responses are neither unique nor unusual (Patterson et al, 2010). Formative assessment with clickers offers a low-stakes opportunity for student discussion regarding the correct response and challenges inherent in the problem or case. Instructors are able to instantaneously gauge overall class comprehension and adjust accordingly. This may prove useful in contexts where sophisticated problem-solving skills are required for student persistence in STEM.

Problem-solving

Over the past several years, all introductory biology courses at UCI have adopted clickers as part of grant-funded research to "help faculty maintain productive research careers while teaching more effectively in large classrooms," (O'Dowd, 2014). The vast majority of the time, we observed clickers being used in the context of problem-solving. Previous literature that associates clickers with positive student outcomes studied courses in which clickers were used in conjunction with case-based questions requiring problem solving, rather than simple content questions, and resulted in higher exam scores (Levesque, 2011) and positive student attitudes (Han & Finkelstein, 2013; Wolter et al., 2011). We intend to collect more detailed data on problem-solving (e.g., the extent to which the instructor explicitly shares scientific thinking and step-by-step strategies to arrive at a solution). We feel problem-solving may be another aspect of instruction associated with increased student achievement.

Peer interaction

Our study affirms other research that clickers provide an opportunity for students to interact with one another and with the instructor, creating a dynamic setting where all students are responding to questions, rather than a select few (Patterson et al., 2010; Mayer et al., 2009). Clickers provide opportunity for collaborative learning. We observed peer interaction in half of the courses using clickers. We intend to investigate further the relationship between clickers and interactive learning, as interactive learning may also be associated with increased student achievement. Feedback, problem solving, and interactive learning in the context of clicker implementation seem to be a catalyst for increased engagement and student achievement.

Student Engagement

It is difficult for STEM instructors to consistently engage students in large gateway courses. Instructor concern regarding student engagement was substantial. Instructors currently using clickers offered considerable detail regarding implementation advantages, including engagement and collaboration with peers on problems, raising the stakes for class attendance and providing a mechanism for students to receive non-threatening formative assessment through feedback. Pedagogical implementation is a particularly important concern. Prior work affirms a positive correlation between professional development for instructors and how students perceive clicker use (Han & Finkelstein, 2013); this perception in turn correlates with how effective clicker use increased student interactions (Peterson et al., 2010). Instructors in our study favored clickers as a means to enhance student engagement. Further study is needed to explore student self-efficacy and motivation, as increased engagement may be course-specific, but increased self-efficacy and motivation may strengthen long-term persistence in STEM.

Our naturalistic, population approach of observing STEM introductory courses revealed that instructors used clickers as a means to improve student engagement. Accordingly, we have modified our observation protocol to capture further details on *how* clickers operate in conjunction with problem solving, feedback, and interactive learning as avenues of not only increased student engagement, but potentially increased self-efficacy and motivation. In summary, clickers are well adapted to improving instruction in large, introductory STEM courses. Like any instructional innovation, implementation success depends on how the tool is embedded within a coherent and effective instructional pedagogy.

REFERENCES

Baillie, C., & Fitzgerald, G. (2010). Motivation and attribution in engineering students. *European Journal of Engineering Education, 25* (2), 145–55.

Blasco-Arcas, L., Buil, I., Hernandez-Ortega, B., & Sese, F. J. (2013). Using clickers in class: The role of interactivity, active collaborative learning and engagement in learning performance. *Computers & Education, 62,* 102–110.

Cain, J., Black, E. P., & Rohr, J. (2009). An audience response system to improve student motivation, attention, and feedback. *American Journal of Pharmaceutical Education, 73* (2), 1–21.

DeGagne, J. C. (2011). The impact of clickers in nursing education: A review of literature. *Nurse Education Today, 31,* 34–40.

Gauci, S. A., Dantas, A. M., Williams, D. A., & Kemm, R. E. (2009). Promoting student centered active learning in lectures with a personal response system. *Advances in Physiology Education, 33,* 60–71.

Grossman, H., & Grossman, S.H. (1994). *Gender issues in education.* Boston: Allyn & Bacon.

Guthrie, R. W., & Carlin, A. (2004). Waking the dead: Using interactive technology to engage passive listeners in the classroom. *Proceedings of the Tenth Americas Conference on Information Systems,* New York, August 2004. http://clickers.ntu .edu.sg/wp-content/uploads/CPSWP_WakindDead082003.pdf

Han, J. H., & Finkelstein, A. (2013). Understanding the effects of professors' pedagogical development with clicker assessment and feedback technologies and the impact on students' engagement and learning in higher education. *Computers & Education, 65,* 64–76.

Hora, M. T., & Ferrare, J. J. (2014). Remeasuring postsecondary teaching: How singular categories of instruction obscure the multiple dimensions of classroom practice. *Journal of College Science Teaching, 43* (3), 36–41. http://www.hhmi .org/scientists/diane-k-odowd

Lansiquot, R.S., Blake, R.A., Liou-Mark, J., & Dreyfuss, A.E. (2011). Interdisciplinary problem-solving to advance STEM success for all students. *Peer Review: Association of American Colleges and Universities,* Summer, pp. 19–22.

Levesque, A. A. (2011). Using clickers to facilitate development of problem-solving skills. *Life Sciences Education, 10,* 406–417.

Lundeberg, M. A. (2008, October). *Case pedagogy in undergraduate STEM: Research we have, research we need.* White Paper commissioned by the Board of Science Education, National Academy of the Sciences.

Mayer, R. E., Stull, A., DeLeeuw, K., Almeroth, K., Bimber, B., Chun, D., Bulger, M., Campbell, J., Knight, A., & Zhang, H. (2009). Clickers in college classrooms: Fostering learning with questioning methods in large lecture classes. *Contemporary Educational Psychology, 34,* 51–57.

Mervis, J. (2010). Better intro courses seen as key to reducing attrition of STEM majors. *Science, 330* (6002), 306.

National Academy of Engineering (NAE). (2005). *Educating the engineer of 2020: Adapting engineering education to the new century*. Washington, DC: The National Academies Press.

Nielsen, N. (Ed.). (2011). *Promising practices in undergraduate science, technology, engineering, and mathematics education: Summary of two workshops*. Washington, DC: National Academies Press.

O'Dowd, D. (2014). *Howard Hughes Medical Institute*. Retrieved from http://www.hhmi.org/scientists/diane-k-odowd

Patterson, B., Kilpatrick, J., & Woebkenberg, E. (2010). Evidence for teaching practice: The impact of clickers in a large classroom environment. *Nurse Education Today, 30*, 603–607.

Ryan, G. W., & Bernard, H. R. (2003). Techniques to identify themes. *Field Methods, 15*(1), 85–109.

Sadker, M. P., & Sadker, D. M. (2003). *Teachers, schools, and society*. Boston: McGraw-Hill.

Sawada, D., Piburn, M. D., Judson, E., Turley, J., Falconer, K., Benford, R. & Bloom, I. (2002). Measuring reform practices in science and mathematics classrooms: The reformed teaching observation protocol. *School Science and Mathematics, 102*, 245–253.

Sheppard, S., Pellegrino, J., & Olds, B. (2008). On becoming a 21st century engineer. *Journal of Engineering Education, 97* (3), 231–34.

Siedman, I. (2006*). Interviewing as qualitative research: A guide to researchers in education and the social sciences*. Teacher's College Press. New York, NY.

Stump, G. S., Hilpert, J. C., Husman, J., Chung, W.T. and Kim, W. (2011). Collaborative learning in engineering students: Gender and achievement. *Journal of Engineering Education, 100*, 475–497. doi: 10.1002/j.2168-9830.2011.tb00023.x

Walkington, C., Arora, P., Ihorn, S., Gordon, J., Walker, M., Abraham, L., & Marder, M. (2012). *Development of the UTeach observation protocol: A classroom observation instrument to evaluate mathematics and science teachers from the UTeach preparation program*. Preprint publication.

Wolter, B. H. K., Lundeberg, M. A., Kang, H., & Herreld, C. F. (2011). Students' perceptions of using personal response systems ("Clicker") with cases in science. *Journal of College Science Teaching, 40*(4), 14–19.

NOTE

This material is based upon work supported by the National Science Foundation under Grant #1256500.

ABOUT THE AUTHORS

Lynn C. Reimer is an NSF Graduate Research Fellow in the School of Education at the University of California, Irvine in Irvine, California.

Amanda Nili is a Research Assistant in the School of Education at the University of California, Irvine in Irvine, California.

Tutrang Nguyen is a Doctoral Student in the School of Education at the University of California, Irvine in Irvine, California.

Mark Warschauer is a Professor of Education and Informatics and the Associate Dean of the School of Education at the University of California, Irvine in Irvine, California.

Thurston Domina is an Associate Professor of Education and Sociology in the School of Education at the University of California, Irvine in Irvine, California.

2

Closing the Loop: A Model for Inter-Institutional Collaboration Through Delivering Formative Assessment in Large, First-Year STEM Classes

Gwendolyn Lawrie, Anthony Wright, Madeleine Schultz,

Tim Dargaville, Roy Tasker, Mark Williams, Simon Bedford,

Glennys O'Brien, and Christopher Thompson

When students start their tertiary studies they move into a new world which differs in many ways from their prior experiences, including the way they were taught, access to faculty, learning environments, class sizes, expectations of independence and time management (Torenbeek, 2011). The first year experience (FYE) has become a pivotal focus for institutional programs that recognize that many students struggle in this transition. Such programs aim to improve student retention in tertiary studies through provision of orientation and mentoring activities. These initiatives have become widespread and are typically informed by key research in the field in terms of transition pedagogies (Kift, 2009; Kift, 2010; Lawrence, 2005) and student engagement and retention (Kuh, 2008; Carini, 2006; Tinto, 1987; Tinto, 2005) (Table 1).

Engagement and retention of students in the science, technology, engineering and mathematics (STEM) disciplines in particular has attracted widespread attention worldwide for several decades. In Australia, the number of students opting to study STEM subjects at high school, or entering STEM fields of study at university has been in decline (Office of the Chief Scientist, 2012). A recent report that benchmarked Australian achievements in STEM disciplines against international data further endorsed the need to enrich STEM education at all levels (Office of the Chief Scientist, 2014).

As students begin their tertiary studies, they are likely to encounter multiple programs, opportunities and activities to support them offered by different stakeholders, including their peers, discipline faculty, their department and their institution. Students who possess a high degree of self-efficacy and/ or confidence, which many school leavers have at the commencement of their tertiary studies, may not perceive that they require support at that stage and indeed a relationship between high self-efficacy and academic success exists

TABLE 1. Key Recommendations for Institutional Initiatives to Improve the FYE and Retention

Context	Recommendation	Source
First Year Transition Pedagogy	First year experience program strategies: • Curriculum that engages students in learning. • Proactive and timely access to learning and life support. • Intentionally fostering a sense of belonging. • Sustainable academic faculty-professional partnerships.	Kift et al. (2010)
Practices	A model for action for retention in terms of the impact of: • Ppedagogy • Assessment • Faculty development • Learning communities • Part-time faculty	Tinto (1987)
Strategies	Orientation objectives: • Familiarize students with the university. • Develop students' sense of purpose and direction. • Facilitate students' engagement. • Promote and enhance students' learning.	Pitkethly and Prosser (2001)
Engagement	Scales to measure student success related to student engagement: • Level of academic challenge • Active and collaborative learning • Student-faculty interaction • Enriching educational experiences • Supportive campus environment • Reading and writing • Quality of relationships • Institutional emphasis on good practices • Higher-order thinking • Student-faculty interaction concerning coursework • Integration of diversity into coursework	Carini et al. (2006)

(Friedlander et al, 2007; Chemers et al, 2001; McKenzie & Schweitzer, 2001). For many students, as they progress in their studies, they encounter assessment across multiple units of study and it is often at this point that many begin to lose confidence or the ability to successfully manage their time (Rausch & Hamilton, 2006). Indeed, many students are the first in their family to enter tertiary study and hence do not possess the required cultural capital in regard to the expectations of the tertiary experience (Tinto, 1987).

Initiatives to monitor student engagement and completion of assessment tasks in lower-division courses, through learning management systems, are becoming prevalent in many institutions. These can enable early detection of students who are at risk of not completing their studies successfully. The problem is that once the issue has been identified, there is still a substantial reliance on the individual student to self-remediate in their studies. Self-regulation is unlikely to be a skill that they have been taught to apply during high school or that they can transfer. While such systems are useful, faculty are often better placed to identify any individual students who have disengaged or drifted away from course activities and to support their re-engagement. A strengthened relationship between student and faculty in relation to the coursework is one of the recommendations to improve retention (Table 1). However, an individual unit (course) coordinator or instructor may not have either the capacity or strategies to identify and support students at risk so a range of models of institutional-level support for faculty are required. In this study, we report the outcomes of a multi-institutional collaborative initiative to provide faculty with strategies and resources to diagnose and respond to students who are poorly prepared for their studies in a core STEM discipline, chemistry. The principles and exemplars of practice are transferrable to any STEM discipline.

RATIONALE

Developing sustainable practices at the unit level

While programs and support processes are likely to be available at the institutional level, there is often disconnection with practices at the disciplinary unit (course) level, particularly in very large enrolment STEM courses. A unit of study (course) is generally designed and implemented by discipline faculty who are often unaware of the individual issues that their students are facing, and are focused on measuring disciplinary learning outcomes. Pedagogical strategies, such as inclusive practice to address student diversity (Lawrence, 2005), can be challenging for faculty to implement and, without professional development or support, they may be overwhelmed by the institutional goals in terms of recommended practices and pedagogies (Table 1). One goal that can be achieved through strategic initiatives at the discipline unit (course) level addresses the academic orientation objective of enhancing students' learning by: "taking into account the variation in skills and experience of students, and where necessary, raising their skills and knowledge to a basic level by addressing deficiencies and enhancing skills already present" (Pitkethly & Prosser, 2001, p192). This objective requires that faculty establish what the existing knowledge and skills of students are when they start their studies. This assessment activity aligns with the shift in institutional practices to measure and benchmark student achievement as part of accreditation, and the project reported here provides a mechanism to attain this goal.

In recent years, accreditation has become increasingly important to both governments and tertiary institutions, and it influences institutional assessment of student learning. In Australia, the Australian Learning and Teaching Council implemented the Learning and Teaching Assessment Standards (LTAS) project in 2011, which catalyzed the development of discipline-based statements of threshold learning outcomes (TLOs) for many disciplines. These are the minimum levels of achievement that a student must demonstrate when they graduate from their program of study. The science TLOs include understanding science, scientific knowledge, inquiry and problem-solving, communication and personal and professional responsibility (Jones et al., 2011). Evaluation of how each program of study measures students' achievement against each TLO involves the difficult task of mapping the form and weighting of assessment across multiple courses/units within a degree, but can provide insights into how we can align and improve assessment (Schultz et al, 2013). Thus, accreditation has the potential to drive assessment of student learning outcomes.

Providing students with information on the extent of their preparedness for studies as they start their tertiary studies serves three purposes: (1) it provides a "baseline" to judge a student's progression in their learning outcomes; (2) it provides faculty with a profile of the students in their class to which they can respond in teaching; and (3) it provides an opportunity to supply formative feedback to students on which they may act. The last of these is a persistent theme in recommendations at both the institutional and program levels, which recognize the potential role and importance of the provision of diagnostic feedback to students as part of transitional pedagogies (Yorke, 2001; McInnes, James, & McNaught, 1995). Common themes arising from investigation of first-year student attrition are large classes, poor time-management skills and lack of educational direction, all of which may be addressed, in part, through the provision of useful formative feedback. While known to be a critical factor in students' engagement in their learning, there are few studies that report large-scale initiatives to deliver formative feedback to first-year students. Formative feedback can only be delivered as part of formative assessment, which opens a can of worms because of tensions between faculty spending time formulating and delivering assessment and feedback, and student engagement in assessment that is not traditionally awarded course marks (Nicol & MacFarlane-Dick, 2006). As part of a recently completed Australian Office for Learning and Teaching (OLT)

Innovation and Development Grant, the authors, a team of chemistry faculty, located at five separate institutions, have collaborated to explore mechanisms for diagnosing student conceptions in chemistry and providing formative feedback in parallel with learning resources that support students in self-regulated study (Lawrie et al, 2013). These institutions, all located in metropolitan sites, represent the three largest states in Australia and over 5,000 commencing undergraduate students in chemistry courses/units. Thus, this study provided a unique multi-institutional lens into the opportunities and challenges that face faculty in a first-year STEM environment.

EXEMPLARS

During the first year of the project, the diagnostic tool (a concept inventory) was developed and delivered to students as they commenced their first semester of first-year chemistry at four of the participating institutions (N=1,287). This enabled exploration of the validity and reliability of the instrument. The project team experienced a very high degree of consensus and shared vision for goals and activities of the project for its entire duration. However, through development of the associated learning modules, it became evident that the faculty in this project preferred to adapt these collaboratively developed teaching resources to fit their own context and students' needs. With hindsight it is not surprising that a "one-size-fits-all" approach did not evolve from the project, but instead a range of exemplars for implementation arose. The icons shown in Figure 1 emphasize the variability in teaching environments that faculty worked in, and summarize the range of tools and contexts that existed.

This diversity in teaching practice highlighted the potential difficulty in the translation of any institutionally-driven pedagogical reform from the perspective of the faculty. Indeed, a single diagnostic instrument, aimed at evaluating the core chemistry conceptions possessed by students who had completed high school chemistry was not sufficient for the project. A second instrument was required to provide feedback to students who had not completed upper high school chemistry but had experienced the foundations in junior science. Both instruments consist of clusters of questions for each conceptual area being investigated. Faculty opted to link the diagnostic questions to the structure of their course and hence delivered the subset clusters of questions when the corresponding concept was being discussed. Researchers who develop concept inventory instruments might argue that this removes the validity and reliability of these questions, but this is driven by practice and the goal of delivering formative feedback (Lawrie et al, 2015).

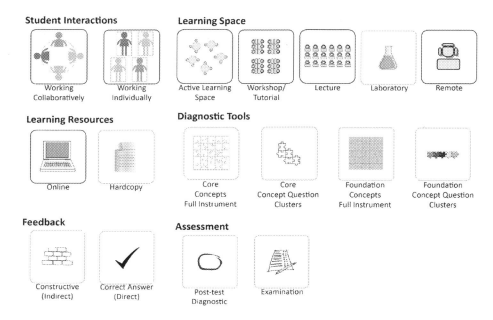

FIGURE 1. Schematic summary of the range of tools, strategies and learning environments that faculty adopted in diagnosing student prior understanding of chemistry, delivering supporting resources for learning, and assessing learning outcomes.

Exemplars of the ways in which the faculty combined these elements of teaching practice are provided in Figure 2. The variables expanded beyond those illustrated in Figure 1 to encompass:

- How feedback was delivered (by faculty, through the learning management system) and its format
- The learning management system (LMS), which impacted on the delivery of online tools and resources and how the instructors assessed whether students had gained in their conceptual understanding

Formative feedback to support student learning is optimal when the timing and format are perceived to be useful by students (Carless et al., 2010; Hattie & Timperlay, 2007). Individual faculty are best situated to judge when, how and where feedback should be delivered to their students. The timing and delivery of feedback, illustrated in a schematic way in each of the exemplars in Figure 2, was highly variable because it was tailored to each context.

It may be tempting to look across these exemplars and conclude that the individual faculty were working independently of each other in their own

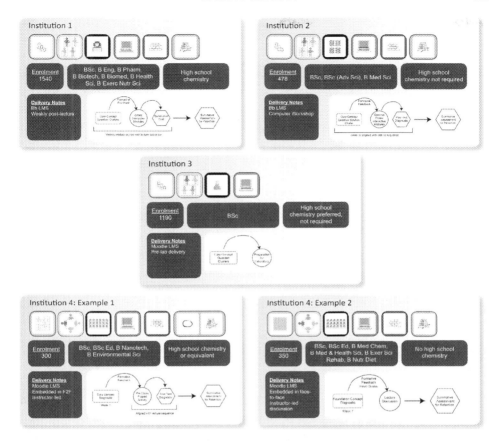

FIGURE 2. Exemplars showing how faculty combined the core tools and resources of this project into their practices. These are structured to provide a snapshot of each learning context in terms of the identity of the student cohort and the classroom context.

contexts, and if this were the case, that would make this project similar to many other teaching innovations at the course/unit level in single institutions. However, an underlying foundation was established during the process of collaboration in the development of the instruments, the application of these common tools for diagnosing alternate conceptions, the development of open resources that students can access in response to formative feedback and shared evaluation strategies that led to a shared awareness of the challenges of engaging first-year chemistry students in their learning. This foundation and awareness improved the quality of all the outputs of the project. The involvement of multiple institutions in this study generated a qualitative and quantitative data set that will translate into multiple recommendations for practice and benchmarking information that will inform institutional reforms of the first-year experience for STEM disciplines. The principles, strategies and outcomes of this project represent more than resources to be shared for adoption or adaptation by faculty, they involve a transformation in beliefs and practices, which are effective change strategies in STEM education (Henderson et al., 2011).

In the context of national accreditation of tertiary institutions, the data from this project provides a profile of the conceptions possessed by students entering their tertiary studies across three Australian states using a single assessment tool. This data can inform the process of developing chemistry TLOs or statements around the minimum levels of understanding that students may achieve by the end of their first-year studies and so contributes to the overall process of curricular reform. Implementation of similar diagnostic assessment tools for the provision of formative feedback to students in other STEM disciplines will both inform the TLOs for those disciplines and further enhance transitional pedagogies.

Institutional reform to support the first-year experience is likely to be more successful if communities of academics attempting to innovate in their practice are fostered and supported. The project reported here demonstrates how the shared experiences and contexts between faculty placed in parallel roles in their respective institutions enabled a deeper understanding and richer examples of how the first-year experience could be enhanced through formative assessment and feedback. A recommendation from this study therefore is that higher education funding agencies support collaborative teams of faculty within a single discipline across multiple institutions as a strategic route to making visible the barriers that often halt individual practice. Shared perspectives, experiences and strategies amongst faculty provide the opportunity to identify different routes to a common destination. As noted by Vincent Tinto in 2005 when considering how the theory of student retention could be translated into action, "Two areas, among many, that are ripe for exploration are the effects of classroom practice upon student learning and persistence and the impact of institutional investment in faculty and staff development programs on those outcomes" (Tinto, 2005). While there is still scope for research to be completed addressing these areas, there is substantial room for growth at the institutional level.

CONCLUSION

This study has highlighted the impact that can be achieved by a critical mass of faculty, working collaboratively across multiple institutions, when their activities align with broader institutional goals in working to increase retention of students in their first-year experience. Assessment for accreditation represents one of many external drivers for institutional reform in teaching and learning practices, and faculty may often perceive that they are subjected to a "top-down" approach to any implementation of change (Cummings et al., 2005). The activities in this project have been driven by faculty who were acting as leaders of change in an institutional context; their initiatives have supported several thousand students in becoming more independent and self-regulated learners, skills that will integrate across all their studies. The resulting alignment between "top-down" transition programs, "bottom-up" student-driven expectations of support and "middle"-active faculty, implementing transferable learning approaches through formative feedback, can develop connections that are more likely to enable sustainable transformations in practice (Brinkhurst et al., 2011).

ACKNOWLEDGMENTS

Support for this activity has been provided by the Australian Government Office for Learning and Teaching (ID12-2277). The views expressed in this activity do not necessarily reflect the views of the Australian Government Office for Learning and Teaching. The significant contributions of project team members Chantal Bailey, Hayden Dickson, Aaron Micallef and Md Abdullah Al Mamun to the development of tools and resources are gratefully acknowledged.

REFERENCES

Brinkhurst, M., Rose, P., Maurice, G., & Ackerman, J.D. (2011). Achieving campus sustainability: Top-down, bottom-up, or neither? *International Journal of Sustainability in Higher Education*, 12, 338–45. DOI: 10.1108/14676371111168304

Carini, R.M., Kuh, G.D., & Klein, S.P. (2006). Student engagement and learning: Testing the linkages. *Research in Higher Education*, 47(1) 1–32. DOI: 10.1007/s11162-005-8150-9

Carless, D., Salter, D., Yang, M., & Lam, J. (2010). Developing sustainable feedback practices. *Studies in Higher Education*, 36, 395–407. DOI: 10.1080/03075071003642449

Chemers, M.M., Hu, L.-T., & Garcia, B.F. (2001). Academic self-efficacy and first-year college student performance and adjustment. *Journal of Educational Psychology*, 93(1) 55–64. DOI: 10.1037/0022-0663.93.1.55

Cummings, R., Phillips, R., Tillbrook, R., & Lowe, K. (2005). Middle-out approaches to the reform of university teaching and learning: Champions striding between the "top-down" and "bottom-up" approaches. *International Review of Research in Open and Distance Learning*, 6, 1–18.

Friedlander, L.J., Reid, G.J., Shupak, N., & Cribbie, R. (2007). Social support, self-esteem, and stress as predictors of adjustment to university among first-year undergraduates. *Journal of College Student Development*, 48(3) 259–74. DOI: 10.1353/csd.2007.0024

Hattie, J., & Timperley, H. (2007). The power of feedback. *Review of Educational Research*, 77, 81–112. DOI: 10.3102/003465430298487

Henderson, C., Beach, A., Finkelstein, N. (2011). Facilitating change in undergraduate STEM instructional practices: An analytic review of the literature. *Journal of Research in Science Teaching*, 48, 952–984. DOI: 10.1002/tea.20439

Jones, S., Yates, B., & Kelder, J. (2011). Learning and teaching academic standards (p. 11). *Threshold learning outcomes for science: Australian learning and teaching council.* Retrieved April 26, 2014 from: http://www.olt.gov.au/system/files/resources/altc_standards_SCIENCE_240811_v3_0.pdf

Kift, S. (2009). *Articulating a transition pedagogy to scaffold and to enhance the first year student learning experience in Australian higher education.* ALTC Fellowship Report. Retrieved March 11, 2011 from: http://www.altc.edu.au/resource-first-year-learning-experience-kift-2009

Kift, S., Nelson, K., & Clarke, J. (2010). Transition pedagogy: A third generation approach to FYE—A case study of policy and practice for the higher education sector. *International Journal of the First Year in Higher Education*, 1(1), 1–20.

Kuh, G.D., Cruce, T.M., Shoup, R., Kinzie, J. & Gonyea, R.M. (2008). Unmasking the effects of student engagement on first-year college grades and persistence. *Journal of Higher Education*, 79(5) 540–63. DOI: 10.1353/jhe.0.0019

Lawrence, J. (2005). *Addressing diversity in higher education: Two models for facilitating student engagement and mastery.* Paper presented at the HERDSA Conference 2005: Higher Education in a Changing World, Sydney, Australia. Retrieved January 3, 2015 from: http://conference.herdsa.org.au/2005/pdf/refereed/paper_300.pdf

Lawrie, G., Wright, A, Schultz, M., Dargaville, T., O.Brien, G., Bedford, S., Williams, M., Tasker, R., Dickson, H., & Thompson, C. (2013). Using formative feedback to identify and support first-year chemistry students with missing or misconceptions. *International Journal of the First Year in Higher Education*, 4(2) 111–6.

Lawrie, G., Schultz, M., & Wright, A. (2015). Developing a practitioner's tool to explore students' conceptions as they transition into tertiary chemistry: A case study of a longitudinal journey. Submitted to *Research in Science Education* (Feb 2015).

McInnes, C., James, R., & McNaught, C. (1995). *First year on campus: Diversity in the initial experiences of Australian undergraduates.* Melbourne: Centre for the Study of Higher Education, University of Melbourne.

McKenzie, K. & Schweitzer, R. (2001). Who succeeds at university? Factors predicting academic performance in first year Australian university students. *Higher Education Research & Development*, 20(1) 21–33. DOI: 10.1080/07924360120043621

Nicol, D. & Macfarlane-Dick, D. (2006). Formative assessment and self-regulated learning: A model and seven principles of good feedback practice. *Studies in Higher Education*, 31, 199–218. DOI: 10.1080/03075070600572090

Office of the Chief Scientist. (2012). *Mathematics, engineering & science in the national interest.* Canberra: Australian Government. Retrieved September 28, 2014 from: http://www.chiefscientist.gov.au/wp-content/uploads/Office-of-the-Chief-Scientist-MES-Report-8-May-2012.pdf

Office of the Chief Scientist. (2014). *Benchmarking Australian science, technology, engineering and mathematics.* Canberra: Australian Government. Retrieved January 19, 2015 from: http://www.chiefscientist.gov.au/wpcontent/uploads/BenchmarkingAustralianSTEM_Web_Nov2014.pdf

Pitkethly, A. & Prosser, M. (2001). The first year experience project: A model for university-wide change. *Higher Education Research & Development.* 20(2), 185–198. DOI: 10.1080/758483470

Rausch, J. L. & Hamilton, M. W. (2006). Goals and distractions: Explanations of early attrition from traditional university freshmen. *The Qualitative Report* 11(2) 317–334.

Schultz, M., Crow, J. M., & O'Brien, G. (2013). Outcomes of the chemistry discipline network mapping exercises: Are the threshold learning outcomes met? *International Journal of Innovation in Science and Mathematics Education.* 21(1) 81–91.

Tinto, V. (1987). From theory to action: Exploring the institutional conditions for student retention. Smart, J.C. (Ed), *Higher Education: Handbook of Theory and Research Volume 25.* New York: Springer. 51–89. DOI: 10.1007/978-90-481-8598-6_2

Tinto, V. (2006/2007). Research and practice of student retention: What next? *Journal of College Student Retention*, 8(1) 1–19. DOI: 10.2190/4ynu-4tmb-22dj-an4w

Torenbeek, M., Jansen, E., & Hofman, A. (2011). The relationship between first-year achievement and the pedagogical-didactical fit between secondary school and university. *Educational Studies*, 37(5) 557–568. DOI: 10.1080/03055698.2010.539780

Yorke, M. (2001). Formative assessment and its relevance to retention. *Higher Education Research & Development.* 20(2) 115–126. DOI: 10.1080/07294360120064385

ABOUT THE AUTHORS

Gwendolyn Lawrie is an Associate Professor in the School of Chemistry and Molecular Biosciences at the University of Queensland in St. Lucia, Queensland, Australia.

Anthony Wright is a Lecturer in the School of Education at the University of Queensland in St. Lucia, Queensland, Australia.

Madeleine Schultz is a Senior Lecturer in the School of Chemistry, Physics, and Mechanical Engineering at Queensland University of Technology in Brisbane, Queensland, Australia.

Tim Dargaville is a Senior Lecturer in the School of Chemistry, Physics, and Mechanical Engineering at Queensland University of Technology in Brisbane, Queensland, Australia.

Roy Tasker is a Professor in the School of Chemistry at the University of Western Sydney in Penrith, New South Wales, Australia.

Mark Williams is a Senior Lecturer and Academic Course Advisor in the School of Chemistry at the University of Western Sydney in Penrith, New South Wales, Australia.

Simon Bedford is a Lecturer in the School of Chemistry at the University of Wollongong in Wollongong, New South Wales, Australia.

Glennys O'Brien is a Senior Lecturer in the School of Chemistry at the University of Wollongong in Wollongong, New South Wales, Australia.

Christopher Thompson is a Lecturer in the School of Chemistry at Monash University in Clayton, Victoria, Australia.

3

Describing Instructional Practice and Climate: Two New Instruments

Emily M. Walter, Andrea L. Beach, Charles Henderson, and Cody T. Williams

IDENTIFICATION OF CHALLENGE

Most faculty have knowledge of evidence-based instructional practices and access to the resources to carry them out. Despite this, efforts to transform postsecondary instruction have met with only modest success (e.g., American Association for the Advancement of Science [AAAS], 2013). Institutional environments and structures may be one of the underlying barriers to changing instruction (Beach, Henderson, & Finkelstein, 2012; Henderson, Beach, & Finkelstein, 2011). One measure of an institutional environment is climate. Climate is a more immediately accessible and malleable construct than organizational culture, as it can be changed through policy or other administrative and organization-member actions. As such, climate is a productive conceptual frame to apply in research that attempts to inform policy and practice change initiatives (Schneider, Ehrhart, & Macey, 2013).

However, in order to measure the impact of change initiatives, it is paramount to have reliable and valid methods to measure climate and instructional practice (AAAS, 2013). The goal of this research study was to design and validate instruments that elicit (a) organizational climate for instructional improvement and (b) postsecondary instructional practices. The resulting surveys, SCII and PIPS, are reliable, interdisciplinary, and can collect data quickly from a large number of participants. In this paper, we share these research tools, explain our development and data collection processes, highlight preliminary results, and provide suggestions for use of the instruments.

RESEARCH STUDY

As part of a larger project on postsecondary instructional change, we have developed two instruments to elicit climate and instructional practices in higher education settings. In this section, we describe background literature,

411

conceptual frameworks, item development, scales, and validation of our surveys. We follow with a discussion of preliminary results and implications. The results we present in this chapter represent our thinking as of the *21st Century Transforming Institutions* conference in October 2014. We encourage interested individuals to contact our research team for the most relevant publications associated with this project.

Research Tool 1—Survey of Climate for Instructional Improvement (SCII)

Background

Climate can be described as either an individual (psychological) construct or as a property of an organization (Kozlowski & Klein, 2000) when individual perceptions are aggregated to the group level and consensus can be demonstrated (Dansereau & Alluto, 1990; James, Demaree & Wolf, 1993; James & Jones, 1974; Kozlowski & Hults, 1987). Since our research project focused on the influence of climate on postsecondary instructional practices, we chose to explore the institutional environment through the lens of organizational climate. This choice limits potentially idiosyncratic data and explores different questions than the work relating teaching practices and self-efficacy (e.g., Tschannen-Moran & Johnson, 2011).

Organizational climate is defined as the shared perceptions of organization members about elements of the organization. These elements influence individual attitudes and behaviors and include patterns of relationships, atmosphere, and organizational structures (Peterson & Spencer, 1990; Schneider, 1975, Schneider & Reichers, 1983; Schneider et al., 2013). Climate can operate on many different organizational levels (Kozlowski & Klein, 2000) and therefore is most useful when focused on a specific outcome—i.e., climate *for* something (Schneider, 1975). In our case, we were interested in *climate for instructional improvement*, which we define as the action or process of making changes in teaching with the goal of achieving the best possible learning outcomes. This change-making process includes the introduction or continued use of evidence-based instructional strategies, technologies, and/or curriculum.

Conceptual framework

We first examined the literature for theoretical and conceptual frameworks from which to develop the climate survey. The framework of faculty work elements identified by Gappa, Austin, and Trice (2007) was eventually chosen for its alignment with the aspects of climate that we were interested in. This framework consists of three aspects of faculty work experience (academic freedom and autonomy, collegiality, professional growth) and three characteristics of academic organizations (resources, rewards, leadership). An important strength of this framework for our purposes was that it aligned with related literature on workplace "climate for change" (Bouckenooghe, Devos, & Van den Broeck, 2009), the nature of academic work and workplaces (Massy, Wilger, & Colbeck, 1994), departmental teaching climate (Beach, 2002; Knorek, 2012), and leadership for teaching (Ramsden, Prosser, Trigwell, & Martin, 2007).

We identified seven components of climate for instructional improvement that could potentially be measured through survey by combining the Gappa et al. framework with related literature (Table 1). These seven components include: resources (Beach, 2002; Gappa et al., 2007, Knorek, 2012), rewards (Beach, 2002; Gappa et al., 2007; Knorek, 2012), professional development (Beach, 2002; Gappa et al., 2007), leadership (Beach, 2002; Bouckenooghe et al., 2009; Gappa et al., 2007; Ramsden et al., 2007), collegiality (Beach, 2002; Gappa et al., 2007; Massy et al., 1994), academic freedom and autonomy (Gappa et al., 2007), and general attitudes about students and teaching (Beach, 2002; Ramsden et al., 2007).

Item development

Items for the SCII were developed based on existing surveys when possible (Bouckenooghe et al., 2009; Hurtado, Eagan, Pryor, Whang, & Tran, 2011; Knorek, 2012; Ramsden et al., 2007) and self-generated when necessary. We sought to refer to group, rather than individual, perceptions as items were generated and revised, so that organization-level perceptions were properly represented (Glick, 1985). This approach involved changing the referent of existing items from the individual to the organizational level (e.g., "the instructors in my department think" rather than "I think"). We also revised existing items to refer to "instructors" instead of "faculty" and changed terms like "tenure" to "continued employment" since full-time, part-time, graduate student instructors were surveyed.

Scale

We purposefully chose a six-point Likert style scale for SCII that uses the following response options: strongly agree, agree, somewhat agree, somewhat disagree, disagree and strongly disagree. Six-point agree-disagree scales are considered preferable to 4-point scales, as they generate better variance (Bass, Cascio, & O'Connor, 1974). There is no neutral point on the scale, as forcing agreement or disagreement avoids an increase in participants claiming "no opinion" when they actually have one (Bishop, 1987; Johns, 2005).

TABLE 1. Operational Definitions and Sources of Organizational Climate Components Used to Develop Items on the SCII.

Component	Definition	Concept Source	Definition Source
Rewards	Recognition of teaching excellence through awards or job security measures.	Beach, 2002 Knorek, 2012	Self-generated
Resources	Tools necessary for instructional improvement, including funding, office space, equipment, and support services.	Gappa et al., 2007 Beach, 2002	Gappa et al., 2007 (modified)
Professional Development	Opportunities that enable instructors to broaden their knowledge, abilities, and skills to address challenges, concerns, and needs, and to find deeper satisfaction in their work.	Gappa et al., 2007 Beach, 2002 Knorek, 2012	Gappa et al., 2007, p. 280
Collegiality	Opportunities for instructors to feel they belong to a mutually respectful community of colleagues who value their contributions, and to feel concern for their colleagues' well-being.	Massy et al., 1994 Gappa et al., 2007 Bouckenooghe et al., 2009	Gappa et al., 2007, p. 305
Academic Freedom and Autonomy	Right of all instructors to teach without undue institutional interference, including freedom in course content and instructional practices.	Gappa et al., 2007	Gappa et al., 2007, p. 140–141 (modified)
Leadership	Policies, actions, or expectations established by the formal leader of the department that communicate the value of teaching and instructional improvement.	Beach, 2002 Bouckenooghe et al., 2009	Self-generated
Shared perceptions about Students and Teaching	Shared perceptions of the individuals in a department regarding student characteristics and instructional practices that may influence improvements in teaching.	Beach, 2002 Ramsden et al., 2007 Hurtado et al., 2011	Self-generated

Research Tool 2—Postsecondary Instructional Practices Survey (PIPS)

Background

There are multiple ways to measure the teaching practices of postsecondary instructors, including self-report surveys and observational protocols. We see surveys as a preferable method, since observational protocols (e.g., RTOP, Piburn, Sawada, Falconer, Turley, Benford, & Bloom, 2000; TDOP, Hora, Oleson, & Ferrare, 2012) require training and expertise, are expensive and difficult to implement at scale, and risk reliability issues.

Although 10 surveys of instructional practices were summarized in a recent AAAS report (AAAS, 2013), none were designed to elicit teaching practices (and only teaching practices) from an interdisciplinary group of postsecondary instructors. Most existing instruments are designed for use in a particular discipline: physics and engineering (Borrego, Cutler, Prince, Henderson, & Froyd, 2013; Brawner, Felder, Allen, & Brent, 2002; Dancy & Henderson, 2010), chemistry and biology (Marbach-Ad Schaefer-Zimmer, Orgler, Benson, & Thompson, 2012), geosciences (MacDonald, Manduca, Mogk, & Tewksbury, 2005), or statistics (Zieffler, Park, Garfield, delMas, & Bjornsdottir, 2012). Other instruments elicit teaching beliefs or goals for student learning, and not actual teaching practice (e.g., ATI; Trigwell & Prosser, 2004). The remaining surveys are interdisciplinary and elicit teaching practices, but elicit a very wide range of faculty practices beyond teaching. These include the FSSE (Center for Postsecondary Research, 2012), HERI (Hurtado, Eagan, Pryor, Whang, & Tran, 2011), and NSOPF (National Center for Educational Statistics, 2004). Two of these are only available on a proprietary basis (NSOPF, HERI).

Seeking an interdisciplinary, non-proprietary, and succinct survey of postsecondary instructional practices, we designed a new instrument. The resulting survey, PIPS, is designed to be easy-to-use, non-evaluative, and collect data quickly from a large number of participants.

Conceptual framework

In absence of an appropriate instrument, we turned to the empirical and theoretical literature about evidence-based teaching practices. There is no conceptual model of instructional practice despite excellent literature reviews describing research on instructional practices (e.g., Pascarella & Terenzini, 1991; 2005). Without a model from which to develop instructional practice items, we shaped the dimensions of our instrument by finding themes among (a) developed instruments, (b) teaching observation protocols and (c) patterns in research on instructional practice. We compiled 153 items by combining all available questions and literature patterns from two published instruments (FSSE, ATI), two observational protocols (RTOP, TDOP), and comprehensive literature reviews (Iverson, 2011; Meltzer & Thornton, 2012; Pascarella & Terenzini, 1991; 2005).

From an initial set of 153 questions, we reduced the number of questions by removing redundant items, items that did not refer to actual teaching practices (i.e., items that elicited beliefs about teaching or intent to teach in a given manner), and lists of generalized practices (e.g., "lecture", "lecture with demonstration", "multiple choice tests"). The final set of 24 items was categorized into four components (Table 2), revised for clarity and to reduce the potential of eliciting socially acceptable responses.

Intended context

PIPS items are designed for respondents to describe teaching the largest enrollment, lowest level course they have taught in the last two years. We believe this setting is one of the most challenging in which to use evidence-based instructional strategies in comparison to smaller enrollment, higher level courses. This setting is also of most concern to researchers and others involved with instructional change (AAAS, 2013).

TABLE 2. Operational Definitions and Sources of Instructional Practice Concepts Used to Develop Items on the PIPS

Component	Definition	Definition Source
Instructor-student interactions	Practices that influence the classroom relationship between the instructor and students (e.g., the role of the instructor in class sessions).	Self-generated
Student-content interactions	Practices that influence how students interact with course concepts (e.g., reflection activities, connecting concepts to students' lives).	Self-generated
Student-student interactions	Practices that influence the classroom interactions among students. These approaches include classroom discourse, small group work, and other collaborative approaches.	Self-generated
Assessment	Practices that provide feedback to students and the instructor on what, how much, and how well students are learning (Angelo & Cross, 1993). Assessment practices include what is assessed, how often students are assessed, how instructors use assessment data, and grading.	Angelo and Cross, 1993, p. 4 (modified)

Scale

PIPS uses a 5-point Likert style scale as recommended by Bass, Cascio, & O'Conner (1974), with options including: not at all descriptive, minimally descriptive, somewhat descriptive, mostly descriptive and very descriptive of my teaching. There is no neutral point on the scale in order to generate more variability in the data (Bishop, 1987; Johns, 2005).

Field testing

Face validity

An instrument has face validity if, from the perspective of participants, it appears to have relevance and measures its intended subject. This requires developers to use clear and concise language, avoid jargon, and write items to the education and reading level of the participants (DeLamater, Miles, & Collett, 2014). We pilot tested the PIPS and SCII in their entirety with a representative sample of instructors in order to achieve face validity with an interdisciplinary group of instructors. We refined items based on the feedback of these individuals prior to implementing the instruments at scale. The reader can note some of our wording changes in our previous sections on *Item Development* as relevant to the SCII and PIPS.

Content validity

Content validity requires surveys to properly represent aspects of the subject of interest (e.g,. teaching practices). A panel of subject matter experts was used to access the content validity of both SCII and PIPS (as recommended by Anastasi & Urbina, 1997). As with the pilot testing with postsecondary instructors, this process allowed for items to be evaluated for clarity and revised. New items were added, several were removed, and the structure and operational definition of each component was further developed.

Construct validity

This refers to the degree an instrument is consistent with theory (Coons, Rao, Keininger, & Hays, 2000); this is often achieved through confirmatory and/or exploratory factor analyses (Thompson & Daniel, 1996). We completed an iterative process of confirmatory and exploratory factor analyses to refine the constructs (see *Analyses*). The constructs presented in this chapter represent our thinking as of the *21st Century Transforming Institutions* conference in October 2014. As such, the constructs herein should be seen as tentative, as we are in the process of publishing on the psychometric development of each instrument.

Implementation and analysis

We collected pilot data from 889 postsecondary instructors at four institutions in the United States (Table 3). Two of these institutions (A and B) completed both PIPS and SCII, and the other institutions completed only PIPS (C and D).

Analysis followed Floyd and Widaman's (1995) recommendations for instrument development and refinement. We first ran exploratory factor analyses (EFA) using maximum-likelihood extraction with Promax rotation to identify dimensions of climate and teaching practice. We made note of items that consistently loaded together across institutions, since instructional practices and climate had the potential to manifest differently at different institutions.

We subsequently ran confirmatory factor analyses (CFA) using *SPSS AMOS 22.0* to create structural equation models based on our a priori categorization of the items and the results of the exploratory factor analyses. We refined the models based on item modification indexes and regression loadings produced by *AMOS* to reach an acceptable chi-squared/df value below 5.0, a CFI near 0.90, and RMSEA below 0.10 (Byrne, 2013). Using the SCII and PIPS constructs that emerged from the modeling process, we created individual construct scores by adding the sum of the items in each construct. Construct scores were generated only if a participant completed all of the items contributing to the construct. We

TABLE 3. Demographic and Sample Size Information for the Surveyed Institutions

	Institution A	Institution B	Institution C	Institution D
N	214	164	87	424
Departments Surveyed	13	9	10	40
Data Sources	PIPS; SCII	PIPS; SCII	PIPS	PIPS
Disciplines	STEM and Applied Sciences	STEM	Biological Sciences	All Departments
Instructors Surveyed	Full- and part-time faculty	Full- and part-time faculty	Full-time faculty only	Full- and part-time faculty
U.S. Region	Great Lakes	Mid-Atlantic	South Atlantic	Mountain West
Control	Public	Public	Public	Public
Carnegie Classification	Research university; High research activity	Research university; Very high research activity	Research university; Very high research activity	Masters College or University (larger program)
Student Population	25K	28K	34K	22K

lastly ran statistical comparisons among mean construct scores for each institution and among departments within an institution.

RESULTS

This section includes instrument reliability scores, a list of the constructs for each instrument, and select differences in institutional and department construct means. We do not include all findings to meet length requirements. In addition, we remind the reader that the constructs presented in this chapter are representative of our thinking in October 2014, and may not represent the current and finalized constructs for each instrument.

Reliability and Construct Structure

In this chapter we present the October 2014 versions of the instruments as presented at the 21st Century Transforming Institutions conference. These may not represent the final published version of these instruments. In the October 2014 versions, the SCII had 26 items within six constructs and PIPS had 20 items within five constructs. Both instruments had high internal reliability (a > 0.8) and could not be improved with removal of additional items (Table 4).

Climate for Instructional Improvement Means by Institution and Department

Climate for instruction improvement as elicited by SCII factored into six distinct constructs in our EFA and CFA analyses. In the order of their contribution to overall variance (Table 5), the constructs include leadership (six items), collegiality (six items), resources (three items), professional development (PD, three items), autonomy (three items), and respect (five items) (see *Appendix*).

Climate construct means significantly differed between Institutions A and B for each construct ($p < .0001$), with the exception of professional development ($p = 0.944$, Table 5). Climate means also significantly differed among departments within each institution. However, these differences were rarely significant in post-hoc comparisons. One notable exception is the significant difference in the mean leadership scores between the Mathematics Department and Industrial and Manufacturing Engineering Department at Institution A (Figure 1).

Significant differences in climate means by institution are detailed in Table 5. We also present a graph of departmental means for one of the constructs that shows instructional clusters of department means (Figure 1). In this case, we chose a plot of the leadership construct as it contributed most to overall variance (44.51% for this sample).

TABLE 4. Reliability Statistics for the October 2014 Versions of the Survey of Climate Instructional Improvement (SCII) Survey and the Postsecondary Instructional Practices Survey (PIPS)

	Survey of Climate for Instructional Improvement (SCII)	Postsecondary Instructional Practices Survey (PIPS)
Number of Items	26	20
Constructs	6	5
N	300	661
Reliability (a)	0.943	0.812

TABLE 5. Mean Climate Construct Scores by Construct and Institution, as Measured by the Survey of Climate for Instructional Improvement (SCII).

	Respect	Autonomy	PD	Resources	Collegiality	Leadership
# Items	5	3	3	3	6	6
Institution A M (SD)	2.69 (1.01)	2.75 (0.87)	3.74 (1.06)	3.08 (1.01)	2.97 (0.92)	2.65 (0.99)
Institution B M (SD)	4.25 (0.91)	4.14 (0.67)	3.75 (0.94)	4.19 (0.98)	4.03 (0.95)	4.05 (0.97)
t-test p-value	****	****	0.944	****	****	****

Scale. 1 = strongly disagree; 2 = disagree; 3 = somewhat disagree; 4 = somewhat agree; 5 = agree; 6 = strongly agree.

Note. **** = $p < .0001$

Instructional Practices by Institution and Department

Instructional practices factored into five distinct constructs by our EFA and CFA analyses. In the order of their contribution to overall variance (Table 6), the constructs include: instructor-student interactions (four items), student-student interactions (four items), student-content interactions, formative assessment (four items), and summative assessment (four items). PIPS items organized by construct are provided in the *Appendix*.

The instructional practice construct means significantly differed among Institutions A, B, C and D for each construct ($p < .01$, Table 6). Instructional practice means also significantly differed among departments within each institution. However, these differences were rarely significant in post-hoc comparisons. One notable exception is a significant difference in the mean leadership scores between the Mathematics Department and Industrial and Manufacturing Engineering Department at Institution A (Figure 1).

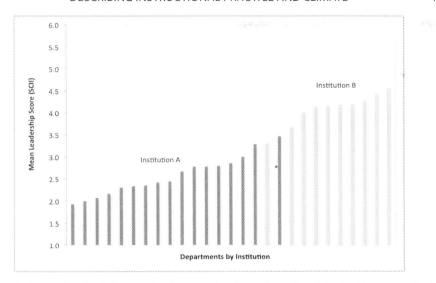

FIGURE 1. Mean leadership construct scores by department and institution as collected by the Survey of Climate for Instructional Improvement (SCII). Departments are listed in order of lowest to highest mean leadership score. Scale. 1 = strongly disagree; 2 = disagree; 3 = somewhat disagree; 4 = somewhat agree; 5 = agree; 6 = strongly agree.

Significant differences in climate means by institution are detailed in Table 5. We also present a figure that displays institutional clusters for mean department teaching practice scores (Figure 2). As with the climate constructs by department, we chose to create the figure for the construct that contributed most to overall variance (student-student interactions, 22.83% for this sample).

LESSONS LEARNED AND TRANSFERABILITY

Understanding and measuring differences in climate and teaching practices in higher education settings enables users to identify levers for improving teaching, thereby better planning future change initiatives. Our research documents support for instruments that can differentiate among elements of climate and instructional practices of postsecondary instructors. The instruments are reliable, easy-to-use, and can quickly collect data from a large number of participants. Furthermore, the instruments are designed modularly so that they can be used together or separately to understand the current situation and/or document changes over time through repeated measurements.

TABLE 6. Postsecondary Instructional Practices Survey (PIPS) Mean Scores by Construct and Institution

Summative Assessment	Formative Assessment	Student-Content Interactions	Student-Student Interactions	Instructor-Student Interactions	
4	4	4	4	4	# Items
2.23 (0.83)[d]	2.48 (0.91)[f]	1.67 (1.10)	2.36 (0.76)[e]	2.72 (0.98)[e]	Institution A M (SD)
2.09 (0.83)[d]	2.70 (0.70)[b]	1.61 (1.14)	2.55 (0.70)[e]	2.97 (0.73)[b]	Institution B M (SD)
1.62 (1.15)[a]	2.17 (1.17)[c]	1.26 (1.29)	2.21 (0.97)	2.45 (0.92)	Institution C M (SD)
2.77 (0.72)[a]	2.85 (0.67)[b]	2.55 (0.98)[a]	2.09 (0.85)[c]	2.25 (0.82)[c]	Institution D M (SD)

Scale. 0 = not at all like my teaching; 1 = minimally descriptive of my teaching, 2 = somewhat descriptive of my teaching, 3 = mostly descriptive my teaching, 4 = very descriptive of my teaching.
Note. [a] Significantly different than the other three institutions ($p < .05$), [b] Significantly higher ($p < .05$) than the two lowest scoring institutions, [c] Significantly lower ($p < .05$) than the two highest scoring institutions, [d] Significantly different ($p < .05$) than the lowest and highest scoring institution, [e] Significantly higher ($p < .05$) than the lowest scoring institution, [f] Significantly lower ($p < .05$) than the highest scoring institution.

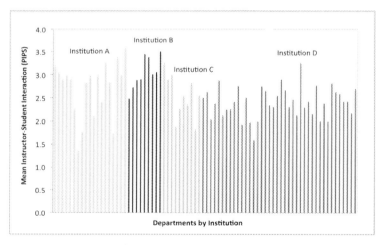

FIGURE 2. Mean instructor-student interaction scores by department and institution as collected by the Postsecondary Instructional Practices Survey (PIPS).
Scale. 0 = not at all like my teaching; 1 = minimally descriptive of my teaching, 2 = somewhat descriptive of my teaching, 3 = mostly descriptive my teaching, 4 = very descriptive of my teaching.

Unique Features of Our Instruments

Although at least 10 surveys of instructional practices (AAAS, 2013) are available, none are designed to elicit teaching practices (and only teaching practices) for an interdisciplinary group of postsecondary instructors. The survey is also non-evaluative, designed for respondents to score practices as descriptive of their teaching without judging the quality of these practices. Furthermore, PIPS is concise, non-proprietary, and designed with clear and consistent item scales.

The SCII is unlike any other instrument available. Although other instruments elicit different elements of climate including organizational climate for change (Bouckenoghe et al., 2009) and faculty teaching climate (particularly rewards and resources; Knorek, 2012), SCII is built in alignment with the essential elements of faculty work described by Gappa et al. (2007). Our results not only provide empirical support for the factors described by Gappa et al. (2007), but it also elicits constructs that could serve as levers for change in planned initiatives.

Identifying Differences with SCII and PIPS

Although not presented in detail in this paper, our findings align with those identified by other instruments. Practices in the instructor-student interaction construct were more descriptive of male instructors than female instructors. This construct includes practices such as "students sitting and taking notes" and "teaching with the assumption that students have little incoming knowledge." Henderson, Dancy, and Niewiadomska-Bugaj (2012) and Kuh, Laird, and Umbach, (2004) likewise found women using fewer instructional practices of this nature.

We also found rank-based differences in teaching practices and in perceptions of department climate similar to those in the literature. For example, part-time instructors reported less flexibility in their teaching methods and fewer teaching resources than their tenure-track counterparts (e.g. Gappa & Leslie, 1993). Graduate students were also less likely to claim assessment practices (both formative and summative) were descriptive of their teaching, perhaps due to a lack of autonomy to develop these assessment practices.

Unique to our study are institutional clusters in teaching practices and organizational climate for instructional improvement (e.g. Figure 1). These clusters may indicate that some elements are more normative at the institution level than the department level, with the exception of certain disciplines. Institution A, which is less research intensive than Institutions B and C by Carnegie classification, reported using more instructor-student interactions. We also found a significant negative correlation ($p < 0.01$) between traditional teaching practices and evidence-based teaching practices, which supports the logical argument that use of one relates to less use of the other.

Future Work

One of our next steps will be to complete hierarchical linear models to understand the sources of variance within the data. This will identify contributions to variance at levels higher than the individual, including department and institution-level variance.

We will also be triangulating the results of our instructional practices survey with teaching observation data (collected using the TDOP) and interviews with instructors. These observations will provide additional support for our constructs and help gain further insight into their organizational climate and undergraduate instructional practices.

Access to the instruments

The instruments are available in their full pilot versions and with items organized into constructs from our website: http://homepages.wmich.edu/~chenders/Publications/. We request that if you plan to use the instruments, please use them in their entirety and please share the data with our research team for further refinement of the instruments.

REFERENCES

American Association for the Advancement of Science (AAAS). (2013). *Measuring STEM teaching practices: A report from a national meeting on the measurement of undergraduate science, technology, engineering, and mathematics (STEM) teaching.* Washington, DC: Author.

Anastasi, A., & Urbina, S. (1997). *Psychological testing* (7th ed.). Upper Saddle River, NJ: Prentice Hall.

Bass, B. M., Cascio, W. F., & O'Connor, E. J. (1974). Magnitude estimations of expressions of frequency and amount. *Journal of Applied Psychology, 59,* 313–320.

Beach, A. L. (2002). *Strategies to improve college teaching: The role of different levels of influence on faculty instructional practices.* Unpublished doctoral dissertation, Michigan State University, East Lansing, MI.

Beach, A. L., Henderson, C., & Finkelstein, N. (2012). Facilitating change in undergraduate STEM education. *Change: The Magazine of Higher Learning, 44*(6), 52–59. doi: 10.1080/00091383.2012.728955

Bishop, G. F. (1987). Experiments with the middle response alternative in survey questions. *Public Opinion Quarterly, 51,* 220–232.

Borrego, M., Cutler, S., Prince, M., Henderson, C., & Froyd, J. (2013). Fidelity of implementation of Research-Based Instructional Strategies (RBIS) in engineering science courses. *Journal of Engineering Education, 102*(3).

Bouckenooghe, D., Devos, G., & Van den Broeck, H. (2009). Organizational change questionnaire—Climate of change, processes, and readiness: Development of a new instrument. *The Journal of Psychology, 143*, 559–599. doi: 10.1080/00223980903218216

Brawner, C. E., Felder, R. M., Allen, R., & Brent, R. (2002). A survey of faculty teaching practices and involvement in faculty development activities. *Journal of Engineering Education–Washington, 91*, 393–396.

Byrne, B. M. (2013). *Structural equation modeling with AMOS: Basic concepts, applications, and programming*, 2nd Ed. New York, NY: Routledge.

Dancy, M., & Henderson, C. (2007). Barriers to the use of research-based instructional strategies: The influence of both individual and situational characteristics. *Physical Review Special Topics—Physics Education Research, 3*, 1–14. doi: 10.1103/PhysRevSTPER.3.020102

Dansereau, F., & Alluto, J. A. (1990). Level-of-analysis issues in climate and culture research. In B. Schneider (Ed.), *Organizational climate and culture* (pp. 193–236). San Francisco, CA: Jossey-Bass.

DeLamater, J. D., Myers, D. J., & Collett, J. L. (2014). *Social psychology* (8th ed.). Boulder, CO: Westview Press.

Floyd, F. J., & Widaman, K. F. (1995). Factor analysis in the development and refinement of clinical assessment instruments. *Psychological Assessment, 7*, 286–299.

Gappa, J. M., & Leslie, D. W. (1993). *The invisible faculty: Improving the status of part-timers in higher education*. San Francisco, CA: Jossey-Bass.

Gappa, J. M., Austin, A. E., & Trice, A. G. (Eds.). (2007). *Rethinking faculty work: Higher education's strategic imperative*. San Francisco, CA: Jossey-Bass.

Glick, W. H. (1985). Conceptualizing and measuring organizational and psychological climate: Pitfalls in multilevel research. *Academic Management Review, 10*, 601–616.

Henderson, C., Beach, A. L., & Finkelstein, N. (2011). Facilitating change in undergraduate STEM instructional practices: An analytic review of the literature. *Journal of Research in Science Teaching, 48*, 952–984. doi: 10.1002/tea.20439

Henderson, C., Beach, A. L., & Finkelstein, N. (2012). Promoting high quality teaching practices in higher education: Lessons learned from the USA. In W. Bienkowski, J. C. Brada & G. Stanley (Eds.), *The university in the age of globalization*. New York, NY: Palgrave Macmillan.

Henderson, C., Dancy, M., & Niewiadomska-Bugaj, M. (2012). Use of research-based instructional strategies in introductory physics: Where do faculty leave the innovation-decision process? *Physical Review Special Topics—Physics Education Research, 8*(2), 020104. doi:10.1103/PhysRevSTPER.8.020104

Hora, M. T., Oleson, A., & Ferrare, J. J. (2012). *Teaching Dimensions Observation Protocol (TDOP) user's manual.* Madison, WI: Wisconsin Center for Education Research, University of Wisconsin-Madison.

Hurtado, S., Eagan, K., Pryor, J. H., Whang, H., & Tran, S. (2011). *Undergraduate teaching faculty: The 2010–11 HERI faculty survey.* Los Angeles, CA: Higher Education Research Institute.

Iverson, H. L. (2011). *Undergraduate physics course innovations and their impact on student learning.* PhD Dissertation, University of Colorado, Boulder, CO.

James, L. R., & Jones, A. P. (1974). Organizational climate: A review of theory and research. *Psychological Bulletin, 81,* 1096–1112.

James, L. R., Damaree, R. G., & Wolf, G. (1993). R_{WG}: An assessment of within-group inter-rater agreement. *Journal of Applied Psychology, 78,* 306–309.

Johns, R. (2005). One size doesn't fit all: Selecting response scales for attitude items. *Journal of Elections, Public Opinion, & Parties, 15,* 237–264. doi: 10.1080/13689880500178849

Knorek, J. K. (2012). *Faculty teaching climate: Scale construction and initial validation.* Unpublished PhD dissertation, University of Illinois, Urbana, IL.

Kozlowski, S. W. J., & Hults, B. M. (1987). An exploration of climates for technical updating and performance. *Personnel Psychology, 40,* 539–563.

Kozlowski, S. W. J., & Klein, K. J. (2000). A levels approach to theory and research in organizations: Contextual, temporal, and emergent processes. In K. J. Klein & S. W. J. Kozlowski (Eds.), *Multilevel theory, research, and methods in organizations* (pp. 3–90). San-Francisco, CA: Jossey-Bass.

Kuh, G. D., Laird, T. F. N., & Umbach, P. D. (2004). Aligning faculty activities and student behavior: Realizing the promise of greater expectations. *Liberal Education, 90,* 24.

MacDonald, R. H., Manduca, C. A., Mogk, D. W., & Tewksbury, B. J. (2005). Teaching methods in undergraduate geoscience courses: Results of the 2004 "On the Cutting Edge Survey" of U.S. faculty. *Journal of Geoscience Education, 53,* 237–252.

Marbach-Ad, G., Schaefer-Zimmer, K. L., Orgler, M., Benson, S., & Thompson, K. V. (2012). *Surveying research university faculty, graduate students and undergraduates: Skills and practices important for science majors.* Paper presented at the annual meeting of the American Educational Research Association (AERA), Vancouver, Canada.

Massy, W. F., Wilger, A. K., & Colbeck, C. (1994). Overcoming "hollowed" collegiality. *Change, 26,* 10e20.

National Center for Education Statistics. (2004). *National study of postsecondary faculty (NSOPF).* Washington, DC: Author. Retrieved from http://nces.ed.gov/surveys/nsopf/

Pascarella, E. T., & Terenzini, P. T. (1991). *How college affects students.* San Francisco, CA: Jossey-Bass.

Pascarella, E. T., & Terenzini, P. T. (2005). *How college affects students. Volume 2. A third decade of research*. San Francisco, CA: Jossey-Bass.

Peterson, M. W., & Spencer, M. G. (1990). Understanding academic culture and climate. In W. G. Tierney (Ed.), *Assessing academic climates and cultures: New directions for institutional research, No. 68* (pp. 3–18). San Francisco, CA: Jossey-Bass.

Piburn, M., Sawada, D., Falconer, K., Turley, J., Benford, R., & Bloom, I. (2000). *Reformed Teaching Observation Protocol (RTOP)*. Tempe, AZ: Arizona Collaborative for Excellence in the Preparation of Teachers.

Prince, M., Borrego, M., Henderson, C., Cutler, S., & Froyd, J. E. (2013). Use of research-based instructional strategies in core chemical engineering courses. *Chemical Engineering Education, 47*, 27–37.

Ramsden, P., Prosser, M., Trigwell, K., & Martin, E. (2007). University teachers' experiences of academic leadership and their approaches to teaching. *Learning and Instruction, 17*, 140–155. doi: 10.1016/j.learninstruc.2007.01.004

Schneider, B. (1975). Organizational climates: An essay. *Personnel Psychology, 56*, 211–217.

Schneider, B., & Reichers, A. E. (1983). On the etiology of climates. *Personnel Psychology, 36*, 19–39.

Schneider, B., Ehrhart, M. G., & Macey, W. H. (2013). Organizational climate and culture. *Annual Review of Psychology, 64*, 361–388. doi:10.1146/annurev-psych-113011-143809

Trigwell, K., & Prosser, M. (2004). Development and use of the Approaches to Teaching Inventory. *Educational Psychology Review, 16*, 409–424.

Tschannen-Moran, M., & Johnson, D. (2011). Exploring literacy teachers' self-efficacy beliefs: Potential sources at play. *Teaching and Teacher Education, 27*, 751–761.

Walczyk, J. J., & Ramsey, L. L. (2003). Use of learner-centered instruction in college science and mathematics classrooms. *Journal of Research in Science Teaching, 40*, 566–584. doi: 10.1002/tea.10098

Zieffler, A., Park, J., Garfield, J., delMas, R., & Bjornsdottir, A. (2012). The statistics teaching inventory: A survey on statistics teaching classroom practices and beliefs. *Journal of Statistics Education, 20*, 1–29.B

ABOUT THE AUTHORS

Emily M. Walter is an Assistant Professor of Biology at California State University in Fresno, CA. She was also a post-doctoral researcher at the Center for Research on Instructional Change in Postsecondary Education (CRICPE) at Western Michigan University in Kalamazoo, MI.

Andrea L. Beach is the Co-Director of the Center for Research on Instructional Change in Postsecondary Education (CRICPE) and the Director of Faculty Development in the Department of Educational Leadership, Research, and Technology at Western Michigan University in Kalamazoo, Michigan.

Charles Henderson is the Co-Director of the Center for Research on Instructional Change in Postsecondary Education (CRICPE) and Professor of the Department of Physics and the Mallinson Institution for Science Education at Western Michigan University in Kalamazoo, Michigan.

Cody T. Williams is a Doctoral Candidate in the Mallinson Institution for Science Education and the Center for Research on Instructional Change in Postsecondary Education (CRICPE) at Western Michigan University in Kalamazoo, Michigan.

4

The Roles of Data in Promoting Institutional Commitment to Undergraduate STEM Reform: The AAU STEM Initiative Experience

James Fairweather, Josh Trapani and Karen Paulson

One barrier to improving post-secondary teaching and learning in undergraduate STEM, indeed in all disciplines, is the conflicting views about relevant types and uses of data in reform efforts. Most of the literature on STEM teaching and learning focuses on individual faculty members and the performance of their students in the classroom (National Research Council [NRC], 2012). This micro-level focus is consistent with the findings of Seymour and Hewitt (Seymour & Hewitt, 1997) who found that student experiences in the classroom had large (and often adverse) effects on retention in the major. The intent of this micro-level focus is to redress the shortcomings of traditional instructor-centered instruction and to promote active student engagement in their learning. This approach has the advantage of fine-tuning strategies to meet specific instructional goals and settings using widely available data about effective evidence-based instruction (NRC, 2012). Yet this approach fails to address wider aspects of the reform process, which must take into account the departmental and institutional environments, as well as those of the academic disciplines. Moreover, reforms that rely on idiosyncratic strategies, while appealing to the individual teacher, may not easily lead to aggregate measures of program or departmental effectiveness.

In contrast, decision-makers at the federal, state, foundation, and institutional levels (and at the college and departmental levels, for that matter) often ask questions about overall effectiveness of instructional programs and strategies such as: Did the program work? Was it worth the cost? How does it compare with other programs? Here, the focus is on ways to seek common information across disparate units (departments, colleges, institutions) to make summative judgments about effects, often for decisions about resource allocation. Because the more idiosyncratic micro-level classroom information meant to improve instruction does not easily aggregate into measures of overall effectiveness, decision-makers often are left with cross-department and cross-institution measures that can oversimplify the context and effects of the reforms. Even seemingly straightforward calculations such as "retention in the major" are made complex when institutions have distinct definitions of the major in question, when student formal selection of major can vary both within and across institutions, and when the answer varies depending on how many years of student progress is monitored. Structural differences in how institutions organize their academic disciplines also can make causal inferences about institutional actions invalid. For example, for phenomena likely affected by administrative actions at the college level, it is difficult to compare cross-institutional effects when some institutions place all humanities, social sciences, and natural sciences into a single college and others organize them into three distinct colleges. Common measures have the advantage of a seemingly uniform metric; that apparent advantage can dissipate if complex phenomena are forced into generic, often ill-fitting categories.

In the AAU Initiative, we have attempted to develop and work with both types of data. We are interested in helping the eight project sites (and indeed, all AAU institutions) improve their teaching and learning by adopting evidence-based best practices. We are, however, also interested in finding out what worked and what might be transferred successfully to other institutions. For the overall AAU Initiative to work, we had to find ways to defuse the potential conflict between locally useful classroom-level information and broader measures of program effects (sometimes this conflict happened within a single institution!). In this paper, we discuss a parallel data collection strategy using two sources of information. The first source is qualitative information collected on visits to the eight AAU project sites, including examples of how student performance data have been used as evidence in departmental adoption of curricular reforms. The second source, quantitative in nature, is a cross-site instrument on faculty teaching practices and attitudes. The survey was sent to all AAU institutions. Responses were received from faculty members from both project and non-project AAU institutions. The non-project institutions represented a wide variety in their current involvement in STEM reforms. Throughout, we focused on AAU's obligation to determine the overall effectiveness of the project while allowing for local variation in project activities, cultures, and structures.

SITE-SPECIFIC QUALITATIVE DATA

AAU carried out visits to each project site during the first year of implementation. These visits were meant to identify challenges and possible solutions to implementing project activities. Table 1 shows the questions asked of key individuals, including project leaders, department chairs and deans, and provosts. The latter groups are considered keys to future dissemination and institutionalization efforts. These implementation data helped project sites and AAU identify a baseline for pre-/post- comparisons of project initiatives. Site visits also started the process of building trust between AAU and project sites—especially that AAU staff were on site to better understand the local institutional environment, rather than to evaluate project work—and beginning cross-institution collaboration.

The site visits made apparent the need for more than aggregate cross-institution data in assessing overall project effects. The substantial historical, cultural, and structural differences between institutions and the wide variety of project objectives made it clear to AAU staff that part of the assessment of overall effectiveness must rely on a meta-analysis of locally relevant information. Accordingly, AAU and local project leaders agreed that annual project reports would address evaluation of implementation and effects including both common cross-institution data (discussed in the next section) and idiosyncratic data relevant for local use. Such local data would include student-level analyses where relevant. AAU staff then would carry out a meta-analysis of the reports to determine effect sizes for some of the more complex measures of cultural change, e.g., changes in promotion and tenure criteria.

The site visits also revealed an important and unanticipated use of data in local decision-making. Most literature on STEM teaching and learning assumes that the evidence of effectiveness is best directed to the teacher (or teaching assistant) who then can use the information to improve instructional outcomes. Although we found this assumption to have merit, we also identified an important political element in curricular decisions. As one example, we found that data about the performance of students in *subsequent courses* convinced the departmental faculty as a whole to approve a reformed sequence in chemistry. The focus was not on within-class improvement but on how well the students performed later in the curriculum. These data are crucial to studies of institutionalization and are also likely to be idiosyncratic to each institution (or department).

TABLE 1. Site Visit Protocol

INTERVIEW PROTOCOL FOR PROJECT TEAM LEADERSHIP
We wish to understand the project at a deeper level, help situate and align local activities with the national initiative, demonstrate AAU's support, and address questions.   • What is the plan for implementation?   • What is the current progress toward implementing the project?   • With the launch, have they confronted unanticipated challenges or opportunities? Have changes occurred to the plan/scope of work? Why? How are they adapting?   • What activities, types of support, and the like are the institution, college, and department providing to help the project succeed?

TABLE 1. Site Visit Protocol (continued)

INTERVIEW PROTOCOL FOR RELEVANT DEPARTMENT CHAIRS AND DEANS
- Please tell us about yourself and about your role and responsibilities in the AAU STEM Initiative. Looking to understand:
  - What is your personal belief about the importance of reforms in undergraduate STEM educational reform?
  - What is your buy-in/commitment to their campus project?
- How do you perceive faculty attitudes toward using evidence-based instructional practices?
- Has the AAU STEM Initiative provided a new forum for conversations about teaching and learning?
- What is your sense of broad-based faculty support within the departments for the project?
- Can you tell us about any changes in the department's program and in how courses are taught?
- What kind of data does the department have/gather about the teaching practice of individual faculty members? How does that relate to promotion/tenure?
- What is the status of teaching and learning infrastructure (e.g., facilities, technology) in terms of facilitating the use of evidence-based teaching practices?

INTERVIEW PROTOCOL FOR PROVOSTS
- What is the current campus climate for change in undergraduate STEM teaching and learning?
- Considering departments are the locus for change in the AAU Initiative, what are institutional efforts to support changes to teaching within the STEM departments?

CROSS-INSTITUTION QUANTITATIVE DATA

At the federal level, where AAU does most of its work as a higher education association, there is a strong push for institutions to provide data in a standardized, comparable format for both enhanced accountability and consumer information purposes. A similar push for standardized data reporting exists in many states. AAU, in its discussions with funders, policymakers, and others, needs to report on the progress made by the Undergraduate STEM Initiative across multiple institutions. This is a challenge, in part, because of the considerable variation among AAU members in their enactment of making systemic improvement to undergraduate STEM teaching and learning. AAU's 62 member universities represent different points on a continuum in enacting systemic change in undergraduate STEM teaching and learning. AAU members also vary in size (enrollment, number of faculty members), geographic region, institutional control (public/private), and student demographics, among other factors. AAU's eight project sites were deliberately chosen to represent this diversity. Additionally, the project sites are focusing on a variety of pedagogical issues across a range of disciplines.

AAU's objective was to collect a set of baseline data from institutions and then measure progress over time through subsequent collections. We required data reporting from project sites. We also encouraged all AAU universities to participate. Our hope is that these data will fulfill the needs of the association, while also providing value-added to individual institutions. We wish to collect information that accounts for differences without "punishing" institutions that are not as far along, providing disincentives to improvement, or forcing institutions into one-size-fits-all actions. To the extent possible, our intention in developing measures was to use existing reliable and valid instruments to collect information. Finally, within the constraints of these objectives, we wanted to minimize the reporting burden.

To begin developing baseline measures, AAU convened a working group of experts on metrics and evaluation in July, 2013. The group included administrators, institutional researchers, and faculty scholars, including some not affiliated with AAU campuses. Many of the eight project sites were represented in this group. We sought guidance on what to collect and how to collect it, as well as more knowledge about existing instruments and data collection efforts, and other issues such as dealing with Institutional Review Boards (IRBs).

Following this meeting, AAU project staff developed a set of research questions, mapped to the levels of the AAU Framework for Systemic Change in Undergraduate STEM Teaching and Learning (AAU, 2015) and matched—to the extent possible—with existing instruments (see Table 2). The working group helped hone the questions and identify potentially useful instruments, and provided advice on the feasibility of collecting certain kinds of information. Through conversations with the working group and others, we were able to define the kinds of common information it made sense for us to collect. For example, data on student learning outcomes are important for sites to judge the effectiveness of instructional reforms, but given the wide range of activities and types of students involved in the project, sites are not easily aggregated across institutions. Our solution was to leave the collection of student learning outcomes to each institution using a form that made sense in the context of their own objectives.

We proposed collecting information on the faculty status and rank of STEM course instructors in departments participating in the initiative, linked to specific courses and enrollment by year. We also decided to collect information on physical infrastructure (using a portion of the PULSE Vision & Change Rubric) (Taking the Pulse, 2013) and to ask for written descriptions of the role of teaching in promotion and tenure by project leads and department chairs. By far, the most challenging topics were instructor practices and attitudes. Answering many of the research questions required an understanding of what instructors were doing in the classroom, as well as their opinions about active learning,

TABLE 2. AAU Research Questions used as Guidelines for Selecting Baseline Data

Pedagogy—*Pedagogy refers to the method and practice of teaching. Much, but certainly not all, of pedagogy occurs in the classroom, and the main actors in changing pedagogical practices are faculty and students.*

- What type of instructional staff and faculty teach STEM courses, and at which level (under or upper division of undergraduates)? How large are those courses?
- What instructional practices are the faculty members who teach STEM courses using in the classroom? And how many students are exposed to these practices?
- What are faculty attitudes toward using evidence-based instructional practices?

Scaffolding—*The notion of scaffolding refers to the supports, including a sense of community, necessary to first incubate and then sustain evidence-based teaching.*

- What opportunities for professional development related to instruction are open to faculty, and to what extent are they taking advantage of these opportunities?
- What departmental and campus resources exist to support faculty in efforts to improve their instruction, and to what extent are faculty utilizing these resources?
- What are administrators' (department chairs, deans, senior administrators) attitudes towards use of evidence-based instructional practices and the importance of teaching?

Cultural Change—*Sustainable change requires cultural change, and faculty members live in at least two cultures: an institutional culture and a disciplinary culture.*

- What role does teaching play in promotion and tenure decisions in the relevant departments or schools at the university?
- What is the status of teaching and learning infrastructure (e.g., facilities, technology) in terms of facilitating the use of evidence-based teaching practices?

Student Outcomes—*While not a section of the framework, we are interested in the effects of projects on student outcomes like learning, progress, and retention. While changes in student outcomes can be attributed to multiple factors, it is important to consider, to the extent possible, the role of faculty teaching practices. AAU considers these data specific to individual project sites and important to local evaluation of the reform efforts.*

- How are students doing in STEM courses in terms of progression/retention/completion?
- How are students doing in STEM courses in terms of learning?
- What are student attitudes toward the use of evidence-based instructional practices?

about the availability and use of professional development activities, and how they felt administrators valued teaching. Although several instructor survey instruments were available (e.g., Bay View Alliance, BVA, 2015a; National Survey of Postsecondary Faculty) or were under development, no tool that encompassed both attitudes and practices, to our knowledge, had been widely used. AAU project staff assembled an instrument from existing tools, drawing especially from the BVA (2015b) (see Table 3 for an example of items in the faculty survey).

There was considerable concern from the project sites about administering an instructor survey. Some arguments focused on burden: It would be costly in time and effort to obtain IRB approval and administer the survey, and instructors—especially faculty members—were already subject to "survey fatigue" and not likely to respond. Other arguments focused on methodology: Instructor self-reports of classroom behaviors were less reliable than collecting such information through direct observation or video, and existing survey devices had not been sufficiently validated. Many individual issues were raised with particular survey questions. Institutions seemed not to agree that such data collection would provide them with value-added information. The political challenge was to convince project sites that AAU's role was not punitive. AAU's goal was (and is) to help ensure that the initiative was given the best chance to work effectively. This goal required that AAU actively monitor project progress (including linking sites that had similar problems and solutions) rather than taking a passive stance and leaving feedback to the end of the initiative.

We addressed these issues in several ways. First, we responded to criticisms of the survey by simplifying and completely revising the device. We discussed alternate questions with several scholars of teaching and learning. We also worked to align the survey questions much more closely with the AAU Framework.

TABLE 3. AAU'S Instructor Survey

The survey requested information from all faculty members whose departments were participating in the STEM Initiative:

- **Instructor information**: Such as institution, department, rank.
- **Classroom practices**: Instructors were asked to rate how descriptive various statements were of their own practice, such as whether they connect class activities to course learning goals, structure class so that students regularly talk with one another about course concepts, and require students to work together in small groups.
- **Attitudes towards teaching**: Instructors were asked to indicate their level of agreement with statements like: "It is important to provide relevant, real-life examples of the concept you are teaching" and "Learning can be facilitated through the use of social interaction among students."
- **Professional development related to teaching**: Instructors were asked to rate the availability of, and their participation in, various types of on- and off-campus professional development activities.
- **Institutional environment for teaching**: Instructors were asked to indicate their level of agreement with statements about the attitudes of other instructors, department chairs, and campus administrators toward teaching, as well as their perception of how important a role teaching played in annual and salary reviews and promotion and tenure.

Second, we wrote a description of how AAU planned to use the data. We specified that: 1) AAU would use these data to provide requested information to our funders, 2) AAU would use these data in aggregated form to help inform national conversations that we participate in, including with federal policymakers and other national associations, 3) AAU would not use these data to benchmark or compare institutions directly to one another (though we promised to provide each campus with a report-back of their responses compared to aggregate statistics for the entire sample), and 4) AAU urged institutions to make it clear to instructors that responses would not be used to evaluate job performance (the responses were sent directly to AAU thereby preserving local confidentiality).

Third, we expanded conversations with the project sites, both individually and collectively. We held a conference call with project team leads to discuss concerns and make clear our objectives. For instance, we made the point that, while self-reported instructional behavior can be limited, other methods of observation were far too costly and not necessary for AAU's purposes. Finally, we also discussed particular aspects of the survey individually with campuses that expressed concerns. The site visits were useful in this process in gaining local acceptance of survey data collection.

These conversations helped project site teams better understand our rationale for crafting the baseline data request, and helped reduce misperceptions or suspicions about how the data would be used. The final data request was sent to project sites in February 2014. Although longer than originally anticipated, this process helped to develop mutual understanding with project sites and increased their participation in the survey data collection.

Early findings show that these efforts paid off. Although many project participants were skeptical that there would be a usable response rate, we achieved a response rate of 37% overall across the eight project sites, which was higher than we expected (though lower than the 50% we hoped for). Some project participants also believed there would be insufficient variance in responses to many items; our preliminary analyses show substantial item variation.

SUMMARY

In the AAU Initiative for Undergraduate STEM education, we have found a key to successful use of data in reforms is to identify clearly which data should be collected across project sites and which data are best collected locally using a meta-analytic approach to combine results. Project sites must feel empowered to collect information relevant to their local needs while allowing for the possibility of sharing information across sites and forming overall assessments of effects. Beyond the role of data in assessment is the importance of data in helping foster a collaborative network across institutions. Our conversations with AAU project leaders and with non-project site personnel showed us how back-and-forth communication about data on effectiveness can foster a larger collaborative network crucial to disseminating reforms beyond the project sites.

REFERENCES

AAU. (2015). The AAU undergraduate STEM initiative. Retrieved from https://stemedhub.org/groups/aau.

Bay View Alliance. (2015a) The Bayview Alliance. Retrieved from http://bayview alliance.org.

Bay View Alliance. (2015b). Bayview Alliance survey. Retrieved from https://kan-sasedu.qualtrics.com/SE/?SID=SV_0ilkxZuB2tN4mnb.

National Research Council. (2012). *Discipline-based educational research: Understanding and improving learning in undergraduate science and engineering.* Washington, DC: National Academy Press.

Seymour, E., and Hewitt, N. (1997). *Talking about leaving: Why undergraduates leave the sciences.* Oxford: Westview Press.

Taking the PULSE Working Group. (2013). *PULSE vision & change rubrics v.1.0.* Retrieved from http://www.pulsecommunity.org/page/pulse-and-vision-change -v-c.

ABOUT THE AUTHORS

James Fairweather is a Professor of the Educational Administration Department at Michigan State University in East Lansing, Michigan.

Josh Trapani is the Director of Policy Analysis at the Association of American Universities in Washington, D.C.

Karen Paulson is a Senior Associate at the National Center for Higher Education Management Systems (NCHEMS) in Boulder, Colorado.

SECTION F
Synthesis of Common Themes

Over the course of the two Transforming Institutions conferences in 2011 and 2014, the challenges involved in aligning undergraduate STEM education with what we know about how people learn, and how to draw students in and support their success, have been explored from multiple perspectives. In this section we examine the larger picture these perspectives provide. Seymour and her colleagues begin to address the difficult question of how far we have come along a trajectory of making proven practices the new normal, and how we can assess progress. They further explore resistance to change, and what factors support and sustain change. Slakey and Gobstein present a reflection on the contents of this volume, both in terms of the trends manifested and the future directions the work described here suggests.

1

The Reformers' Tale: Determining Progress in Improving Undergraduate STEM Education

Elaine Seymour and Catherine L. Fry

In this chapter, we pose two related questions about change in science, technology, engineering, and mathematics (STEM) education at the undergraduate level: (1) What do we know about the extent and nature of uptake of proven STEM education practices? (2) What indicators or measures of "uptake" or "scale-up" have been used to assess these? To address these questions, we draw upon published work that estimates the extent of change in STEM education, and discuss the indicators by which assessments of progress are reached. We also present responses to these questions offered in interviews with an invited panel of 18 expert witnesses. Interviewees included project directors, principal investigators, evaluators, and funding officers of multiple STEM education reform grants. All panel members have been highly involved in efforts to improve quality and access in STEM undergraduate education for between 15 and 30 years. Some also have extensive experience as scholars of teaching and learning, STEM educators, workshop and institute organizers, and advisors to other STEM education reform projects.[1]

---

1. An earlier account of these discussions with 10 interviewees appeared in Seymour, E., DeWelde, K., and C. Fry (2011). A further eight interviews were subsequently added. Projects represented by the panel include: Center for the Advancement of Engineering Education (CAEE), Center for the Integration of Research, Teaching, and Learning (CIRTL), Process-Oriented Guided Inquiry Learning (POGIL), Peer-Led Team Learning (PLTL), Science Education for New Civic Engagements and Responsibilities (SENCER), The Reinvention Center, two National Learning Communities projects, multiple Cooperative Learning initiatives and workshop programs, Women in Science and Engineering (WISE), Recognition Awards for the Integration of Research and Education (RAIRE), the EXCEL Engineering Coalition, the Systemic Reform Projects in Chemistry, and Multi-Initiative National Dissemination (MID) workshops, Curriculum for the Bioregion, Mobilizing STEM Education for a Sustainable Future (MSE), the Student Assessment of their Learning Gains online instruments (SALG), doctoral programs in STEM education at Purdue University and the University of Michigan, the Washington Center for Improving the Quality of Undergraduate Education, and several other university-based Teaching and Learning (T&L) Centers.

THE QUESTION IN CONTEXT

Since the mid-1980s, researchers have pointed to problematic outcomes in undergraduate STEM education in terms of declining enrollment, field-switching rates, and persistently low participation by women and students of color (reviewed in Seymour and Hewitt, 1997 and Seymour, 2002). By the late 1990s, inadequate teaching methods and curriculum content issues had been identified as major contributors to these problems. This diagnosis was acknowledged in a series of reports by national commissions and panels, each of which called for fundamental changes in STEM teaching and learning methods (e.g., National Science Foundation, 1996; Boyer Commission on Educating Undergraduates in the Research University, 1998; National Research Council, 2003a, 2003b, and 2007). Beginning in the early 1990s, these reports also offered to an emergent community of education reformers a set of targets for the improvement of STEM undergraduate education. The National Science Foundation (NSF) and many private foundations responded by funding a variety of experimental approaches that were undertaken by groups of faculty, both in single institutions and multi-institutional coalitions. Over the last two decades, such initiatives have collectively created a body of thematic, contextual, research-grounded curriculum and learning materials, an array of classroom-tested, active, interactive, and inquiry-based pedagogies, and learning assessment methods that explore students' depth of understanding, and ability to apply, extend, and transfer their knowledge. This expanding body of tested and adapted methods and materials has been disseminated to widening circles of faculty through online resources and communities, journal articles and conference presentations, and workshops that offer hands-on exposure to learning theories, research findings, and their classroom applications.

However, some follow-up studies have reported either modest or disappointing progress (The Reinvention Center, 2002; National Research Council, 2010) or have strengthened the urgency of their calls for improvement in STEM education (President's Council of Advisors on Science and Technology, 2010, 2012; National Science Board, 2010). Research articles also echo the impression of unsatisfactory progress (cf., Handelsman et al., 2004; DeHaan, 2005; Dancy and Henderson, 2008; Fairweather, 2008; Baldwin, 2009; Dancy & Henderson, 2010). In their review of research literature between 1980 and 2008, Cox et al. (2010) point to slow progress in fostering learning-centered teaching approaches among faculty, and posit that top-down teaching-focused policies have had "but trivial effects on faculty members' perceptions or behaviors" (p. 20).

While this work conveys a consensus that promising teaching and learning innovations are not being taken up at a satisfactory level or pace, the evidentiary base for these claims is not always clear. We therefore surveyed the research literature to extract what could be said from evidence about the nature and extent of "uptake" or "scale-up" of research-grounded teaching materials and methods. The research evidence focuses either upon institutionally supported adoption by whole institutions, large groups within them, or cross-institution initiatives; or upon individual instructor uptake that may or may not be supported by their respective institutions. The concern of funders is increasingly to see adoption of research-based instructional strategies (RBISs) sustained by departments or institutions, and debate increasingly centers on how this may be brought about (e.g., Wieman, Perkins, & Gilbert, 2010; Anderson et al., 2011). A similar discourse, focused on how best to move from individual uptake to institution-based reform, was also discernable in the interviewees' observations. We begin with a categorized summary of published work that offers evidence both of individual instructor uptake and of adoption at various institutional levels.

EVIDENCE OF UPTAKE OR SCALE-UP OF RESEARCH-GROUNDED IMPROVEMENTS IN STEM EDUCATION BY EXTENT OF INSTITUTIONAL REACH

Whole Institutions

One medical school has overhauled its entire curriculum and instituted active learning and other research-based pedagogies throughout all classes (Elizondo-Montemayor et al., 2008).

Within a Single Institution: Whole Departments or Colleges

There are multiple accounts of whole departments or colleges that have changed their curriculum to incorporate active learning and other research-based education innovations. These include, departments of chemistry (Ege et al., 1997; Coppola et al., 1997), physics (Dori and Belcher, 2005), biology (Ono et al., 2007), and engineering (Pundak and Rozner, 2008; Merton et al., 2009). In an evaluation of the Rochester Institute of Technology's Chester F. Carlson Center for Imaging Science, Pow (2013) reports development and implementation (since 2010) of a freshman year curriculum of linked courses with a pedagogical framework emphasizing experiential and project-based learning and collaborative design projects in multidisciplinary teams. Student gains include a preference for active learning and active pursuit of research opportunities.

Within Single Institutions, Within or Across Departments

Barlett and Rappaport (2009) report on the cross-departmental effects of faculty development programs at two universities that were focused on integrating environmental and sustainability content across the curriculum. In both institutions, course curricula were modified or developed in line with this goal, interdisciplinary teaching became more common, faculty networks grew as they worked on implementation, and the number of collaborations on grant proposals increased. Similarly, in a study of single course transformation in one university, Chasteen et al. (2010) found that innovations were planned, implemented, and sustained with a high degree of fidelity over time and across instructors. What may be significant about these three cases is the high degree of agreement among faculty about the value and goals of these initiatives.

Across Institutions, Sub-Departmental

By 2008, the SCALE-UP (Student-Centered Active Learning for Undergraduate Programs) approach had been adopted at more than 50 institutions (Beichner and Saul, 2003; Beichner et al., 2007; Beichner, 2008) and was found to be spreading across disciplines (Oliver-Hoyo et al., 2004; Biechner et al. 2007). In 2014, Foote and her colleagues estimated that SCALE-UP style instruction had spread to over a dozen disciplines in more than 314 departments located in at least 189 higher education institutions in 21 countries. In the USA, 63% of reported departmental use of SCALE-UP methods occurred outside of physics (the originating discipline), with 20% of usage in non-STEM disciplines. The researchers assessed that, in physics, use of SCALE-UP methods may be approaching a tipping point between adoption by more adventurous early users and the mainstream majority.

Another indicator of institutionally-supported change is the growth of *learning community* programs. In a meta-analysis of 110 single-institution assessments, Taylor and colleagues reported in 2003 that over 1,000 institutions offered some form of learning community programs, many of which were of longstanding and were regularly offered. They are established in all types of institutions—research universities, engineering schools, regional public institutions, liberal arts colleges, and community colleges.

Discipline-Based Initiatives

Since its beginnings in the early 1990s, the STEM reform effort has largely been organized through its component disciplines, with initiatives in engineering, chemistry, physics and mathematics among its early leaders—a movement that deserves a history in its own right. However, some highlights that suggest forward movement include the development of "*discipline-based education research*"[2] (DBER), defined as "an emerging, interdisciplinary research enterprise that combines the expertise of scientists and engineers with methods and theories that explain learning" (National Research Council, 2012, p1.). It is expressed as growth in the publication of articles, monographs, and books on educational scholarship authored by STEM education practitioners and researchers; publication of articles on effective use of particular innovations; citation of these sources in the presentations and proposals of others; inclusion of these articles and citations in faculty portfolios and their acceptance for promotion and tenure purposes. More journals now publish articles on STEM education issues, and new (including online) journals have arisen to carry them, and some reform groups (e.g., the Council on Undergraduate Research) have developed their own journals.

Among other evidences of disciplinary engagement are development and use of *concept inventories*. These have grown, both in the number of STEM disciplines that have developed them, and in the range of institutions in which they are used. In 2008, Libarkin cited 23 distinct concept inventories developed in an array of STEM disciplines; there are now 39 in physics alone.[3] Another indicator of disciplinary engagement is evident in the development and growth of doctoral programs in STEM education, with an increasing number of doctoral theses focused on STEM education topics.

Awareness and Uptake by Individual STEM Instructors

Surveys of faculty across institutions report *higher levels of faculty awareness of teaching alternatives*, some increased inclination to use them, but mixed results for uptake. Beginning in 2007 with a survey of faculty in Louisiana, Walczyk, Ramsey and Zha reported that few faculty had been trained in teaching methods, but those who had were more likely to use that information to improve teaching, and also to consider teaching an important part of their professional identity. However, Dancy and Henderson's (2010) survey data for physics faculty indicate that, despite awareness of particular well-grounded methods, and an apparent willingness to try them, there is a considerable gap between knowledge and use. For example, 64% of their sample knew about peer instruction (Mazur, 1997) but only 29% were using it; 49% knew about cooperative group problem-solving, but only 14% used the method; 48% were aware of just-in-time teaching but only 8% put it into practice.

---

2. The salient characteristics and scope of DBER were articulated in 2012 by the National Research Council Committee on the Status, Contributions, and Future Directions of Discipline-Based Education Research Report.

3. cf., https://www.physport.org/assessments/

Also in 2010, from their study of engineering departments, Borrego, Froyd, and Hall report high levels of awareness of innovations (82%), but much lower levels of adoption (47%). The result from two further studies of engineering faculty in 2013 by the same research team (Froyd et al., and Prince et al.) underscore this point. Among electrical and computer engineering faculty, awareness of RBISs was very high, but their use of particular strategies varied from 10% to 70%. Discontinuation rates (ranging from 25% to 76%) after an initial trial were significant in explaining lower usage rates. A sample of chemical engineering faculty showed levels of awareness of 12 named RBISs of over 80% for all but two strategies. However, awareness outpaced adoption for every strategy, often with large gaps between awareness and adoption, and significant rates of discontinuation after an initial trial period. Studies by Henderson and colleagues (2012) and the Gates Foundation (FTI Consulting, 2015) also point to discontinuation (and also inappropriate use) of RBISs as the largest contributors to the "knowledge-practice gap."

Recently, a group of scholars have sharpened our understanding of the nature, as well as the extent, of change in teaching practices by the use of classroom observation techniques. For example, Budd and colleagues' (2013) observational study of teaching practices in 66 introductory geology classes at different institutions, report that approximately one-third of the classes used traditional, lecture-based methods, another third employed student-centered learning practices, and the remaining one-third of classes were in transitional states between the two. Hora's (2013) deployment of the Teaching Dimensions Observation Protocol (TDOP) provided a fine-grained analysis of classroom sessions, specifically, the degree to which methods that directly engage students in their own learning were used in conjunction with periods of lecturing. Such methods, Hora argues, free research and debate about classroom change from reliance on surveys based on self-report and on reductionist definitions of what "lecturing" means in practice. For his sample of 156 instructors in three large public research universities in four STEM fields, Hora found that 61% lectured without student interactions for periods of 20 minutes or less, 23% for periods between 21 and 40 minutes, and 16% lectured for over 40 minutes. The majority of instructors were, thus, not engaged in "straight lecturing" without any student interaction; rather they used various forms of interactive activity between short lecture periods.

That said, the latest Higher Education Research Institute (HERI) faculty survey (Hurtado et al., 2012) based on self-reported survey data from 23,824 full-time faculty in 417 postsecondary institutions reported "extensive lecturing" in 63% of courses, class discussion in 61%, cooperative learning in 47% and "using student inquiry to drive learning" in 37% of their courses. Hora wryly observes that the 63% of the HERI faculty sample who self-reported their regular teaching method as "extensive lecturing" is very close to his finding of lecturing with pre-made PowerPoint visuals in 64% of all of the two-minute intervals logged via the TDOP. Two other studies of self-reported classroom pedagogies support the view that "lecturing", although still widespread, now includes a wider array of active and interactive methods that augment straight-forward content delivery. Henderson and Dancy (2009) found that 29% of their sample of 722 physics instructors used peer instruction (Mazur, 1997), 14% used interactive lectures, and 14% cooperative group problem-solving. Simi-larly, in their 2013 survey of electrical engineering faculty, Prince et al. (op.cit.) found that 60% used a variety of active teaching methods, 35% problem-based learning, and 15% peer instruction.

WEIGHING THE EVIDENCE

This body of published research offers a picture of uptake of different levels and types. The distinction between evidence of faculty awareness and uptake of RBISs and that of institution-based adoption is important. As we and others have long argued (e.g., Seymour 2002), departmental or institutional support make it more likely that faculty who become aware of strategies that make their lectures more active and interactive—and of ways to document improvements in their students' learning—will both try them and persist in their use.

The largest body of evidence catalogues the national and international suc-cess of a well-disseminated model (SCALE-UP) across multiple disciplines. It also includes: examples of institutional, departmental, and cross-disciplinary curriculum overhaul within a framework of relevant research-grounded ped-agogy; some success in getting usage of particular teaching methods that is institutionally supported; some uptake by whole departments; and examples of the spread of particular pedagogies at sub-department levels by individuals and groups of faculty. We also learn that dissemination and uptake are effec-tively enabled by interpersonal means, including active faculty networks and learning communities, and that growth in these activities are, in themselves, indicators of a changing culture around teaching and learning. We note also the possibility of institution-wide transformations and cross-departmental collaborations in contexts where faculty coalesce around widely valued end goals. From 2007 to the present, researchers also document growing levels of faculty awareness of alternative ways to teach, and some inclination to try them. However, they also observe gaps between awareness and adoption of new teaching approaches, and difficulties in sustaining adoption beyond ini-tial trial periods.

While this patchwork of clues does not indicate a strong national response to the call for change, it does not seem to justify the consensus expressed in post-millennium reports and articles that little or nothing has happened to improve undergraduate STEM education. Rather, we get a sense that we do not have enough information to warrant such a judgment, along with signs that change is in the air, and, in some quarters, is happening. The picture created, especially by the most recent studies, is one of change that is partial, but nevertheless underway.

Why then has a dismal consensus developed that projects little or no change in STEM education? We posit a number of reasons for this, the first being the power of serial repetition of a view originating from a prestigious source to create a shared perspective which may or may not have an independent reality. A second reason may be that we have restricted our gaze for indicators of change to published research. What is little represented in this body of work is evidence (particularly longitudinal evidence) from evaluations of funded change initiatives. One reason for this omission is that articles based on evaluation data are rarely accepted for publication by disciplinary journals. Rather, most evaluation reports are sent only to funders who neither make them publically available, nor synthesize results across like projects to discern patterns of change. There may also be other less-documented indicators of change, for example, the trend in disciplinary society meetings towards sessions in which educational innovations are discussed. These sessions typically showcase results from the classroom experiments of individuals and small groups, and offer workshops in the use of particular pedagogies. Such sessions are themselves the tip of an iceberg of scholarly faculty activity to describe and document their teaching and learning experiments—some in the short-run; some over time.

INDICATORS OF CHANGE EVIDENT AMONG PANEL MEMBERS

With the intention of making a wider sweep for indications of change in STEM education, we sent four questions to our 18 panel members and asked them to discuss their responses in an open-ended telephone interview of one hour or more. We discuss here the panelists' responses to our two opening questions: What do we know about the extent and nature of uptake of proven STEM education practices, and what indicators or measures of "uptake" or "scale-up" have been used to assess these? We discuss in the subsequent chapter, entitled, "Why doesn't knowing change anything?" the panelists' responses to the second two questions: What are the sources and nature of constraints upon and resistance to change; what leverages change and what sustains it?

The interviewer (Seymour) took shorthand notes of the interviews which were then transcribed. The resulting text data were coded and grouped into themes, which are presented below. Some observations by panelists are offered as quotations throughout the following account.

Definitions

We asked panel members to clarify their understanding of the terms "innovation," "uptake," and "scale-up." "Innovations" were seen as methods or materials that were still being tested. A variety of qualifying adjectives (e.g., "research-grounded," "scientific," "proven," "quality," "good") were used to reference classroom resources that were proven and well-used. However, panelists reported no agreed way to reference all non-traditional teaching methods and materials, and understood why faculty at large might use "innovation" to describe anything unfamiliar. They took "uptake" to mean use or adaptation of any classroom resource by individuals or faculty groups, usually in a single discipline, and in a few courses or sections. Uptake was essentially informal—the result of individual faculty choices, the product of promotional efforts by project networks, or the efforts of leaders in particular institutions. Host institutions might not know this was happening. "Scale-up" was interpreted as adoption of a resource into many courses in one or several departments or colleges (e.g., engineering), across one or several institutions, or in whole divisions or institutions. "Scale-up" implied formal decisions to adopt something as "a reform," following the promotional efforts of both formal and informal leaders. For whole institutions this implied policy decisions. Reforms, or scaled-up "innovations" were essentially *public, formal, and durable*. One project director described the resulting normative shift as the feeling that "everyone is doing this."

How did panel members assess the extent and nature of uptake of proven STEM education methods and materials?

We asked panel members to offer estimates of both uptake and scale-up, whether from their direct experience of projects, or from their wider experience of national reform efforts.

Positive overall estimates, largely defined as "uptake"

Many combinations of research-grounded teaching methods were reported to be in widespread use—POGIL, PLTL, just-in-time teaching, calibrated peer review, etc. These and other innovations (e.g., learning communities, small group learning) were reported as having gained a foothold in many institutions and disciplines.[4] Further growth was thought to depend on good professional

---

4. This assessment matches Henderson and colleagues findings, using a DELPHI approach, also with a panel of experts. (Personal communication). For a list of "Well propagated instructional strategies," consult http://www.increasetheimpact.com/resources

development. Other markers of uptake included: greater awareness of the scholarship of teaching and learning and its implications for teaching; more disciplinary discussion of education issues, notably at meetings; more publications available on what works better or worse; and an overall sense of an upward trend: "We haven't slipped back as far as we used to. The high energy is still there."

Observations on the nature of uptake

Change was largely reported to be underway in lower level classes, and most uptake at the individual level because we, as yet, lack strategies for institutional change. Pedagogical innovations with the greatest uptake were easy-to-incorporate technical additions to traditional pedagogy (e.g., clickers) that do not require fundamental change in how faculty think about their pedagogy. Also, more often adopted are innovations that are "concrete," and those that can be grasped without much explanation.

Examples of high-level uptake that are evident in many institutions

Over 1,000 institutions were reported to offer some form of learning community program, many of which are longstanding and regularly offered. They are established in all types of institutions (research universities, engineering schools, regional public institutions, liberal arts colleges, and community colleges). In contrast with individual class uptake, they require teams of faculty and administrators to set up and run them (cf. also Taylor et al., 2003). Other resources cited as widely used in many institutions were: POGIL ("In 2009, 800–1,000 people were doing something they identify as POGIL"); PLTL (2008 director's survey yielded 300 users); various forms of small group learning, SENCER courses, and increasing use of the Student Assessment of their Learning Gains (SALG) online instrument. Professional development programs are established in many institutions, e.g., DELTA program (University of Wisconsin) which is offered annually and attracts good participation, and the CAEE engineering teaching portfolio program for graduate students. The number of teaching and learning centers (T&LCs) and engineering education centers continues to grow, although estimates of faculty usage vary. The CAEE director reported their "CELT" model is successfully used in other centers and departments. However, other panelists cited good T&LCs with low faculty use of their services.

Departmental, divisional, and whole institution scale-up

All project directors on the panel offered examples of departmental, sub-departmental (e.g., for all introductory courses), and institutional uptake of curriculum or pedagogical reforms, some over long periods. Examples offered in our interviews were the iceberg tip. They (and other informants whom they cited) had many more examples to offer. The panel pointed to the value of collecting and categorizing these in a public national registry that was regularly updated. The strongest example of success in a nationwide reform effort was agreed to be achieving higher rates of enrollment, persistence to graduation, and entry to doctoral programs, of women in STEM disciplines. Success was mainly attributed to sustained effort and funding, good publicity, and to both on-campus and nationally-organized women's groups, e.g., WISE, Women in Engineering Programs and Advocates Networks (WEPAN), the Society of Women Engineers (SWE), and the National Center for Women & Informational Technology (NCWIT), rather than to pedagogical or curriculum improvements—a view also expressed by Fox, Sonnert, & Nikiforova (2011). The panel agreed that improvement in the representation among graduating STEM majors by students of color (other than Asians) remains low.

Caveats and factors affecting progress

Nationwide uptake was seen as variable and uneven. Some institutions (e.g., community colleges and liberal arts colleges) are more flexible in their teaching structures and have cultures that are more open to change. Some saw greater uptake in both engineering and physics. "Big ideas" like adoption of the scholarship of teaching and learning, or learning through diversity, are more appealing to senior administrators than to faculty who value resources that they can use immediately. There is also a time-lag for uptake. For example, the PLTL director viewed progress in getting PLTL into department curricula as "the fruit of seeds sown 8–10 years ago." Some long-time project directors reported that, in their *own* institutions, there were no signs of uptake of their resource, or value shifts towards other reforms: "Without institutional framework and college-level support, innovation all falls on the individual professor." In presentations, some directors reported getting the same questions about their initiative that they got 10–12 years ago. They ascribed this to faculty being unaware of evidence of efficacy in the research literature that they and others have produced. Notwithstanding the examples of uptake that they offered, they also observed that the effects of many initiatives have not spread much beyond the immediate group of innovators.

Establishing the extent of STEM improvement

That STEM faculty teaching is unchanged was seen as the dominant NSF view. Panelists proposed a nation-wide inquiry to establish the extent of uptake of established resources, and of formal adoption (scale-up) in particular institutions. Every project director cited institutions, schools, and departments that had formally adopted their projects' products and felt that others could generate such a list. They unanimously supported a formal effort to collect, categorize, publish, and regularly update this information.

What indicators or measures did panel members use to reach their assessments?

We asked the panel members to explain what formal or informal indicators they, or their project evaluators, had used to get a sense of progress, both for the resources they had developed, and for the STEM education reform effort overall. Although all of their projects were formally evaluated, and examples of their measurement methods were offered, their responses indicated that panel members relied more on what they had directly observed than on formal change measures in current use. That said, they were acutely aware of the need for good measures of change, described some of their problems with existing measures, and explained what kinds of measures they would like to have[5].

Indicators of institutional uptake of innovations

A critical indicator of institutional buy-in to reform efforts cited by most panelists was that an institution was providing money and resources to sustain them. They offered examples of institutions that had: created resources (some with endowments), such as teaching and learning centers, with new staff positions to service them; continued staff positions required to service innovations beyond the end of external grant support; and developed professional development programs for faculty, post-docs, and/or teaching assistants (see also Seymour, 2005).

Within larger initiatives, communities and networks have developed, and are attracting new recruits

Community activities in larger initiatives that were taken to indicate growth and sustainability were: regional networks of learning communities, consortia of participating institutions that hold meetings and conferences for faculty and administrators; the engagement of deans and provosts through these meetings; growing attendance figures in education sessions at national conferences; summer institutes that have run for multiple years, and that successfully promote ideas, research, and know-how, build networks and work teams; reform communities that are created and maintained by online networks; new recruits continue to be attracted, engaged, and join the project's dissemination efforts;

---

5. Described in Seymour, DeWelde, and Fry (2011)

project leaders receive many invitations to talk about their own work both in the US and internationally—an activity that builds awareness and prestige.

Greater awareness and acceptance of scholarship of teaching and learning and of research-based teaching methods

Expressed as direct experiences: Panel members reported that they encounter more awareness of teaching and learning research at all levels of academe. Aggressive expressions of skepticism about student-centered forms of instruction that were common in presentations and workshops 15 years ago have disappeared. (Panelists commented that, even if this reflects political correctness, such behavior is no longer legitimated.) They meet more college administrators who are aware of the kinds of change needed to improve student learning. Institutional documents (e.g., mission statements), and the language they employ, indicate some reform ideas have become "givens."

Positive shifts observed in faculty attitudes toward their teaching role

Shifts in orientation toward students as learners were recounted at summer institutes in faculty presentations. Capturing such "conversions" or "moral shifts" is an important change indicator because, as panelists observed, faculty who have made these shifts in perspective do not return to traditional teaching methods (see also Mazur, 2009).

Behavioral shifts noted include faculty attention to their teaching outcomes

Panelists noted faculty adapting or developing learning assessment instruments that reflect their changed content or pedagogy. They also noted growth in faculty use of student feedback tools, notably the SALG online instrument, to solicit student feedback "in real time," as well as at course end. Growth in the use of the SALG website and instruments, since its inception in 1997, is documented on its website, www.salgsite.org, and in a series of reports (2009 to 2014) by the SALG Development Group to its funder, the NSF. The SALG site now includes 6,850 published instruments, administered multiple times by 12,500 instructors and responded to by 279,000 students. Many projects depend upon SALG instruments (including URSSA, the Undergraduate Research Student Self-Assessment instrument) to validate their learning outcomes, and their results have been reported in over 400 peer-reviewed papers in academic journals.

Spontaneously–offered reports of use and promotion of innovations

Project directors learn about uptake of their innovations when people spontaneously contact them. They meet strangers at meetings who describe uptake of their innovation. These encounters also indicate that the adapters themselves are enabling further uptake: "I no longer feel I have to be out there convincing people because there's enough knowledge of it out there, and it's other people who do the convincing on the basis of their own practical experiences."

Institutional shifts in formal criteria for hiring, promotion, and tenure

In some institutions, the formal recruitment criteria now include familiarity with particular teaching methods and with the scholarship of teaching and learning. However, panelists also noted resistance to this, citing examples of senior administrator interventions to correct decisions made without reference to the new formal criteria.

Innovation spreads into new disciplines or from the sciences into other disciplines

Undergraduate research has spread into social science and humanities in multiple institutions (see also, Laursen, Hunter, Seymour, Thiry, and Melton, 2010). POGIL was recently taken up in mathematics and biology; the SALG instrument and website, originally created for STEM faculty, is now used in all disciplines and in learning contexts outside the classroom (e.g., workshops, conferences, libraries, and museums).

Finally, the panelists shared the view of the research community, discussed earlier, that important indicators of change are represented by *the growth of published and presented scholarship on teaching and learning*, as evidenced in disciplinary meetings that showcase them and journals that carry them, and by a gradual institutional acceptance of Boyer's (1998) call to take account of such work for tenure and promotion purposes.

DISCUSSION AND CONCLUSIONS

Notwithstanding their own commitment to STEM education reform, the panelists' assessments of how much progress has been made were not rosy-tinted. They described a wide spectrum of uptake and scale-up with both encouraging and discouraging dimensions. All of the projects represented by our panel used formal evaluation methods.[6] However, panel members' accounts of evaluation questions they had asked, the measures that they had used, and the results these had yielded were clearly only a selection of all the methods they deployed in reaching their assessments. Their comprehensive and nuanced approach to the assessment of progress, grounded in experience and observation over time, utilizes more markers and criteria for change than are typically included in formal

---

6. The evaluation methods and findings reported by panel members for the projects they represent are discussed in Seymour, DeWelde, and Fry, 2011.

evaluation or research studies. Although some indicators could not easily be formalized, many could be included in evaluation designs.

The panelists also argued (with examples and reference to documented data) that there is evidence waiting to be gathered that can establish more accurately and completely than hitherto the extent and nature of uptake of research-grounded teaching and learning methods. They identified project directors and evaluators as primary sources for uncollected information, whether singly, or, more significantly, across funded projects. Some are clearly aware of, and some could already provide, categorized listings of institutional and departmental implementation of reformed curricula and pedagogy, and instances of high levels of implementation that fall short of formal adoption. Others indicated that they would need additional funding to conduct such inquiries, but see themselves as well-placed and motivated to undertake this work. The category of "high-level uptake" of any resource both within an institution and across institutions (i.e., without formal adoption) appears a worthwhile target for inquiry its own right.

On the basis of the information that we distilled from published sources, and gathered directly from our panel members and indirectly from their sources of formally and informally gathered evidence, it is at least plausible to argue that the widely cited view of the National Research Council (NRC)'s report (2003b)—that little change has been made in the learning experiences of undergraduates in introductory science courses—may be insufficiently grounded. Although the NRC report offers no evidence in support of its claim, we note that it is widely cited in other work. The NRC assessment may or may not be correct. However, the credence given to one highly respected, albeit unsupported source, may have created a research myth. It is also possible that the extent of progress may be only partially documented in published work because they, as yet, reflect an overly narrow set of indicators. That said, insofar as funders view this claim as an accurate estimate of change, it has consequences for funding decisions affecting the reform effort:

> It is not important whether or not the interpretation is "correct" . . .
> if men define situations as real, they are real in their consequences
> (Thomas, 1928, p.572).

What are the conclusions that the panel members have collectively reached about progress in STEM education reform? They offer many kinds of evidence—both documented and observed—that change is happening: some as outcomes, some as shifts in professional attitudes and values, some in the distribution of awareness and knowledge, and some as identification of stages in change as a process. They imply that what they and co-reformers observe could (given tools and funding) be captured by more formal inquiries to build a many-faceted picture of the processes and outcomes of change. They argue strongly that data already gathered by many initiatives could, and should, be mined, synthesized, and made available. They suggest new lines of research needed to establish the extent of uptake and scale-up on a national basis, and offer themselves and their fellow PIs, project directors and evaluators as a collective resource in developing research questions and methods.

Even with an admitted under-sampling of the available experts, and perhaps of available literature, we think that our panel members have collectively offered sufficient testimony of observed improvements in STEM education to lift the reform community and its funders out of the Slough of Despond, to use John Bunyan's reference from *The Pilgrim's Progress*, into which it may recently have fallen for want of some well-founded good news. That said, we clearly need a series of inquiries that are both multi-dimensional and national in scope in order to ground the actual extent and nature of what has and has not been achieved in the reform of STEM education.

REFERENCES

Anderson, W. A., Banerjee, U., Drennan, C. L., Elgin, S. C. R., Epstein, I. R., Handelsman, J., . . . Warner, I. M. (2011). Changing the culture of science education at research universities. *Science, 331,* 152–153.

Baldwin, R. G. (2009). The climate for undergraduate teaching and learning in STEM fields. *New Directions for Teaching and Learning,* (117), 9–18.

Barlett, P. F., and Rappaport, A. (2009). Long-term impacts of faculty development programs: The experience of TELI and Piedmont. *College Teaching, 57*(2), 73–82.

Barzun, J. (1991). *Begin here: Forgotten conditions of teaching and learning.* Chicago: University of Chicago Press.

Beichner, R. J. (2008). The SCALE-UP project: A student-centered active learning environment for undergraduate programs. Commissioned Paper for the *National Academies Workshop: Evidence on Promising Practices in Undergraduate Science, Technology, Engineering, and Mathematics (STEM) Education.* Retrieved July 12, 2015 from http://sites.nationalacademies.org/dbasse/bose/dbasse_080106

Beichner, R. J., and Saul, J. M. (2003). Introduction to the SCALE-UP (student-centered activities for large enrollment undergraduate programs) project. *Proceedings of the International School of Physics "Enrico Fermi"* (pp. 1–17). Varenna, Italy.

Beichner, R. J., Saul, J. M., Abbott, D. S., Morse, J. J., Deardorff, D. L., Allain, R. J., Bonham, S. W., Dancy, M. H., and Risley, J. S. (2007). The student-centered activities for large enrollment undergraduate programs (SCALE-UP). In Redish,

E. F., and Cooney, P. J. (Eds.), *Research-based Reform of University Physics*. College Park, MD: American Association of Physics Teachers. Retrieved Nov. 2, 2010 from http://www.compadre.org/PER/per_reviews/volume1.cfm

Borrego, M., Froyd, J. E., and Hall, T. S. (2010). Diffusion of engineering education innovations: A survey of awareness and adoption rates in U. S. engineering departments. *Journal of Engineering Education, 99*(3), 185–207.

Boyer Commission on Educating Undergraduates in the Research University. (1998). *Reinventing undergraduate education: A blueprint for America's research universities*. Menlo Park, CA: Carnegie Foundation for the Advancement of Teaching. Retrieved July 12, 2015 from: http://eric.ed.gov/?id=ED424840

Budd, D. A., Kraft, K. J. V. D. H., Mcconnell, D. A., & Vislova, T. (2013). Characterizing teaching in introductory geology courses : Measuring classroom practices. *Journal of Geoscience Education, 61*, 461–475.

Chasteen, S. V., Pepper, R. E., Pollock, S. J., and Perkins, K. K. (2010). But does it last? Sustaining a research-based curriculum in upper-division electricity & magnetism. *Physics Education Research at Colorado (PERC) Proceedings*. Retrieved Nov. 3, 2010 from http://www.compadre.org/PER/items/detail.cfm?ID=10321

Coppola, B. P., Ege, S. N., and Lawton, R. G. (1997). The University of Michigan undergraduate chemistry curriculum 2: Instructional strategies and assessment. *Journal of Chemical Education, 74*(1), 84–94.

Cox, B.E., MacIntosh, K.L., Reason, R.D. & Terenzini, P.T. (2011). A culture of teaching: Policy, perception, and practice in higher education. *Res High Educ, 52* (8), 808–829.

Dancy, M. H., and Henderson, C. (2008). Barriers and promises in STEM reform. Commissioned paper for the *National Academies Workshop: Evidence on Promising Practices in Undergraduate Science, Technology, Engineering, and Mathematics (STEM) Education*. Retrieved July 12, 2015 from http://sites.nationalacademies.org/dbasse/bose/dbasse_080106

Dancy, M., and Henderson, C. (2010). Pedagogical practices and instructional change of physics faculty. *American Journal of Physics, 78*(10), 1056–1063.

DeHaan, R. L. (2005). The impending revolution in undergraduate science education. *Journal of Science Education and Technology, 14*(2), 253–269.

Dori, Y. J., and Belcher, J. (2005). How does technology-enabled active learning affect undergraduate students' understanding of electromagnetism concepts? *Technology, 14*(2), 243–279.

Ege, S. N., Coppola, B. P., and Lawton, R. G. (1997). The University of Michigan undergraduate chemistry curriculum 1: Philosophy, curriculum, and the nature of change. *Journal of Chemical Education, 74*(1), 74–83.

Elizondo-Montemayor, L., Hernández-Escobar, C., Ayala-Aguirre, F., and Aguilar, G. M. (2008). Building a sense of ownership to facilitate change: The new curriculum. *International Journal of Leadership in Education, 11*(1), 83–102.

Fairweather, J. (2008). Linking evidence and promising practices in science, technology, engineering, and mathematics (STEM) undergraduate education: A status report for the National Academies National Research Council Board on Science Education. Commissioned paper for the *National Academies Workshop: Evidence on Promising Practices in Undergraduate Science, Technology, Engineering, and Mathematics (STEM) Education*. Retrieved July 12, 2015 from http://sites.nationalacademies.org/dbasse/bose/dbasse_080106

Foote, K. T., Neumeyer, X., Henderson, C., Dancy, M. H., & Beichner, R. J. (2014). Diffusion of research-based instructional strategies: The case of SCALE-UP. *International Journal of STEM Education, 1*(1), 10.

Fox, M. F., Sonnert, G., & Nikiforova, I. (2011). Programs for undergraduate women in science and engineering: Issues, problems, and solutions. *Gender & Society, 25*(5), 589–615.

Froyd, J. E., Borrego, M., Cutler, S., Henderson, C., & Prince, M. J. (2013). Estimates of use of research-based instructional strategies in core electrical or computer engineering courses. *IEEE Transactions on Education, 56*(4), 393–399.

FTI Consulting. (2015). *U.S. postsecondary faculty in 2015: Diversity in people, goals and methods, but focused on students*. Retrieved from http://postsecondary .gatesfoundation.org/wp-content/uploads/2015/02/US-Postsecondary -Faculty-in-2015.pdf

Gafney, L. and Varma-Nelson, P. (2008). *Peer-led team learning: Evaluation, dissemination, and institutionalization of a college level initiative*. Dordrecht, Netherlands: Springer.

Handelsman, J., Ebert-May, D., Beichner, R., Bruns, P., Chang, A., DeHaan, R. L., Gentile, J., Lauffer, S., Stewart, J., Tilghman, S.M., and Wood, W. B. (2004). Scientific teaching. *Science,* (304), 521–522.

Henderson, C. R. & Dancy, M. H. (2009). Impact of physics education research on the teaching of introductory quantitative physics in the United States. *Physical Review Special Topics—Physics Education Research, 5*, 020107.

Henderson, C., Dancy, M., & Niewiadomska-Bugaj, M. (2012). The use of research-based instructional strategies in introductory physics: Where do faculty leave the innovation-decision process? *Physical Review Special Topics—Physics Education Research, 8* (2), 020104.

Hora, M.T. (2013). *Exploring the use of Teaching Dimensions Observation Protocol to develop fine-grained measures of interactive teaching in undergraduate science classrooms*. WCER Working Paper No. 2013-6.

Hurtado, S., Eagan, K., Pryor, J.H., Whang, H. & Tran, S. (2012). *Undergraduate teaching faculty: The 2010–2011 HERI faculty survey*. Los Angeles, CA: Higher Education Research Institute, UCLA.

Laursen, S. L., Hunter, A.-B., Seymour, E., Thiry, H., and Melton, G., (2010). *Undergraduate research in the sciences: Engaging students in real science*. San Francisco, CA: Jossey-Bass.

Libarkin, J. (2008). Concept inventories in higher education science. Commissioned paper for the *National Academies Workshop: Evidence on Promising Practices in Undergraduate Science, Technology, Engineering, and Mathematics (STEM) Education*. Retrieved July 12, 2015 from http://sites.national academies.org/dbasse/bose/dbasse_080106

Mazur, E. (1997). *Peer instruction: A user's manual*. Upper Saddle River, NJ: Prentice Hall.

Mazur, E. (2009). Confessions of a Converted Lecturer: Eric Mazur [Video file]. Retrieved April 1, 2010 (http://www.youtube.com/watch?v=WwslBPj8GgI).

Merton, P., Froyd, J. E., Clark, M. C., and Richardson, J. (2009). A case study of relationships between organizational culture and curricular change in engineering education. *Innovative Higher Education, 34*(4), 219–233.

National Research Council. (2003a). Evaluating and improving undergraduate teaching in science, technology, engineering, and mathematics. In Fox, M. A., and Hackerman, N. (Eds.), *Report of the Committee on Recognizing, Evaluating, Rewarding, and Developing Excellence in Teaching of Undergraduate Science, Mathematics, Engineering, and Technology, Center for Education, Division of Behavioral and Social Sciences and Education*. Washington, DC: National Academies Press.

National Research Council. (2003b). *Improving undergraduate instruction in science, technology, engineering, and mathematics: Report of a workshop*. McCray, R. A., DeHaan, R. L., and Schuck, J. A. (Eds.), Center for Education, Division of Behavioral and Social Sciences and Education. Washington, DC: National Academies Press.

National Research Council. (2007). *Rising above the gathering storm: Energizing and employing America for a brighter economic future*. Committee on Science, Engineering, and Public Policy (COSEPUP). Washington, DC: National Academies Press.

National Research Council. (2010). *Rising above the gathering storm, revisited: Rapidly approaching category 5*. Washington, DC: National Academies Press.

National Research Council. (2012). *Committee on the Status, Contributions, and Future Directions of Discipline-Based Education Research Board on Science Education Division of Behavioral and Social Sciences and Education*. Singer, S.R., Nielsen, N.R., and H. A. Schweingruber (Eds.). Washington, DC: National Academies Press.

National Science Board. (2010). *Preparing the next generation of STEM innovators: Identifying and developing our nation's human capital*. Arlington, VA: NSF Publication NSB1033.

National Science Foundation. (1996). *Shaping the future: New expectations for undergraduate education in science, mathematics, engineering, and technology.* Arlington, VA: NSF Publication 96-139.

Oliver-Hoyo, M. T., Allen, D., Hunt, W. F., Hutson, J., and Pitts, A. (2004). Effects of an active learning environment: Teaching innovations at a research I institution. *Journal of Chemical Education, 81*(3), 441–448.

Ono, J. K., Casem, M. L., Hoese, B., Houtman, A., Kandel, J.,and McClanahan, E. (2007). Development of faculty collaboratives to assess achievement of student learning outcomes in critical thinking in biology core courses. In Deeds, D. and Callen, B. (Eds.), *Proceedings of the National STEM Assessment Conference, October 12–21, 2006* (Vol. 104, pp. 209–218). Open Water Media, Inc.

President's Council of Advisors on Science and Technology. (2010). *Prepare and inspire: K–12 education in science, technology, engineering, and mathematics (STEM) for America's future.* Retrieved Nov. 3, 2010 from http://www.white house.gov/sites/default/files/microsites/ostp/pcast-stem-ed-final.pdf

President's Council of Advisors on Science and Technology. (2012). Engage to excel: Producing one million additional college graduates with degrees in science, technology, engineering, and mathematics. Washington, DC: Executive Office of the President.

Pow, J. J. (2013). Using a non-traditional pedagogy in STEM disciplines: Implications for faculty. White paper from 2013 Hawaii University International Conferences, Education & Technology, June 10–12, 2013. Honolulu, Hawaii.

Prince, M., Borrego, M., Henderson, C., Cutler, S., & Froyd, J. (2013). Use of research-based instructional strategies in core chemical engineering courses. *Chemical Engineering Education, 47*(1), 27–37.

Pundak, D., and Rozner, S. (2008). Empowering engineering college staff to adopt active learning methods. *Journal of Science Education and Technology, 17*(2), 152–163.

Seymour, E. (2002). Tracking the process of change in U.S. undergraduate education in science, mathematics, engineering, and technology. *Science Education,* (86): 79–105.

Seymour, E., DeWelde, K., and Fry, C. (2011). Determining progress in improving undergraduate STEM education: The reformers' tale. A white paper commissioned for the forum *Characterizing the Impact and Diffusion of Engineering Education Innovations,* February 7–8, 2011, New Orleans, LA. Retrieved on July 12, 2015 from https://www.nae.edu/File.aspx?id=36664

Seymour E. and Hewitt, N. M. (1997). *Talking about leaving: Why undergraduates leave the sciences.* Boulder, CO: Westview Press.

Seymour, E., with Melton, G., Wiese, D. J., and L. Pedersen-Gallegos. (2005). *Partners in innovation: Teaching assistants in college science courses.* Lanham, MD: Rowman and Littlefield Publishers, Inc.

Taylor, K., with Moore, W. S., MacGregor, J. and Lindblad, J. (2003). *Learning community research and assessment: What we know now.* Olympia, WA: The Evergreen State College, Washington Center for Improving the Quality of Undergraduate Education, in cooperation with the American Association for Higher Education.

The Reinvention Center. (2002). *Reinventing undergraduate education: Three years after the Boyer Report.* Miami, FL: The University of Miami.

Thomas, W. I. (1928). *The child in America.* New York: Knopf.

Walczyk, J. J., Ramsey, L. L., and Zha, P. (2007). Obstacles to instructional innovation according to college science and mathematics faculty. *Journal of Research in Science Teaching, 44*(1), 85–106.

Wieman, C., Perkins, K., & Gilbert, S. (2010). Transforming science education at large research universities: A case study in progress. *Change: The Magazine of Higher Learning, 42*(2), 7–14.

ABOUT THE AUTHORS

Elaine Seymour is the Director Emerita and Research Associate, Ethnography & Evaluation Research at the Center for the Advancement of Research and Teaching in the Social Sciences, and a Research Fellow for the Center for STEM Learning at the University of Colorado, Boulder in Boulder, Colorado.

Catherine L. Fry is the Education Manager at the American Society for Pharmacology and Experimental Therapeutics in Bethesda, Maryland.

2

Why Doesn't Knowing Change Anything?
Constraints and Resistance, Leverage
and Sustainability

Elaine Seymour and Kris De Welde

In the prior chapter, "The Reformers' Tale," we drew on both published research and the responses of our invited panel of seasoned STEM education reformers[1] to address two questions: (1) What do we know about the extent and nature of "uptake" of proven STEM education practices? (2) What indicators or measures of "uptake" or "scale-up" have been used to assess these? In this chapter, we draw on panelists' responses to three further questions: (1) What are the sources and nature of constraints upon, and resistance to, educational improvements in engineering and the sciences? (2) What can leverage change? and (3) What sustains change? Panel members viewed change processes as operating at structural, cultural, and personal-professional levels in highly connected ways. They discussed what kinds of leverage can be applied at these levels, and offered examples of where these have been used with success. Observations from panelists' direct experience are discussed with reference to the work of scholars who have also addressed these issues, including our own work in this field.

CHANGING INSTITUTIONS: POSSIBILITIES AND LIMITATIONS

Institutes of higher education (IHEs) are inherently less nimble than business organizations. Historically, they have had fewer external drivers, and respond less directly to their customers and markets. The organizational structure at IHEs has evolved to preserve customary ways of carrying out formal tasks, even when some processes and their outcomes are acknowledged to be dysfunctional. These are evident, for example, in the design of lecture halls, the criteria for departmental funding, and for faculty rewards, tenure, and promotion. Institutional structures privilege traditional teaching methods and do not easily accommodate new ones. This makes the resulting inertia hard to break, involving as it does both structural and cultural shifts. Indeed, resistance to

---

1. See "The Reformers' Tale" for composition of the panel, interview methods, and analysis.

change in its educational functioning may be seen as normal in higher education institutions.

Aware of an unsuccessful history of experiments with both grassroots-only initiatives and top-down led or mandated changes, the panel members were unanimous that successful teaching reform requires combined top-down and bottom-up approaches. From the top, it requires institutional commitment to the value of research-based instructional strategies (RBISs), shifts in the distribution of funding and rewards, and changes in organizational and physical structures. The panelists described the reform community's (and their funders') disheartening experiences of seeking to "institutionalize" their RBISs without sufficient top-down buy-in and practical support.

In light of this history, the reform community works to convince chairs, deans, provosts, and college presidents of the importance to their mission of high quality teaching and the scholarship of teaching and learning (SoTL). Panel members gave examples of senior administrators who had come to see change as in their institution's best interest—"as something for the president to brag about." And they broadly agreed that the effort to get senior administrators on board has been successful: "The idea that research excellence isn't enough anymore has taken hold." They cited as evidence the successful spread of institution-sponsored teaching and learning centers, undergraduate research (UR) programs, high-school- to-college bridge programs, and women-in-science programs. They also observed that awareness of both successes and failures can motivate institutional leaders to respond. Innovations shown to improve student retention and completion rates can prompt a combined top-down/bottom-up effort. Conversely, dismal drop, withdrawal, or failure rates in key courses can prompt a department to make a trial run of alternative teaching methods. However, the panel cautioned that it is departmental, not institutional, leaders that have most power to determine matters of curriculum and pedagogy. Institutional leverage over these matters is indirect and marginal, not central (Seymour, 2001). What then do institutions have the power to do?

Establishing official criteria for the faculty rewards system

Panel members unanimously identified existing institutional rewards systems as the main structural deterrent to change for faculty who are otherwise disposed to rethink their teaching. As one panelist observed:

> What is not true is that faculty don't care about students. Rather, the strategies that would improve their pedagogy are not as yet embedded in faculty positions and rewards.

The panel gave examples of institutions that have extended their criteria for hiring, promotion, and tenure to include evidence of teaching effectiveness and the scholarship of teaching and learning. Such changes reflect Boyer's (1990) proposition that achievements in research and teaching should be judged by parallel criteria. However, they also gave examples of resistance to revised rewards criteria by tenure committees that continued to operate by traditional standards. Project directors cited cases in their own institutions where they or others had intervened to insist that new formal criteria be followed. One project director observed, "Some tenure processes have stepped up to the 21st century; others have a firm grasp of their rear-view mirror."

The extent to which institutions have revised their rewards criteria, the numbers of faculty who have benefitted from this, and losses among faculty denied promotions or tenure because their teaching achievements and scholarship were discounted, all remain to be documented. However, based on their direct knowledge, the panel was optimistic that the institutional climate for classroom reformers was improving. Program directors cited tenure successes among their program participants, and saw it as a growing trend for new faculty to negotiate career paths that include their innovative teaching and SoTL. However, the panel saw this as a stronger imperative for institutional leaders than for departments whose criteria for hiring and career development largely remain traditional. Again, we lack a national picture of these significant indicators of change.

Aside from institution-level progress in restructuring formal rewards criteria, project directors described the success of financial awards as a means to encourage individual faculty to take up innovations—for example, competitive grants for design, implementation, or evaluation of classroom change experiments. Faculty respond well, even to modest rewards, because they showcase good work and confer status. Also seen as effective were institutional awards to departments for sustained teaching improvements that resulted in desired and documented student outcomes. Panelists viewed this as a promising way for institutional leaders to leverage departmental support for change in curriculum and pedagogy.

Other institutional contributions to educational reform

Sustaining faculty deployment of an innovation after external funding has ceased can be secured by providing administrative support, and funding faculty lines or staff positions that service and support research-based instructional strategies. As one panelist observed, "That's what institutionalization means." Such structural supports were seen as critical in institutionalizing and sustaining successful projects. Administrative and physical structures may need to be rethought to accommodate new ways of teaching. These may include provision of service staff, classroom redesign, and the addition of technical teaching aids. As one panelist observed: "The purposive deployment of money, jobs, and resources is critical if you want to see improvements last."

Senior administrators can also insist on conformity where departments have agreed to revise their curriculum, pedagogy, or learning assessment processes. Strategic use of central resources also plays a role in sustaining change: for example, where institutional research staff monitor the outcomes of an institutionally adopted reform initiative.

Institutions have considerable power to encourage faculty uptake of new curriculum or pedagogy by changing their faculty time allocation policies. Provision for release time to allow innovators to do their work is critical to their motivation and chances of success. This is especially important for the principal investigators (PIs) of reform efforts who are teaching faculty. As several project directors pointed out, few of their number have appointments that allow sufficient time to organize reform efforts, particularly those that are multi-institutional in scope. Any multi-institution project's administrative efficiency is undermined from the outset unless its faculty PI receives appropriate time allowances from the host institution. The practice of funding part-time PIs to run complex change initiatives may be an inappropriate transposition to education reform of the research funding model. Some panelists spoke from painful experience about this:

> "All change project PIs are amateurs who have to learn on the job everything they need to know about how to make things happen and keep them going."

> "The leaders of most new projects don't know much about the theoretical grounding of the remedies they choose to address a problem, or how to evaluate what they are doing. Some of them don't even realize that other people have worked on this before."

> "The burn-out rate is high among PIs trying to run a complex change project. It's even higher in projects that involve multiple institutions. This is a serious barrier to survival or scale-up."

Finally, panelists pointed to the significance of the power and influence of senior administrators and disciplinary leaders who champion education reform. Seymour (2001) pointed to the significance of "radicalized seniors" in publicly promoting educational improvements, legitimating their uptake, protecting younger faculty reformers from negative consequences, and using their influence to leverage change at national, institutional, departmental, and disciplinary levels. Panelists cited examples of Nobel laureates, college presidents, provosts, and deans who have spearheaded reforms in particular institutions or on the national stage. Panelists were unanimous that the work of proactive and vocal institutional and disciplinary leaders was invaluable to durable institutional and nationwide change. They cited as examples of such leadership:

- Seven engineering deans who created the Excellence in Engineering Education (EXCEL)[2] coalition.
- A group of senior women who insisted that their institution reverse a decision not to continue to fund its Women in Science Education (WISE) program.
- Senior women administrators in one ADVANCE program[3] who insisted that all departments include suitably qualified and experienced, locally available women in their hiring searches, and refused to consider hiring proposals without evidence of this.
- Some long-time RBIS developers and change initiative leaders who have now risen to senior positions in their institutions with seats on finance, hiring, promotion and tenure committees are positioned to give direct assistance to change initiatives.

Role of professional development in teaching and learning

Institutions can provide funding and resources for teaching and learning centers (T&LCs) and for professional development programs for faculty, graduate students, post-docs, and graduate teaching assistants (TAs). They can also actively leverage their use and status. Panelists were unanimous in proposing that the single most important strategy for durable, nationwide STEM education reform is development of institution-wide professional education programs that ground the pedagogical knowledge and skills of current and future faculty in SoTL research. In the interim, where faculty innovators teach large classes with the help of graduate TAs, such programs are critical in securing the active

---

2. EXCEL was the first of a series of coalitions funded by the NSF to promote improved teaching across particular university disciplines.

3. The goal of the NSF's ADVANCE program is to increase the representation and advancement of women in academic science and engineering careers, thereby contributing to the development of a more diverse science and engineering workforce. ADVANCE encourages institutions of higher education and the broader science, technology, engineering and mathematics (STEM) community, including professional societies, to address aspects of STEM academic cultures and institutional structures that may differentially affect women faculty and academic administrators.

collaboration and professional socialization of graduate assistants. In our evaluation studies, we found that research-based education programs for TAs can also ward off disapproval, even sabotage, of an instructor's RBISs, by TAs who have internalized the values and methods of more traditional instructors (Seymour, 2005).

Panelists also saw the potential of institutionally supported T&LCs to promote change through the practical assistance they can provide to instructors (for example, with learning assessment methods). However, creating such centers is not enough. Panelists cited examples of well-run centers that faculty do not use. One panelist urged, "Institutions must promote them, make them accessible, and reward their use." Another concurred: "We can't wait for faculty to discover them." Centers may be underused where their staff are post-doctoral appointees, or have doctoral degrees in education but not in STEM disciplines. To be respected, center staff need high skill sets and status derived from disciplinary research. Institutions can enable this with part-time appointments for disciplinary faculty.

Outside of institutions, and predating T&LCs, regional, disciplinary based workshops have, for two decades, been the main conduits of knowledge and know-how developed by larger initiatives. Project Kaleidoscope is probably the best- known example. There is strong evidence that workshops foster cross-institutional reform by drawing in, educating and enabling recruits, and incorporating them into the change effort (Andrews, 1997; Hilsen and Wadsworth, 2002; Connolly and Millar, 2006). Project directors with long experience as workshop organizers observed that, to work optimally, they must be of sufficient duration, offer repeated exposure in a progressive sequence, provide support for new reformers in their home departments, use "old hands" as facilitators, and build facilitator capacity among newer recruits:

> "Workshops give strength, motivation, and skills to people who don't feel a connection to their department colleagues."

> "They offer portals to people who are like-minded, people that they wouldn't ordinarily meet—people in different disciplines, senior people."

Panel members observed that the continuing role of reform-associated workshops was thrown into doubt because of difficulties in finding adequate funding. However, professional and disciplinary societies have stepped into this role by mounting workshops as part of annual meetings.

ENGAGING FACULTY

The role of internalized disciplinary and academic cultures in resistance to change

As we asked in the previous chapter, "Why do faculty not use more of the sound innovative instructional practices that are available to them?" To which, one project director added: "Why don't faculty apply to their teaching the same standards and care that they use in their own research?" We cannot attempt a comprehensive answer to either question here. However, we offer some insights from panel members with long experience of seeking ways to address them.

First, one cannot assume that faculty members are free agents. Their professional choices are shaped by socialization in graduate school and by values, norms, and beliefs transmitted through their discipline and reinforced by their departments—cultural forces that are strong enough to withstand change efforts (Seymour, 2001). In addition, the broader culture has, over the later 20th century, lost much of its traditional respect for teachers (Seymour, 2006). One panelist cited Barzun's (1991) observation that academe has lost its sense of teaching as "a calling." Another described the academy's de-professionalization of teaching: Unlike other professions, the academy does not formally educate its entrants for this aspect of their professional role (cf., Seymour, 2007). Learning how to teach has to be retrofitted. As one panelist observed:

> "Where the cultural norm—that teaching is a "talent"—is dominant, instructor education may be viewed as ineffective for the untalented and unnecessary for everyone else."

It is against this culture that education reformers press institutions to provide mandatory research-grounded teaching, and learning education for graduate students and instructors.

Some panel members themselves subscribed to a commonly expressed view of professional rights by which the norm of "academic freedom" is extended to the belief that faculty are the best judges of how to teach any class. However, they also identified this as the source of a tendency to adapt rather than adopt RBISs. In journal accounts, presentations at meetings, and live encounters with adapters of pedagogical methods that they have developed, panel members observed that users commonly miss out essential elements or violate the innovation's central principles. Their observations are supported by Dancy and Henderson's (2010) findings that, in implementing innovations, faculty commonly altered them in ways that made them more "traditional" and, thus, significantly differed from the original research-grounded techniques. As one frustrated developer observed:

"Then they complain that the method doesn't work and stop using it. If they used it the way it was designed to work, it would."

A number of researchers have documented the "discontinuation rates" among faculty who try particular RBISs but then abandon them. (Henderson et al., 2012; Froyd et al., 2013; Prince et al., 2013; FTI Consulting, 2015). They concur with the panelists' experiences that incorrect adaptations of their methods are significant sources of discontinuation.

Might this widespread adapter behavior simply reflect the variability of teaching situations, or is the presumed uniqueness of each classroom a myth? Panelists thought the latter, pointing to the dominant traditional pattern in all research universities of large classes taught in an entirely uniform manner. The adapt-rather-than-adopt pattern creates hybrids that panelists described as "SENCER-like, POGIL-like, PLTL-like, SALG-like",[4] etcetera—which make evaluation of their efficacy and meta-analysis very difficult (Century et al., 2010) Henderson and colleagues (2012) and The Gates Foundation (FTI Consulting, 2015) also report a "knowledge-practice gap" that manifests in inappropriate adaptation of RBISs.

Student resistance

Faculty and their departments may also be locked into a culture that supports dysfunctional teaching norms and practices, notably, tolerating high rates of student failure, and the "normative resistance" exhibited by students in many traditionally taught classes (Seymour, 2005). In large lecture classes especially, students routinely behave in ways that express disinterest, disengagement, and disrespect. They distance themselves from whatever the teacher is doing by "arriving late, sitting at the back, talking, reading non-class material, doing work for other classes, eating, sleeping, and failing to ask or answer questions" (Seymour, op.cit., p. 128). These behaviors are well-known to faculty, and, for those who are hesitant to change their teaching methods, may be used as a rationale to "let sleeping dogs lie." Panelists cited culturally-supported faculty explanations for these widespread forms of student resistance to large lecture classes that deflect blame away from the teacher's methods onto presumed flaws in students as a genus (see also Seymour and Hewitt, 1997; Campbell et al., 2001; Seymour, 2005).

Panel members were asked why the persistence of negative student responses to traditional teaching did not prompt faculty to try other methods? Their answers focused on fear of the consequences. Learning new teaching

---

4. Science Education for New Civic Engagement (SENCER), Process-Oriented Guided Inquiry (POGIL), Peer-Led Team Learning (PLTL), Student Assessment of their Learning Gains (SALG).

methods may carry a greater emotional load of risks, fears, and potential losses than tolerating bad student behavior—particular when colleagues blame the students, not the teacher. Trying unfamiliar methods also raises fears of exposing what they might not know, losing control over class activities, collegial accusations of reducing course "rigor," or failing to cover a normatively-defined canon. Faculty may also fear negative student responses to new teaching methods in institutional course evaluation surveys that (albeit, woefully irrelevant as measures of student gains) are included in their tenure and promotion portfolios as evidence of teaching competence. Indeed, it was partly to address this realistic fear that the Student Assessment of their Learning Gains (SALG) instrument was developed (in 1997) as a web-based course evaluation instrument that focuses exclusively on student learning[5] (Carroll, 2010).

Why would students resist, rather than rejoice, when instructors give more thought to helping students learn? As argued in Seymour (2005), many students enter college having learned mostly by memorization. Little may have been demanded of them by way of inquiry, independent reflection, or application of knowledge. Early college classes, traditionally taught, do not disrupt familiar learning techniques that focus on passive absorption and short-term memorization. However, students find that these methods do not work where teachers demand active engagement with ideas, independence of thought, application of ideas in new situations, and responsibility for learning. Rather, such approaches expose inadequacies in students' traditional learning practices and in their depth of understanding. An implicit contract between students and teachers about what each expects of the other is thus suspended. The resulting discomfort can provoke anxiety, complaints, withdrawal, and both passive and active resistance from students. Innovators can expect student resistance to be most acute early in their introduction of new learning methods. Seymour and Laursen (2005) reported that innovating faculty can typically expect two semesters of resistance from the start of any course requiring students to engage in active, interactive, or inquiry-based learning. Acceptance and appreciation develop as faculty gain experience and confidence in using their new techniques methods, as students (and their TAs) become aware of increased learning gains, and as the approach comes to be seen as "normal" for this course (Seymour, 2005).

Internalized restraints and other issues of departmental change

The panel was agreed that discovering how to leverage change in departments is critical, and "that we have never really known how to do it." Routine resistance to teaching changes is, as panelists noted, too consistent in particular

---

5. See salgsite.org

sub-disciplines (e.g., organic chemistry) to be coincidental. It is best explained in terms of deeply internalized cultural norms about appropriate teaching methods and their purposes (including the practice of "weeding out" some students). Faculty participating in summer institutes or workshops commonly describe their struggle to break loose from such internalized restraints. As one panelist observed:

> "Although our whole program is built on a critique of traditional practice, we avoid direct attacks which only provoke resistance. Instead, we find ways to get participants to see the critique and its personal consequences for themselves."

However, as another panelist commented, "The power of departmental culture and customary practices makes it difficult to see when and how STEM faculty get to exercise choice."

That said, panel members thought that such resistance is lessening because it has become harder to ignore widely circulating knowledge about improved teaching methods and their results. Faculty are more aware of research on how students learn best and why "straight lecturing" is less effective than more active and interactive pedagogies, some of which may be incorporated into a variable-methods lecture class. (See also, Walczyk, Ramsey & Zha, 2007; Dancy & Henderson, 2010; Borrego, Froyd, & Hall, 2010 Froyd et al., 2013; Prince et al., 2013; Hora, 2013.) They also hear colleagues at disciplinary meetings describing improved results from research-grounded teaching. The panel acknowledged the growing role of disciplinary societies in spreading, and thereby legitimating, uptake of good practices. As one panelist observed, "It has become a more conscious moral choice not to try them."

Panelists saw the ground having shifted from arguments about insufficient evidence of the efficacy of new pedagogies to a more defensive rationale that changing their teaching methods would take too much time and effort. Beichner and colleagues (2007), Dancy and Henderson (2008, 2010), and Fairweather (2008) all present evidence of faculty claims that the presumed time investment is a primary reason for not changing their teaching methods. Panelists advocated providing faculty with time for educational development work as an important institutional contribution to improved teaching. However, panelists also treated the argument that "it will take too much time" as a culturally legitimated fend-off in face of increasingly powerful evidence supporting change:

> "It's a rationale that my colleagues know won't be challenged. And it covers their fears of being criticized for being seen as spending too much time on teaching, possibly at the expense of their research."

Panelists advocated securing the buy-in of chairs as a critical element in creating a departmental climate supportive of improved teaching. It is largely chairs that decide who teaches what courses; they can also protect foundational courses from pressures by departments whose majors they service, and mentor early-career faculty interested in teaching improvements. As discussed above, institutions can also leverage change via rewards to departments that demonstrate improved student learning, persistence of majors to graduation, or increased participation by hitherto under-represented student groups.

However, in a climate showing signs of appreciating the importance of improved curriculum and pedagogy in introductory courses, panelists lamented the departmental response of employing adjunct faculty, lecturers, or faculty appointed to specifically to teach foundational courses. One panelist labeled this "as a form of unbundling," that is, the corporate practice of separating and redistributing its organization's component parts. Others commented on the lower departmental status and salaries of faculty "who are hired just to teach" (cf., Fairweather, 2008).

Under present conditions, what may best secure faculty engagement?

Panel members focused on four main strategies. The first, institutional-wide establishment of professional development for instructors and graduate students in research-based pedagogy is discussed above; three others are discussed below.

Building communities and networks

Panelists gave many examples of communities and networks that function to keep like-minded reformers supported, motivated, and engaged. Historically, these have developed in larger initiatives and are sometimes referred to as "communities of practice." They support faculty who might otherwise be isolated, accumulate a reservoir of know-how and resources for members to tap into, and offer the satisfactions of working together—often with colleagues they might not otherwise meet. Much of the daily life of such communities and networks proceeds by electronic means. However, panelists observed from experience that securing funds for meetings, institutes, and workshops "is always time and money well spent." Face-to-face occasions bring together experienced classroom innovators with interested newcomers. Workshops and active working collaborations build the reform effort by attracting, training, motivating, and supporting recruits who become the next wave to carry the reform effort forward. Engagement in community conversations also enables faculty to deepen their understanding of the underlying principles of an innovation and develop ownership of the adaptations they develop.

The significance of communities and networks in enabling the uptake of RBISs is also reported in a number of studies. For example, in a large survey-based study, Foote and colleagues (2014) identify interpersonal networks as significant in leveraging and accelerating uptake of educational innovations, including SCALE-UP. Many more survey respondents reported that they had learned about SCALE-UP in interpersonal ways (through talks, workshops, and colleagues), than through mass media channels. Similarly, Hora (2007) ascribes SCALE-Up's success in large part to its development of STEM and education faculty networks that use collaborative approaches in teacher preparation. A sense of ownership, which the panel identified as critical in sustaining individual reformers, is also identified by scholars as a key component of successful institution-wide change (e.g., Elizondo-Montemayor et al., 2008). Increasingly, disciplinary and professional societies play an active role in developing communities of education research and practice, and providing the venues to sustain working collaborations.

Panelists also offered single institution examples where development of a cross-department community (that might include faculty, administrators, students, advisors, and staff) sustained reform efforts. For example, ADVANCE initiatives were cited as effectively addressing the concerns of women via campus-wide community action, and the diverse composition of such communities was seen as a major contributor to their effectiveness.

Offering faculty resources they can find and use

Panelists reasoned that faculty are more likely to try out high quality materials and methods if they are introduced to them by people that they respect, if they can find them easily, and if they are easy to use. In addition, because adaption rather than adoption is the norm, the core elements and underlying principles of an innovation also need to "make sense" to faculty. Spillane, Reisner & Reimer (2002), Small (2014), and others argue that people are more receptive to new ideas when they align with existing norms and practices. As one project director observed, "the classroom innovations that we see most readily adopted are the ones that faculty find easy to relate to."

Fairweather (2008) argues that faculty also need practical information on implementation strategies. One director reported that repeatedly showing colleagues' results obtained by using the learning assessments methods developed by his project spreads their uptake. An evaluator reported that what young faculty took away and used (both immediately and over time) from the project's professional development program were doable teaching strategies and classroom management tips, rather than more complex activities such as education scholarship.

What faculty can use is also related to the time and effort they presume will be needed. Notwithstanding their skepticism about time constraints as an avoidance rationale, panelists agreed that, for reforms to be taken up and become established, they should not demand more time than faculty feel they can give. As Borrego and colleagues (2010) also found among engineering faculty, reforms requiring least coordination and commitment of time were the most likely to be used.

Lastly, getting knowledge and know-how out to faculty requires proactive forms of dissemination. As one panelist observed, "It is critical to synthesize what we know for faculty to find easily and use." Popular search engines, like Google, are more likely to be searched (often at short notice) for teaching materials than the National Digital Library collections (Manduca et al., 2006; McMartin et al., 2008). One panelist summarized the general view that, in addition to published accounts of successful innovations, "the NSF has a lot of evaluation data that should be synthesized and published in a single, easily found place with regular updates." Some panelists pointed to congressional and presidential encouragement to the NSF to become more active in compiling, synthesizing, disseminating, and promoting what they know from evidence to be effective (President's Council of Advisors on Science and Technology Report, 2010).

The role of evidence

Panelists shared the view, also proposed by other scholars, that proving a resource works is necessary, but is not sufficient, by itself, to persuade faculty to use it (Seymour, 2001; Dancy and Henderson, 2008, 2010; Fairweather, 2008; Borrego and colleagues, 2010). We have long known that validation by disciplinary colleagues with high *research* prestige legitimates *educational* innovation (Foertsch et al., 1997). Providing such validation is one important contribution made by "radicalized seniors." Panelists recounted another effective source of persuasion—the personal stories of experienced innovators offered at their summer institutes and workshops. They described the powerful effects on newcomers of engaging in hands-on experiences of new methods, personal accounts by people who use them, and a safe and supportive environment in which to discuss concerns. In combination, such experiences contributed to the many "conversions" they had witnessed. Project evaluators agreed that, once faculty reach an intellectual and personal understanding of what they need to do in their own classrooms, and experience of using new methods with their students, they typically discover that they cannot return to how they previously taught (cf., Mazur, 2009).

What, then, did panel members see as the role of evidence? They argued that, strong evidence of success presented in disciplinary and professional meetings or at project meetings can be a good driver of uptake. They also pointed to increasing awareness of learning research that supports many innovations. However, in line with a number of studies that report a gap between knowledge and practice (e.g., Borrego et al., 2010; Dancy and Henderson, 2010; Froyd et al., 2013; Prince et al., 2013), the panelists observed that faculty awareness of available RBISs far outpaces adoption levels. Greater faculty understanding of the evidence in support of RBISs is a necessary precursor to change in teaching practice. However, it does not necessarily prompt or sustain it. Both Henderson and Dancy (2009) and Norton and colleagues (2005) found that faculty conceptions and beliefs about teaching were "more reformed" than their actual practice. Finally, panelists were doubtful whether any amount of evidence would sway those most opposed to change: "We won't get more change by producing more work that's focused on becoming credible with disciplinary conservatives."

Several panelists pointed out the irony that, *lack of evidence* of the efficacy of lecture-only teaching does not undermine its widespread use. Nor does it curb uptake of practices that are widely believed to be good. As an example, they pointed to the major expansion of undergraduate research (UR) programs funded by large grants. Although both UR funders and directors sought evidence of student gains that they attributed to these experiences—and of ways to measure them—they proceeded without either. Faculty confidence in UR methods was grounded in their own direct observation of student responses and outcomes—a form of validation that the panelists broadly shared. Research-based evidence for such confidence is now available, and Laursen and colleagues (2010) offer a comprehensive review of their own and other studies of UR benefits. Clearly, academics do not necessarily require evidence before making pedagogical choices, but they do have to believe in their value.

BROADER INFLUENCES ON CHANGE

Finally, the panel discussed four types of leverage that have the capacity both to instigate and to sustain change: (1) external pressures, (2) creating a demand for change, (3) transcendent concerns that overcome resistance, and (4) the role of adequate funding.

The significance of external leverage has been noted since the early days of the STEM reform effort (Seymour, 2001). Panelists pointed to the leverage of engineering education reforms by the Accreditation Board for Engineering and Technology (ABET). The earliest of the major coalitions, EXCEL, was grounded in ABET's concern to restore design to the engineering curriculum, a goal which the EXCEL coalition achieved. Panelists observed that the power of accrediting agencies to encourage and endorse change is beginning to be deployed as educational standards gain currency in higher education.

Also recently expounded is the proposition that *change can be generated by creating a demand for it*, for example, by students or employers (Zemsky, 2009). The director of SENCER described an example in which students had successfully lobbied for introduction of SENCER courses in their curriculum.

In documenting the adoption of cross-disciplinary, environmental problem-focused curricula at two universities (Emory and Tufts), Barlett and Rappaport (2009) did not draw the broader inference that institutional and personal resistance to innovation in STEM education may be surmounted by a rallying call with sufficient power to command widespread response. However, this proposition resonated with panel members. They offered further examples of *transcendent concerns* that had overcome departmental and individual resistance and generated cross-institution efforts to address widely shared concerns, notably, global and regional environmental issues. Engineering panel members cited curriculum changes in schools of engineering to address the "Fourteen Grand Challenges awaiting engineering solutions in the 21st century" (as delineated by the National Academy of Engineering, 2008–2014). Other issues cited by panelists as having rallied administrators and faculty into concerted action were regional concerns over job shortages for graduates that prompted both "Rust Belt" and California schools to provide education geared to environmentally focused industries. Similarly, a shared concern to break the patterns of disadvantage that have historically limited access to STEM education and careers for students of color and women of all races and ethnicities was also cited as a catalyst for institution-wide change efforts.

All panel members discussed the proposition that *strategic deployment of money* (whether from institutional or external sources) *can enable or hold back desired educational change.* Directors of projects that were operating successfully across multiple institutions, were concerned how their initiatives would survive or maintain growth, especially where their own institution offered no financial support. On no other issue was the distinction between uptake and scale-up more sharply made, for example:

> "We don't currently have any funds to run the project. If there's any funding, it has shifted to people who are using what we developed."

There was consensus that:

> "Funding must be adequate to the size of the task. It's not enough just to grow communities; you need to support them if you want national outcomes."

Panelists pointed to a long-standing, but unrealistic, expectation by funders that, somehow, projects will become institutionalized, or otherwise sustain themselves. They observed (from experience) that "the typical PI does not have the skills to invent a self-sustaining structure for a project being used in many institutions."

As veterans of many grant proposals, they noted patterns in the solicitations and awards practices of funders that they assessed as counterproductive for scale-up. Among these, were an unstated, but normative, timeline for achieving project goals, and refusal of further funding after a program, however successful, is "thought to have received enough"—a norm that effectively dismantles what has been built over time. As one respondent observed, "Reluctance to keep a good thing going is built into NSF's distribution of funds." They understood that the NSF in particular was struggling with the constraints of its original mission to create new knowledge, rather than to disseminate and promote proven practices, but were generally skeptical that the expectation of project sustainability was grounded in an understanding of *how* widespread change is to be secured. One panelist wondered "if the NSF had a model for scale-up, or were expecting grantees to discover it by trial and error." They stressed that the focus on innovation at the expense of consolidation and growth was self-defeating: "We never get to see what could be achieved if successful innovations were promoted and supported."

The mixed messages they saw in patterns of NSF funding confused program directors: "Some projects with demonstrated success continue to get funding, but they starve and stunt the growth of others." They tried to make sense of this. The composition of review panels, or the guidelines they were instructed to use, were thought to reduce their chances of sustaining successful programs:

> "I question the criteria by which a decent proposal from a project such as this, with widely acknowledged teaching methods and growing faculty uptake, gets rated only as 'fair.'"

Another panelist (with "rotator" experience[6]) wondered if "the rotators system creates loss of institutional memory." Whatever their causes, they described the damaging effects of what seemed irrational withdrawals of funds:

> "Cutting off funding from workshop-based projects that have developed expertise in how to do them well, that are getting good response, attracting new facilitators, and growing the movement, is disastrous. It demoralizes people who have invested a lot and have taken professional

---

6. "Rotators" are disciplinary faculty who are invited to work at the NSF in various capacities for a defined period, then return to their institutional duties.

risks to do this work. It is important to reward people. It's also important not to punish them for doing a good job."

"Innovations are now intrinsically sustained by the concern of individuals and small groups who want to make a difference. I question whether this is enough."

All the directors were expecting more difficult financial days ahead. As one director observed, "I am afraid that large budget cuts will choke off progress." And, in various ways, they were trying to prepare for this. One initiative focused on building faculty capacity in research and scholarship in the expectation that external funds would run out, and individuals and small groups would have to continue the work unaided. Some multi-institutional projects were looking for ways to become financially independent from host institutions that had accepted their grant overhead but did not support them. They were exploring endowment funding, establishment as independent not-for-profit entities, and exploring how to charge fees for some services. One project had shifted to a business model that transfers some of direct costs of summer institutes to institutions that send participants.

We were surprised to discover that long-time directors of large projects who shared the same problems of survival and growth, saw opportunities to discuss collective concerns and ideas with each other as rare. They valued meetings among project directors, evaluators, and seasoned NSF officers to discuss ways to sustain successful initiatives. They also identified questions of how to secure wider uptake of proven methods and materials, and to scale up and sustain successful programs, as research questions deserving of an awards category in their own right.[7]

The role of theory

The work of the reform community is the practical lab where theories of change—often implicit rather than stated—are tested. Writing from an evaluator's perspective in 2001, Seymour characterized the nature of much of the experimentation in STEM education then underway as a-theoretical. Proposals for action commonly jumped from diagnosis to action without if-then rationales to explain why chosen strategies might work. As Borrego and Henderson (2014) observe, projects are still apt to explain their choices of action in terms of

---

7. This wish may now have been partially granted by the launch of two NSF programs—Innovation Corps (i-corps) Teams whose mission is to enable widespread adoption, adaptation, use of resources developed with NSF funding; and i-corps Learning Teams which offer practical guidance to developers in building appropriate organizational and funding structure to independently sustain their work.

a single change strategy. Since 2001, a growing group of theorists have come to the reformers' aid by building a body of change theories and models that can inform their choice of action. Borrego and Henderson review this history and the array of available perspectives that can enable the design of more robust change strategies by making their underlying assumptions explicit. Funders (and their reviewers) have also responded by requiring project proposers to explain the rationales for their proposed choices of action.

Panelists were aware of, and some were collaborating in, the development of theoretical frameworks by which to understand how change happens. However, they assessed, this work as not yet widely realized as praxis—the process by which theory is embodied or enacted. A gap (partly of language) between theorists and practitioners may need to be bridged before theories of change are deployed by lay planners and implementers as a matter of course. However, theorists and reformers share common ground in viewing as paramount the task of determining how change may best be made at departmental, institutional, and national levels.

DISCUSSION AND CONCLUSIONS

As we have described, panelists responded to our opening question—"Why doesn't knowing change anything?"—with an array of both experienced and observed strategies that have taken them from diagnosis (i.e., "knowing") to successful interventions. What the panelists (and many researchers and evaluators) also "know" are the nature and sources of inertia and resistance that constrain freedom of choice for faculty, their departments and institutions, and their students. In a situation of increased awareness of what might be done to improve the quality of STEM education, what the panelists have learned by direct and shared experience is what kinds of leverage can surmount these obstacles and enable faculty to choose and maintain research-based instructional strategies. For example, moving the criteria for faculty rewards towards the Boyer model, has long been identified as essential to free faculty from constraints on their choice of teaching methods. However, panelists also attested to the strategic distribution of teaching achievement rewards to whole departments as an efficient way to leverage both individual and institution-wide change. Panelists prioritized institutional change strategies by their observed effectiveness. Important among these were: strategic institutional deployment of rewards (as above) and also of funding, jobs, and resources (to sustain successful initiatives); institution-wide professional development programs in research-based teaching methods for both faculty and graduate students; building widely inclusive communities and networks within and across institutions; making RBISs easy for instructors to find and use; and linking change goals to shared professional values or transcendent social concerns.

The panelists also saw themselves, their fellow PIs, project directors, and evaluators as an underutilized resource in the effort to ground change efforts in experience over time. What they have observed and documented could be drawn upon to build what one panelist descried as "a battleground map" of strategies that enable and sustain change, and to identify initiatives that are already working but need stronger support. Evaluation findings from change initiatives should, they argued, also be mined, synthesized, and made available. They also understood that the historic mission of their primary funder, the National Science Foundation, has inhibited its direct engagement in sustaining the practices they have helped to develop and test. However, they invited all funders of improvements in STEM education to become active partners in promoting practical answers to the question, "How best do you take successful change strategies, develop them to scale, and sustain them?"

REFERENCES

Andrews, G. J. (1997). Workshop evaluation: Old myths and new wisdom. In Fleming, J. A. (Ed.), *New perspectives on designing and implementing effective workshops,* (pp. 71–85). San Francisco: Jossey-Bass.

Barlett, P. F., and Rappaport, A. (2009). Long-term impacts of faculty development programs: The experience of TELI and Piedmont. *College Teaching, 57*(2), 73–82.

Barzun, J. (1991). *Begin here: Forgotten conditions of teaching and learning.* Chicago: University of Chicago Press.

Beichner, R. J., Saul, J. M., Abbott, D. S., Morse, J. J., Deardorff, D. L., Allain, R. J., Bonham, S. W., Dancy, M. H., and Risley, J. S. (2007). The student-centered activities for large enrollment undergraduate programs (SCALE-UP). In Redish, E. F., and Cooney, P. J. (Eds.), *Research-based reform of university physics.* Retrieved Nov. 2, 2010 from http://www.compadre.org/PER/per_reviews/volume1.cfm

Borrego, M., Froyd, J. E., and Hall, T. S. (2010). Diffusion of engineering education innovations: A survey of awareness and adoption rates in U. S. engineering departments. *Journal of Engineering Education, 99*(3), 185–207.

Borrego, M. and Henderson, C. (2014). Increasing the use of evidence-based teaching in higher education. *Journal of Engineering Education VC ASEE,* 103 (2), 220–252. Retrieved from http://wileyonlinelibrary.com/journal/jee. doi:10.1002/jee.20040

Boyer, E. L. (1990*). Scholarship reconsidered: Priorities of the professoriate.* San Francisco: Jossey-Bass.

Campbell, J., Smith, D., Boulton-Lewis, G., Brownlee, J., Burnett, P. C., Carrington, S., and Purdie, N. (2001). Students' perceptions of teaching and learning: The influence of students' approaches to learning and teachers' approaches to teaching. *Teachers and Teaching: Theory and Practice, 7*(2): 173–187.

Carroll, S.B. (2010). Engaging assessment: Using the SENCER-SALG to improve teaching and learning. In S. Sheardy, *Science education and civic engagement: The SENCER approach.* ACS Symposium Series 1037. Oxford, UK: Oxford University Press.

Century, J., Rudnick, M., and Freeman, C. (2010). A framework for measuring fidelity of implementation: A foundation for shared language and accumulation of knowledge. *American Journal of Evaluation, 31*(2), 199–218.

Connolly, M. R., and Millar, S. B. (2006). Using workshops to improve instruction in STEM courses. *Metropolitan Universities, 17*(4), 53–65.

Dancy, M. H., and Henderson, C. (2008). Barriers and promises in STEM reform. Commissioned paper for the *National Academies Workshop: Evidence on Promising Practices in Undergraduate Science, Technology, Engineering, and Mathematics (STEM) Education.* Retrieved Nov. 2, 2010 from http://www7 .nationalacademies.org/bose/PP_Commissioned_Papers.html

Dancy, M., and Henderson, C. (2010). Pedagogical practices and instructional change of physics faculty. *American Journal of Physics, 78*(10), 1056–1063.

Elizondo-Montemayor, L., Hernández-Escobar, C., Ayala-Aguirre, F., and Aguilar, G. M. (2008). Building a sense of ownership to facilitate change: The new curriculum. *International Journal of Leadership in Education, 11*(1), 83–102.

Fairweather, J. (2008). Linking evidence and promising practices in science, technology, engineering, and mathematics (STEM) undergraduate education: A status report for the National Academies National Research Council Board on Science Education. Commissioned paper for the *National Academies Workshop: Evidence on Promising Practices in Undergraduate Science, Technology, Engineering, and Mathematics (STEM) Education.* Retrieved Nov. 2, 2010 from http://www7.nationalacademies.org/bose/PP_Commissioned_Papers.html

Foertsch, J. A., Millar, S. B., Squire, L. L., and Gunter, R. L. (1997). *Persuading professors: A study of the dissemination of educational reform in research institutions (Vol. 5).* Madison, WI: University of Wisconsin-Madison, LEAD Center.

Foote, K. T., Neumeyer, X., Henderson, C., Dancy, M. H., & Beichner, R. J. (2014). Diffusion of research-based instructional strategies: The case of SCALE-UP. *International Journal of STEM Education, 1*(1), 10.

Froyd, J. E., Borrego, M., Cutler, S., Henderson, C., & Prince, M. J. (2013). Estimates of use of research-based instructional strategies in core electrical or computer engineering courses. *IEEE Transactions on Education, 56*(4), 393–399.

FTI Consulting. (2015). *U.S. postsecondary faculty in 2015: Diversity in people, goals and methods, but focused on students.* Washington, DC: FTI Consulting

Global. Retrieved from http://postsecondary.gatesfoundation.org/wp-content /uploads/2015/02/US-Postsecondary-Faculty-in-2015.pdf

Henderson, B. C., Finkelstein, N., and Beach, A. (2010). Beyond dissemination in college science teaching: An introduction to four core change strategies. *Journal of College Science Teaching, 39*(5), 18–25.

Henderson, C., Dancy, M., & Niewiadomska-Bugaj, M. (2012). The use of research-based instructional strategies in introductory physics: Where do faculty leave the innovation-decision process? *Physical Review Special Topics—Physics Education Research,* 8 (2), 020104.

Higher Education Research Institute at UCLA. (2009). *The American college teacher: National norms for 2007–2008.* Los Angeles, CA: University of California, Los Angeles.

Hilsen, L. R., and Wadsworth, E. C. (2002). Staging successful workshops. In Gillespie, K. H. (Ed.), *A guide to faculty development: Practical advice, examples, and resources.* Bolton, MA: Anker.

Hora, M. T. (2007). Analyzing cultural processes in higher education: STEM and education faculty collaboration in teacher education. Presented at the *Annual Meeting of the American Educational Research Association.* Chicago, Illinois. Retrieved Nov. 2, 2010 from http://scale.mspnet.org/index.cfm/19915

Hora, M.T. (2013). *Exploring the use of Teaching Dimensions Observation Protocol to develop fine-grained measures of interactive teaching in undergraduate science classrooms.* WCER Working Paper No. 2013-6.

Laursen, S. L., Hunter, A.-B., Seymour, E., Thiry, H., and Melton, G., (2010). *Undergraduate research in the sciences: Engaging students in real science.* San Francisco, CA: Jossey-Bass.

Manduca, C. A., Fox, S., & Iverson, E. R. (2006). Digital library as network and community center: A successful model for contribution and use. *D-Lib Magazine, 12*(12). doi:10.1045/december2006-manduca

Mazur, Eric. 2009. *Confessions of a converted lecturer: Eric Mazur* [Video file]. Retrieved April 1, 2010 (http://www.youtube.com/watch?v=WwslBPj8GgI).

McMartin, F., Iverson, E., Wolf, A., Morrill, J., Morgan, G., and Manduca, C. (2008). The use of online digital resources and educational digital libraries in higher education. *International Journal of Digital Libraries, 9,* 65–79.

National Academy of Engineering. (2008–2014). *Grand Challenges in Engineering.* Retrieved from www.engineeringchallenges.org

Norton, L., Richardson, J. T. E., Hartley, J., Newstead, S., and Mayes, J. (2005). Teachers' beliefs and intentions concerning teaching in higher education. *Higher Education,* (50), 537–571.

President's Council of Advisors on Science and Technology. (2010). *Prepare and inspire: K–12 education in science, technology, engineering, and mathematics*

(STEM) for America's future. Retrieved Nov. 3, 2010 from http://www.white house.gov/sites/default/files/microsites/ostp/pcast-stem-ed-final.pdf

Prince, M., Borrego, M., Henderson, C., Cutler, S., & Froyd, J. (2013). Use of research-based instructional strategies in core chemical engineering courses. *Chemical Engineering Education, 47*(1), 27–37.

Seymour, E. (2007). The U.S. experience of reform in science, technology, engineering and mathematics (STEM) undergraduate education. Presented at *Policies and Practices for Academic Enquiry: An International Colloquium,* held at the Marwell Conference Centre, Winchester, UK, 19/21 April, 2007. Retrieved Nov. 3, 2010 from http://portal-live.solent.ac.uk/university/rtconference/rt colloquium_home.aspx.

Seymour E. and Hewitt, N. M. (1997*). Talking about leaving: Why undergraduates leave the sciences.* Boulder, CO: Westview Press.

Seymour, E. (2001). Tracking the process of change in U.S. undergraduate education in science, mathematics, engineering, and technology. *Science Education,* (86): 79–105.

Seymour, E. and Laursen, S. (2005). Student resistance and student learning in undergraduate science classes using active learning pedagogies: A white paper. *Ethnography & Evaluation Research,* University of Colorado at Boulder, February 2005.

Seymour, E., with Melton, G., Wiese, D. J., and L. Pedersen-Gallegos. (2005). *Partners in innovation: Teaching assistants in college science courses.* Lanham, MD: Rowman and Littlefield Publishers, Inc.

Small, A, (2014). In defense of the lecture: A good lecturer doesn't just deliver facts but models how an expert approaches problems. *The Chronicle of Higher Education,*146787, (July 10, 2015), p. 15.

Spillane, J.P., Reisner, B.J., & Reimer, T. (2002). Policy implementation and cognition: Reframing and refocusing implementation research. *Review of educational research, 72*(3), 387–431.

Undergraduate science, mathematics and engineering education: What's working? Hearings before the Research Subcommittee of the House Committee on Science, 109th Cong. 40 (2006) (testimony of Elaine Seymour).

Walczyk, J. J., Ramsey, L. L., and Zha, P. (2007). Obstacles to instructional innovation according to college science and mathematics faculty. *Journal of Research in Science Teaching, 44*(1), 85–106.

Zemsky, R. (2009). *Making reform work: The case for transforming American higher education.* Piscataway, NJ: Rutgers University Press.

ABOUT THE AUTHORS

Elaine Seymour is the Director Emerita and Research Associate, Ethnography & Evaluation Research at the Center for the Advancement of Research and Teaching in the Social Sciences, and a Research Fellow for the Center for STEM Learning at the University of Colorado, Boulder in Boulder, Colorado.

Kris De Welde is the Associate Dean of Undergraduate Studies and an Associate Professor of Sociology at Florida Gulf Coast University in Fort Myers, Florida.

3

Toward a New Normal

Linda Slakey and Howard Gobstein

The preceding chapters in this volume explore issues that drive the need for change in undergraduate STEM education, and models for bringing about institutional transformation. They present a rich array of approaches to this challenge, from concepts in their early stages of implementation to mature projects that provide opportunities for reflection, and from a focus on individual courses and faculty pedagogy to institution-wide frameworks. In this chapter, we view the work from the perspective of time, looking back over recent advances, and reflecting on the findings presented in this volume, and then forward, noting new or growing external demands and developments, and the implications they have for institutional transformation.

We are interested in what it will take to make student-centered practices the new normal. How might we ensure that practices that have been proven to be effective will replace a culturally embedded reliance on the transmission of information via the 50-minute lecture as the principal means of instruction? The accounts in this volume show that we have moved well along a path from having a single course revision seem like a breakthrough to a focus on systemic change. Before reflecting on studies presented at the Transforming Institutions Conferences, we note two factors that support systemic change that are in place now and that weren't present, or were less developed, as little as five years ago.

First, large bodies of evidence for the effectiveness of student-centered approaches to pedagogy have been collected and analyzed. Newcomers who want to learn about the depth and quality of evidence, or those who need to provide convincing arguments, perhaps to campus leadership or funders new to this effort, will find this is much easier to do than was the case a few years ago. Examples include:

- Singer, Nielsen and Schweingruber, *Discipline-Based Education Research: Understanding and Improving Learning in Undergraduate Science and Engineering* (2012).
- President's Council of Advisers on Science and Technology (PCAST,) *Engage to Excel: Producing One Million Additional College Graduates*

with Degrees in Science, Technology, Engineering, and Mathematics (2012).

- Freeman et al. *Active Learning Increases Student Performance in Science, Engineering, and Mathematics* (2014).

Second, research groups and faculty learning communities focused on this effort have begun providing practical guides to designing and implementing change at the course and department level. The first two listed explicitly connect evidence from studies of cognition and memory to pedagogy. Some of the practical conclusions from studies of memory and attention are counterintuitive, and important to designing pedagogy that maximizes the opportunities for learning.

- For a faculty member or department just getting into conversion of lecture-based courses to more student-centered pedagogy, the Carl Wieman Science Education Initiative (CWSEI) has prepared a Course Transformation Guide. CWSEI also maintains a webpage listing a number of additional resources for course transformation. See http://www.cwsei.ubc.ca/resources/index.html
- A faculty learning community at Carnegie Mellon University has produced a guide to pedagogy based on cognitive science. See Ambrose et al., *How Learning Works: Seven Research-Based Principles for Smart Teaching* (2010).
- Lessons learned from cognitive science and tested in the Howard Hughes Medical Institute/National Academy of Sciences Summer Institutes are summarized in Handelsman, Miller, and Pfund, *Scientific Teaching* (2006).
- There is now a companion volume to *Discipline Based Education Research* from the National Research Council, that provides a handbook of practical examples and advice: Kober, *Reaching Students; What Research Says About Effective Instruction in Undergraduate Science and Engineering* (2015).

Turning now to reflections on the work presented in this volume, we see that the development of frameworks for institutional change, and the implementation of these frameworks within networks of institutions, have added substantial visibility and institutional credibility to efforts for reform. Frameworks lay out elements at an organizational level higher than a course or curriculum that need to be in place for complex cultural change to occur. A framework typically has some theoretical underpinning, but is also designed to be a practical guide for action and analysis. Framework projects may contain or lead to rubrics by which a campus or smaller unit can measure progress. Examples explored in this volume include the framework developed in the Keck-funded initiative led by Project Kaleidoscope (PKAL), and the STEM initiative of the Association of American Universities (AAU). In Section A4, writing about the Keck-PKAL initiative, Elrod and Kezar describe a process in which a group of institutions each worked through identification of campus-level goals, planning for change, and implementation, and then collectively developed an understanding of this process that can be shared with other institutions. In a series of four chapters, members of the AAU initiative describe its goals and organizational structure (A3), provide examples of project implementation at the campus level (B1 and 2), and discuss the challenge of collecting evidence across the initiative that would enable assessment of the impact of the project, as well as provide benchmarks about current practice (E4).

Groups working to change the way teaching is viewed and carried out in college and university settings have begun to apply knowledge about organizational change developed in the social sciences and in the business sector. Ferrini-Mundy (A1) emphasizes the importance of grounding change work in a theory of change, and shares the high level constructs that presently guide the approaches the National Science Foundation is taking to funding the transformation of STEM education. Increasingly, we see studies that apply specific models and theoretical constructs from these sectors to the challenge of bringing about change in the practices of academic life. Kezar and Holcombe (A2) explicitly connect research findings on organizational change to the Keck-PKAL initiative, and emphasize a point made in several other chapters: that sustainable change requires action from both the highest levels of leadership and the core workers in an organization. They further emphasize the importance of integrating the multiple dimensions of change in higher education—structural, human resource, political and symbolic.

Reinholz and his colleagues (B4) present examples of two approaches that deal in different ways with the interactions among the levels of the organization, again with the discussion grounded in the organizational change literature. Marker and his colleagues (B8) also test an explicit model of change taken from business, one that features integration of the efforts of different organizational levels. The model they examine echoes one of the empirical findings of the Keck-PKAL initiative, namely that change begins with developing a common understanding of the present circumstances and the desired change.

Potter and his colleagues (B6) also pay substantial attention to sharing information during the planning stages for change, and further emphasize the use of detailed data on student outcomes as a way of grounding plans. The report from Scott Franklin (B3) provides an interesting example of an institutionalization of change that arose from the bottom. Administrative support was needed, but the motivating force came from the desire of a group of faculty to bring about change. The themes of careful attention to student data, and understanding what motivates faculty, feature as well in Bunu-Ncube and her colleagues' report (B7). They found that a relatively modest financial incentive in support of the work was enough to draw in faculty.

Describing one of the more mature projects in this collection, Chasteen and her colleagues (B5) report on a theory of change that utilizes an investment of resources at the department level. Course reform that will bring about sustainable improvements in pedagogy and student learning is achieved in this instance by providing expertise to support faculty in going through a backwards design process. In addition to practical lessons learned about the successful implementation of this model, the author comments that, while the model succeeded without addressing the faculty reward system, it would be strengthened by alignment of the reward system with investment of faculty time in pedagogic reform. The report from Kirkup (B11) provides details of individual interventions at different campuses, and can also be seen as an example of a top-down initiative, since the work on which he reports was done while he was a recipient of a fellowship from the Australian government. The fellowship program itself clearly represents a belief on the part of the leadership that investment in supporting catalysts for change is worthwhile.

Collectively, participants in the 2014 conference presented a rich array of case studies of particular approaches to curricular change. The studies that focus at this level are gathered in Section C, and more examples are embedded in studies for which the focus was on faculty development, or a model for widespread institutional change. Looked at as a whole, they make the critical point that a primary interest in deep understanding of the subject is a very strong motivator for faculty to rethink the content of courses, and in that process they can be more easily drawn into rethinking pedagogy as well. While, at first glance, this may not seem to address our theme of institutional-level change, in fact we really need the kind of grassroots, long-term commitment that arises from love of the subject for sustainable change, and it is well to be reminded of that.

A body of studies (Section D) directly addresses the ways in which faculty can be supported as they take on new teaching approaches. Much of this work draws on research that may be unfamiliar to STEM faculty—including the importance of pedagogical content knowledge, the intentional development of a sense of self-efficacy in both faculty and students, and new approaches to assessment of student mastery of content. There is a rich tradition of methods of faculty development, and the impact of faculty learning communities has been known for decades. Nonetheless, the studies presented here suggest increasing use of both learning communities and centers for faculty development as explicit tools for organizational change, rather than simply supports for individual faculty members. Further, these increases are accompanied by enhanced efforts to understand the barriers to change—as they are perceived by faculty, which is critical feedback for the construction of models for institutional change.

Finally, to ask where we are in a trajectory of institutional transformation, we must be able to track faculty beliefs and practices as well as student learning. The last several years have seen rapid progress in the development of instruments that support quantitative, as well as qualitative, documentation of what is actually happening in classrooms. New surveys and observation protocols are moving through development and validation (E3). Groups of institutions are beginning to ask how they can deal with the challenge of creating large enough data sets to see trends while still maintaining institutional ownership of data and complying with requirements for protection of personal privacy (E2). As particular pedagogic approaches move beyond their innovators to become mainstream, we are beginning to see specific studies of how well their effectiveness travels (E1).

Looking back then over recent advances that support institutional, as well as individual, change, we see a growing number of initiatives that deal explicitly with larger units than a course or curriculum, a rich array of case studies of particular approaches to curricular change, the impact of expanding approaches to faculty development, and advances in metrics for documenting classroom practice. With all these aspects of change studied, how do we create the will and the energy for broader change to occur, for wider transformation of STEM learning in the next five to 10 years? Will these reforms become mainstream, or are they to be interesting sidelines well into the future?

We believe there are opportunities for many of these changes in STEM education to propagate, perhaps in a surprisingly short time. In order for this expansion of efforts to occur, we suggest that change advocates need to take far greater advantage of broader forces acting upon our institutions and faculty. To begin, we must recognize that our institutions are not static—they *are* changing, probably with increasing rapidity, in response to stimuli that affect the opportunity to enhance STEM education.

Perhaps the foremost factor affecting the future of higher education institutions is increasingly constrained finances. Public institutions, particularly, are beset by continued declines in state support. At the Association of American Colleges and Universities meeting in January 2015, the president of Oregon State University quipped that state funding for his public institution decreases every year until the next recession—when it decreases dramatically more. Most institutions also face resistance to increasing tuition, as the cost of college and higher student debt levels have become national issues.

Declining institutional funding could make it more difficult to adequately invest in the refinements necessary to enhance STEM learning, such as: faculty time to reconceive their courses/curricula; support for teaching/learning/STEM centers; reconfiguration of classroom space to enhance more active learning; and the pedagogical development of faculty, instructors and graduate teaching assistants. Cost cutting measures may lead institutions to rely to a greater extent on part-time instructors to teach lower level courses. Part-time instructors often lack the time to learn and work together to enhance their pedagogy. To counter these financial trends, those promoting STEM learning improvement will need to be armed with cost data for particular reforms. The economic motivation for improvement in education is typically framed in terms of benefit to students and society, while the financial impact of reform on educational institutions is analyzed in terms of the costs of the activities undertaken. Rarely is reform perceived as an investment, with analysis of the immediate positive financial impact of improved student learning on the institution. A body of evidence from the National Center for Academic Transformation (2005) shows that individual course redesigns can lead to at least equivalent learning at the same or lower costs. The National Research Council has recently reviewed criteria of productivity in higher education (Sullivan, Mackie, Massy & Sinha, 2012). However, use of routine accounting practices to capture positive financial impacts for the institution of improving instruction is not common.

The picture is further complicated by constraints in federal research funding for universities, which has been flat or less for a number of years. With research proposal success rates hovering in the low double digits for many programs, faculty spend more time writing proposals—perhaps with less success than previously. Research funding constraints have a number of potential effects on the capacity to transform STEM learning, particularly at more research-oriented universities. First, since research funding is the major source of support for graduate students in many fields of science and engineering, there will likely be less support, and eventually, cuts in the numbers of graduate students. Second, witnessing the difficulty experienced by their professors in getting research funding, graduate students, or potential graduate students, will opt for more sustainable and lucrative fields and not pursue careers as future faculty. On the other hand, it may also be possible that constraint on resources for research could result in greater attention to teaching as part of an adjustment in the balance of faculty responsibilities.

Perhaps intuitively at odds with declining university resources is the growing pressure to grant more postsecondary degrees. As the share of the population holding postsecondary degrees in other nations is slowly rising and eclipsing the respective share of the U.S. population, there is alarm that the declining U.S. rank suggests the U.S. workforce will not remain internationally competitive. A corollary to granting more degrees with constrained resources is the growth in demand for accountability and transparency by states and the federal government. Reacting to perceived increases in college costs, governments are demanding greater success rates by students seeking degrees, shorter times to graduation to save students tuition costs, and new measures to track post-degree student success, such as whether they get jobs and how much they earn in their post-graduation employment. In response to these demands, the two major associations for four-year public universities, the Association for Public and Land Grant Universities and the American Association of State Colleges and Universities, organized a joint Project Degree Completion, in which nearly 500 institutions committed to doing their part in reaching the goal of 60 percent of the working age population holding postsecondary degrees by 2025.

Those promoting enhanced STEM learning may be missing a major opportunity to motivate institutional change. It is surprising how the political pressure for more STEM degrees and greater STEM literacy for non-STEM majors has been largely disconnected from academically driven efforts to improve instruction. The White House Office of Science and Technology Policy attempted to create such a link by convening several regional meetings on enhancing degree completion in STEM fields in advance of the second annual White House College Opportunity Summit convened in December 2014. At one of those regional sessions, the chancellor of the University of Colorado, Boulder, noted how student success translates into additional university revenue. He reasoned that, *on the margin*, every additional student who succeeds in courses and becomes an upper classman nets the university additional tuition revenue that would probably have been lost. He cited a figure in the millions that he believes his institution has netted since enhancing STEM student learning.

Another way that STEM education reform advocates can play off the pressure for accountability is through attention to post-graduate student success. As summarized in its powerful opening section—*Why Systemic Reform Can No Longer Wait*—*Achieving Systemic Change: A Sourcebook for Advancing and Funding Undergraduate STEM Education* (Fry, 2014) notes how employment is growing faster in both STEM fields and for those graduates with STEM literacy; and that both jobs for STEM majors and STEM literates pay more than other fields. Industry groups are alarmed that there are too few graduates in a number of STEM fields and urge attention. The Business Higher Education Forum (BHEF) has begun industry-university collaborations across the country in a growing number of STEM-related fields of particular industrial need to foster undergraduate course re-design, greater student internships, enhanced student interest, and a larger number of graduates to feed corporate workforce needs. While in the past there had been intense attention to university-industry collaborations for research, there is now increased focus on potential partnerships for enhancing undergraduate education. BHEF is demonstrating that this might be an untapped opportunity of some potential for those interested in enhancing STEM learning.

Connecting to overall degree completion efforts could have another benefit to those wishing to enhance STEM learning. Much of the pedagogical transformation to enhance STEM learning (active learning, flipped classrooms, etc.) could also enhance learning in other disciplines. Perhaps a larger faculty cohort from a broader disciplinary set would have greater success in pressing for institutional instructional/learning change, and justifying appropriate investments and faculty incentives in return for a higher success rate by students. Indeed, such wider scale pedagogical change might be pulled along by students demanding better instruction overall as they begin to experience better teaching methods in some classes, or their peers share their better experiences. Existing STEM change literature only hints at how STEM transformation might benefit from this broader degree completion push.

One of the strongest drivers of pedagogical change is the rapid change in technology. The STEM change literature reflects this with its incorporation of technology into flipped classrooms and student response systems. But while some have called technology the great disruptor to higher education, how much further change in STEM pedagogy it will prompt is unclear. Will the rise of technology diminish the education dominance of our present brick and mortar university system? Use of technology in education is ubiquitous and has already changed learning dramatically. For example, new pedagogies, new start-ups and major corporations make online courses available to millions of people—for degrees or personal interest. Hybrid courses blend traditional face-to-face instruction with significant online components. Campuses create new facilities (exploratoriums, labs, gymnasiums) for personal learning. Finally, students now arrive at college with technology-rich and totally different learning styles and expectations than their peers of just a few years ago. The STEM education literature reflects this rise of technology, mostly at the classroom level, but does not yet appear to have addressed broader institutional effects, except conjecturally.

Given the centrality of changing faculty culture and incentives in order to transform STEM education, the growing attention, leadership and beginnings of alignment among disciplinary and scientific societies in improving undergraduate education offers substantial potential. These new activities of the societies include increasing opportunities for interested faculty to find support, and unaligned faculty to be stimulated, or perhaps recruited into the improvement movement. While the attention of scientific societies to improving pedagogy in their respective disciplines is aimed at their members and not at the academic institutions in which they live, it is still a key component of institutional transformation, given the importance of the respect of disciplinary peers in faculty culture.

One of the early leaders in education reform has been the American Physical Society, as described in *History of APS in Education* (Popkin, 2012), with a key example of their being an early significant promoter of the learning assistants program initiated at the University of Colorado, Boulder, and spreading the program to some three dozen institutions. Now, the APS is far from alone, as many key efforts have been undertaken in recent years by other faculty membership organizations. *Achieving Systemic Change* (Fry, 2012) contains a particularly good description with links to national efforts, including Vision and Change in Undergraduate Biology Education; SPIN-UP (physics); SCALE-UP (physics, now across disciplines); Project Kaleidoscope faculty summer institutes across disciplines; and CIRTL (graduate education). There is now a nascent attempt by some two dozen scientific societies—the Integration of Strategies that Support Undergraduate Education in STEM (ISSUES) group—to be more intentional and synergistic in improving pedagogy. They have met to review how faculty workshops can improve teaching across the disciplines (Association of American Physics Teachers, 2012). They plan to share information so as to provide links among attendees from the same campus to enable them to support one another and perhaps become the core of a learning community on their campus or in their region.

In response to the third recommendation of the PCAST report (2012) on undergraduate education, *Engage to Excel,* there is a very significant effort growing across major mathematics societies. In Transforming Post-Secondary Education in Mathematics (TPSEmath, n.d.), a group of distinguished mathematicians are sounding the alarm and encouraging colleagues to sign on to change both introductory mathematics for all, and mathematics curricula for majors. In what appears to be a major implementing effort in the works (Mathematical Association of America, n.d.), Common Vision is seeking to transform the first two years of introductory mathematics, bringing in other major mathematics societies, including two of the leaders of TPSE.

In its workshops and meetings focused on developing precepts and collaborative actions for education improvement, the Carnegie Foundation for the Advancement of Teaching routinely reminds attendees that, "Every system is perfectly designed to give the results that it does."[1] Thus, in order to change the results (enhance STEM learning) we have to change the system. We must address the relationships and connections among components of the system, that is, transform our institutions.

With creative leadership and effort, we believe there could be a sufficient confluence of forces to change our system of education and prompt significant adoption of better STEM learning practices. As described throughout this book, with faculty and students at the center of our system, there has been significant growth in knowledge of effective classroom practices; development of a range of models both in pedagogy and broader departmental change; creation of frameworks to guide institutional change; enhanced understanding of more effective faculty training and professional development. We have noted here the activity of disciplinary societies in providing platforms for faculty to share their knowledge; presidents and their associations (AAU and APLU) beginning to incorporate improvements in STEM learning as part of their agendas; and growth in accountability pressures and an increase in broader external stimuli to the system.

To transform our institutions to achieve their potential in providing the best STEM learning for all students, we will need to significantly connect and integrate activities that are occurring at different places in the overall system of education and STEM learning. Those working at the faculty or classroom levels need to be mindful of what is happening in the broader system to enhance their opportunities to grow and sustain the changes that they are bringing about. Leaders working at the overall system level might recognize that all change is local—class by class, faculty member by faculty member. So the challenge of those leading universities, or working with associations (such as AAU or APS), or across associations (TPSE) at the broader system levels is how to empower those creating and studying change locally, and how to integrate their many efforts to create a more significant synergy—a momentum for transformation. How do system leaders identify and energize "mid-system" champions, such as distinguished faculty members and department chairs, to enable and promote the efforts of those working at the classroom or program level, and perhaps grow their efforts one step larger?

Our growth in understanding each component of the system augurs well for increasing take-up of improved education practices, and enhanced STEM

---

1. This statement circulates widely in oral form and in public presentations among leaders of organizational change. It is often attributed to either Tom Northup or Peter Senge, but their earliest recorded use of it is in the mid-'90s. The community working on health care improvement attributes it to Paul Batalden, MD, who was quoted as saying it in the late '80s. It appears on his website at http://www.dartmouth.edu/~cecs/hcild/hcild.html.

learning. But we need to make a much stronger effort to connect activities across different places in the system. Individuals empowered to work across the system—from institution and association heads to department chairs to faculty members—can all increase their focus on building connections. In this way, we can imagine creating a new normal of transformed institutions.

REFERENCES

Ambrose, S., Bridges, M., DiPietro, M., Lovett, M., & Norman, M. (2010). *How learning works: Seven research-based principles for smart teaching.* San Francisco, CA: John Wiley & Sons.

American Association of Physics Teachers. (2012). *The role of scientific societies in STEM faculty workshops.* Available from http://www.aapt.org/Conferences/newfaculty/upload/CSSP_May_3_Report_Final_Final_Version_3-15-13.pdf

Association of Public and Land Grant Universities. (n.d.). *Project degree completion.* Available from http://www.aplu.org/projects-and-initiatives/project-degree-completion/index.html

Business and Higher Education Forum. (n.d.). *The national higher education and workforce initiative.* Available from http://www.bhef.com/our-work/hewi

Carl Wieman Science Education Initiative. (2014). *Course transformation guide.* Boulder, CO: Science Education Initiatives. Available from http://www.cwsei.ubc.ca/resources/files/CourseTransformationGuide_CWSEI_CU-SEI.pdf

Freeman, S., Eddy, S. L., McDonough, M., Smith, M. K., Okoroafor, N., Jordt, H., & Wenderoth, M. P. (2014). Active learning increases student performance in science, engineering, and mathematics. *Proceedings of the National Academy of Sciences 111,* 8410–8415. Available from http://www.pnas.org/content/111/23/8410.full

Fry, C. (Ed.). (2014). *Achieving systemic change: A sourcebook for advancing and funding and funding undergraduate STEM education.* Washington, DC: Association of American Colleges and Universities.

Handelsman, J., Miller, S., & Pfund, C. (2006). *Scientific teaching.* New York, NY: W.H. Freeman.

Kober, N. (2015). *Reaching students: What research says about effective instruction in undergraduate science and engineering.* Washington, DC: The National Academies Press. Available from http://www.nap.edu/catalog/18687/reaching-students-what-research-says-about-effective-instruction-in-undergraduate

Mathematical Association of America. (n.d.). *Common vision.* Available from http://www.maa.org/programs/faculty-and-departments/common-vision

The National Center for Academic Transformation. (2005). *A summary of NCAT program outcomes.* Available from http://www.thencat.org/Program_Outcomes_Summary.html

Popkin, G. (2012). *History of APS involvement in education.* Washington, DC: American Physical Society. Available from http://www.aplu.org/projects-and -initiatives/stem-education/SMTI_Library/history-of-aps-in-education/file

President's Council of Advisers on Science and Technology. (2012). *Engage to excel: Producing one million additional college graduates with degrees in science, technology, engineering, and mathematics.* Available from http://www.whitehouse .gov/sites/default/files/microsites/ostp/pcast-engage-to-excel-final_feb.pdf

Singer, S. R., Nielsen, N. R., & Schweingruber, H. A. (Eds). (2012). *Discipline-based education research: Understanding and improving learning in undergraduate science and engineering.* Washington, DC: The National Academies Press. Available from http://www.nap.edu/catalog.php?record_id=13362

Sullivan, T. A., Mackie, C., Massy, W. F., & Sinha, E. (2012). *Improving measurement of productivity in higher education.* Washington, DC: The National Academies Press. Available from http://www.nap.edu/catalog/13417/improving -measurement-of-productivity-in-higher-education

TPSEmath. (n.d.). *Transforming post-secondary education in mathematics.* Available from http://www.tpsemath.org/

ABOUT THE AUTHORS

Linda Slakey is a Professor Emerita of the Department of Biochemistry and Molecular Biology at the University of Massachusetts, Amherst in Amherst, Massachusetts. She has a consulting practice in Washington, DC, with appointments as Senior Advisor at both AAU and AAC&U.

Howard Gobstein is the Executive Vice President and Co-Project Director of the Science and Mathematics Teaching Imperative at the Association of Public and Land-Grant Universities in Washington, D.C.

SECTION G

Appendices

1

Editor Biographical Information

Gabriela C. Weaver is a Professor in the Department of Chemistry and serves as the Vice Provost for Faculty Development and Director of the Institute for Teaching Excellence and Faculty Development (TEFD), at the University of Massachusetts, Amherst. In that role, she oversees initiatives across the TEFD and represents both the TEFD and the broader university on issues of teaching, learning, and faculty development. Previously, she was a Professor of Chemistry and served as the Jerry and Rosie Semler Director of the Discovery Learning Research Center at Purdue University (2008–2014). She received her BS degree in Chemistry in 1989 from the California Institute of Technology, and her PhD in Chemical Physics in 1994 from the University of Colorado at Boulder. Dr. Weaver carries out research in STEM education at the undergraduate level and faculty development in higher education. She has also carried out work in science education and teacher professional development in K–12 schools. From 2004 through 2011, she directed the NSF-funded Center for Authentic Science Practice in Education (CASPiE), a consortium of 17 universities and two-year colleges that developed innovative approaches to integrate undergraduate research experiences into laboratory course instruction in several disciplines.

Dr. Weaver has co-authored two chemistry textbooks (*Chemistry in Context* from the American Chemical Society and *Chemistry and Chemical Reactivity, 6e.*), as well as numerous book chapters on science education. In 2012, she was named as a Fellow of the American Association for the Advancement of Science (AAAS).

Wilella D. Burgess is the Managing Director of the Discovery Learning Center at Purdue University. She holds degrees in Biology, Earth Sciences, and Ecology from the Pennsylvania State University and has worked as a research scientist in both industry and academia. Over the past 20 years, Ms. Burgess has been instrumental in developing and analyzing the effectiveness of education programs and methodologies aimed at reaching audiences ranging from K–16 to graduate students, professionals, and the general public through a variety of formal and informal venues.

Amy L. Childress is the Center Operations Manager for the Discovery Learning Research Center. She is a graduate of Purdue University, where she earned her BS in Biology/BA in History, MBA in Finance and Strategic Management, and Ph.D. in Educational Studies. Her research examined financial and strategic decision-making within higher education. At Purdue, she has held numerous positions supporting education at the K–12, undergraduate, and graduate student levels. She began her work as a multimedia supervisor for educational software development in the School of Agricultural & Biological Engineering and contributed to more than 20 projects funded through USEPA Region 5. She then served as the Coordinator of Advising for the Krannert School of Management. In her current position, she is responsible for overseeing the research facilities and student internship programs, as well as contributing to grant proposal development and research project coordination.

Linda Slakey is a graduate of Siena Heights College (BS in Chemistry), and the University of Michigan (PhD in Biochemistry). She was appointed to the faculty of the Department of Biochemistry at the University of Massachusetts Amherst in 1973. She was Head of the Department of Biochemistry from 1986 until 1991, and Dean of the College of Natural Sciences and Mathematics (NSM) from 1993 until 2000. From 2000 through 2006, she was Dean of Commonwealth College, the honors college of the University of Massachusetts Amherst. As Dean of NSM and of Commonwealth College, she was active in supporting teaching and learning initiatives throughout the university, with particular attention to engaging undergraduate students in research, to faculty development activities that promote the transition from lecturing to more engaged pedagogies, and to the support of research on how students learn.

From 2006 through 2011, she was Director of the Division of Undergraduate Education at the National Science Foundation. At present, she has a consulting practice in Washington, DC, with appointments as Senior Advisor at both AAU and AAC&U, focused on bringing about a shift in the culture of undergraduate teaching from one in which lecture is an acceptable norm toward one characterized by personal and institutional expectations of more student-centered teaching practices.

2

Transcript of Keynote Address by Freeman Hrabowski III at the 2014 *Transforming Institutions* Conference

Freeman A. Hrabowski III

President, the University of Maryland, Baltimore County, Baltimore, Maryland

Transcription of Keynote Remarks, Delivered October 23, 2014

Transforming Institutions:
21st Century Undergraduate STEM Education Conference,
Indianapolis, Indiana

Good evening! Thank you very much, from the heart. I appreciate it, I really do. I am delighted to be here. I took the time to ask colleagues on my campus about UMBC students who have gone on to graduate school at Purdue and how they are doing. I've gotten some great messages from people. You'll appreciate that our former students are enjoying their experience here.

It's always a challenge to speak after somebody says you gave another speech that was reasonably received because then people say, "Okay, what's he going to do this time?" I want to do a combination of things. One is to talk about some of the research that we're doing on our campus that focuses on transformation. We've taken a lot of time to think about this notion of culture change on campuses from a variety of perspectives, from thinking about issues of race, gender, and income across STEM areas, as well as undergrad, grad, and faculty levels. But we've also taken the time to think about something else, and I was reminded of it earlier today when a speaker talked about the importance of speaking from the heart. I think sometimes when we talk about STEM, we may do the analysis, but we don't take the time to talk about the inspiration that we need, the passion that is required to inspire people to want to do science, to understand why somebody might be willing to wait for years before knowing whether or not things are going to work out. Why would somebody wait until she was 40 before knowing if she'd ever get an NIH R01 grant? What could possibly have someone waiting that long? It shouldn't take that long, of course, but there are reasons for that.

I begin in a way that often surprises people when talking about success in science and institutional culture: I begin with literature. It was Zora Neale Hurston, one of my mother's favorites, who wrote a book entitled *Their Eyes Were Watching God*. The book begins, "Ships at a distance have every man's wish on board. For some they come in with the tide, for others they sail forever on the horizon, never out of sight, never landing, until the watcher turns his head away in resignation, his dreams mocked to death by time. That is the life of men." My mother would say, "And women."

The point Hurston sought to make in the '20s and the '30s, during the Harlem Renaissance, was that you had these two groups of people, people whose dreams would be—for whatever reasons—fulfilled, and then you had people whose dreams were, in the words of Langston Hughes, "forever deferred." My mother's point was this: The difference between those two groups, essentially, was often education. The people who could get some kind of education, who had those values and skills, would be able to get a job, to see some dreams realized. Those who did not saw their dreams somehow deferred.

Why do I bring that up? Because when you think about why you're here, when you think about transforming institutions, when you think about the success of students in general and then success in STEM, you really are talking about people who come to college to see dreams fulfilled. They want to be doctors; they want to be engineers; they want to do something in science or technology, engineering or mathematics.

I grew up loving mathematics. I have always proudly loved mathematics. I grew up in a home where people did math all of the time. We read and we did math. Why? Well, my mother was an English teacher, but later on, there was a curricular innovation called the New Math. My mother was brave enough to get training in the New Math, despite the fear surrounding it on the part of many teachers. As many of you know, if you are old enough to remember the New Math, there was a disconnect between the higher ed community and the K–12 community. Members of the higher education community went in to fix the problem of K–12 mathematics education. We were going to tell K–12 what they needed to do. Quite frankly, the approach we used, in my opinion—from reading about it and from listening to my mother—too often reflected a sense of arrogance on the part of the higher ed community. We thought we had the answers. But we did not understand that we really didn't know children. Teachers know children. We did not understand the fear that people sometimes have of math and science. And as a result, there were problems as the New Math was rolled out.

But my mother decided to be brave, as an English teacher, and go back and learn the New Math. The reason she was successful, and my dad wanted to be supportive of her—he always liked math—was that she loved language.

What she came to understand, what all of you will understand whether you're in engineering, science, math, was that there is a connection between language skills and word problems. The better one can read and think, whether it's for chemistry or physics or engineering, the more clearly one can understand what the problem is at least asking. And you begin to see relationships among words. You begin to understand how you use symbols, develop equations, and begin to solve the problems. And so mother could use her language skills in helping people begin to understand, and she became this master math teacher, along with loving literature. And there I was, the guinea pig. She tried out the New Math problems she was working on with me. I loved it.

We all are the product of our childhood experiences. Working with my mother on math problems as she worked on the New Math was important to me. So was my sudden participation in an even bigger change. In 1963, the Civil Rights Movement came to Birmingham, Alabama, my hometown. And one evening, I was sitting in the back of church not wanting to be there, and I hear this man say, "And if the children participate in this peaceful protest, all of America will understand. Even our children know the difference between right and wrong. They will show that they want a better education, and will get a chance to go to the better schools." I looked up from what I was doing. I had not wanted to be there, but my parents had placated me by letting me do my math problems in the back of the church and doing the other thing I really loved to do—eat. I loved to eat, so I was getting fatter and smarter all the time. So I'm eating my M&Ms with the peanuts—the good kind—and doing my math, and I look up as I'm chewing and say, "Who is this man?" And what was his name? Dr. Martin Luther King, Jr.

When I get home, I say, "I've got to go. He said we can make a difference." Why do I want to go? Because I'm tired of these damn books. We get the books that the white kids discard. Those are the books they give to the colored children. And I know it because I peel back [the covering] and I see that they're old raggedy books. And there's this sense that we are less than the white students. So I want to be at the school where you get the better books and everything. I want to see if these white kids are smarter than I am because I think I'm as smart as anybody else—because, to me, smart means you work hard, smart means you are excited about the ideas, smart means you don't stop until you solve the problem. And I get home and my parents say, "Absolutely not. You cannot go." And I say to them, "You guys are hypocrites. You make me go and listen to the guy. You tell me to do what he says. He says, 'go and march,' and now you say I can't go."

Now, at that time, you did not tell your parents they were hypocrites. Just as faculty may not tell administrators that they're hypocrites, right? Depending on the campus, of course. As it turns out, the next day they told me they didn't want to let me go because they didn't trust the people at the jail if I were arrested. But they did let me go. I did join the Children's March. I did go to jail for five nights. It was a horrible experience, but it was also a rich learning experience. It was horrible because the adults were terrible to the children, but it was rich because it taught me the significance of citizenship. It taught me what Thoreau meant by "civil disobedience." Most important, it taught me about change. It taught me about transformation. Because what Dr. King was saying was exactly what we are saying here today—that the world of tomorrow does not have to be the same as the world of today, that we can look in the mirror and know what's good and what's not and think about the gap between the two, and talk about the difference between them.

This was my childhood experience. Coming out of Birmingham, Dr. King led the March on Washington that fall. And then important legislation followed in 1964 and 1965: Civil Rights, Voting Rights, the Elementary and Secondary Education Act, and the Higher Education Act. And the world changed dramatically. If it had not, I would not be standing here. I would not be president of a university with students from more than 100 countries. My university could not have admitted Black students. We would not be here, as men and women, so many in higher education today. That's how different the world is.

And the first point I would make is this: that often some of the most dramatic transformations occur in such a way that we don't even take the time to reflect and appreciate just how significantly different our lives are as a result of whatever the transformation has been.

Let me make my point. What percent of Americans at age 25 in the mid-'60s had a college degree? What would you think? What percent? [Audience responses: 25%, 10%, 10%, 42%] Anybody else? [20%] It was only 10 percent. Ten percent of Americans in 1964–65 had a college degree. Ten percent. What percent of whites? It was 11 percent. What percent of Blacks? Two percent, not quite three. And we had not started counting the other groups separately. Today, what percent have a college degree? I can tell this is a risk-averse group. Now you cannot talk about transformation if you're risk averse. You've got the second principle? If you're going to talk about transformation, you have to be willing to take risks! Am I right? You have to be willing to be wrong. If there's one thing about my campus, people are willing to take risks. Why? Because we are willing for people to fail, because we understand that on a campus that makes progress, we often learn more from the failure than we do from the success. It's when you fail and then take the time to understand what went wrong that you learn. You bounce back; you don't punish people. You say, "What didn't work?" and you go from there.

So, again, what percent today? We just got to 30 percent. What percent of whites? About 37 percent. What percent of Blacks? Not quite 20 percent, 19 percent. What's the fastest growing group? [Audience response: Hispanics] What percent? We're up to almost 15 percent. What percent of Asian-Americans? I heard an 80, I usually get a 90 from somebody. It is 55 percent. And [the percentage of] Native Americans is much lower than Hispanics. You put it all together, this is the point—two thirds of Americans today over [age] 25 do not have bachelor's degrees. Two thirds. Two thirds do not. In fact, think of it this way: We're saying that 70% of whites, 80% of Blacks, 85% of Hispanics, and a high percentage of Native Americans don't have bachelor's degrees. And while we can say that 55% of Asian Americans have college degrees, there are certain groups of Asian Americans that are really poorly educated, depending on the group.

The next point about transformation, in terms of talking about different groups, is that we must engage in data analysis and be specific—that you cannot generalize about Hispanics, you cannot generalize about women, you cannot generalize about Blacks, about any group, that there are subgroups within any population, whether it's the Asian American population, the Black population, the Hispanic—it depends on the group. That we can say for a number of people with parents from other countries—again, though, depending on the group—that people may be doing better academically, depending on the group. And that there is this need when talking about transformation, when talking about student success, to look at everything, from test scores to academic preparation to the particular department or major of the students. All of this is very important, right? But what is my point? That we've come a long way from 10% to 30%.

Now when I say to my friends of any race—white, Black, Asian or whatever—that so few people have a bachelor's degree, what's the first response? "Oh, that couldn't be true, Freeman. Most of my friends—all of my friends—have a college degree." I get that from all of my friends. Why? Because lawyers are around lawyers, professors are around professors, and doctors are around doctors. And plumbers, who make more money than any of you, are around plumbers. [Laughs] Eat your hearts out! So it just depends on what your profession is, right? And the other part that's really significant is that literally 50% of the Americans who began college—50 million Americans—never graduated with a bachelor's degree. Think about it. Imagine if hospitals said that half the people who came to them were not successful. Think about the metaphor.

And so what am I saying? I'm saying that before we even get to STEM, we need to just talk about success in general. You can't get to STEM before you talk broadly about the issue of student success. Just as when people talk to me about minority students, I say don't start with minority students, don't start with women, don't start with students who are low income. Start with all students. Because here is the point I would make to you with certainty after being an undergrad at Hampton, going to grad school at Illinois, and working for more than 40 years: If you show me a campus that has focused carefully on the academic performance and the social well-being of students in general, I will show you a campus that has looked with care at every group. If they've really looked at the culture for students—not just for one group, but for students, if they've said we care about our students, then they're going to look at every group with care. The challenge that we face is that we tend to be accustomed to having cultures that were shaped decades and decades ago. And if a student is extraordinarily advantaged, for example—and we know the demographics of that population (in technology or engineering, they're men)—that particular student is far better off than students who are either from low-income backgrounds or are of color. We know that. All of the students who've gone on to college—this is just historical perspective—there was the 1965 Higher Education Act that made the difference.

I was speaking and talking about this to the Georgia Association of School Boards some years ago, and I'll never forget, the CEO of a major company in Georgia—wonderful gentleman—said, "Freeman, may I interrupt you for a minute? Because some of the people in the room seem bothered or uncomfortable with your talking about Civil Rights. Because sometimes we tend to think that when you talk about Civil Rights, you're talking about Blacks and women." And then he said, "When you look at me, you know I'm white. You look at me, you can tell I'm well-heeled." Everybody laughed. "Yeah, I'm rich. I'm a CEO. I'm the chairman of the school board in a very wealthy district. So you assume I come from money. Nothing could be further from the truth. My daddy died when I was young. My mother had to be a sharecropper." He said, "My mother saw the little Negro children in the '60s, after that Higher Education Act, after financial aid became available for poor people, she saw the little Negro children getting a chance to go to college, and she said, 'Where are they going?' And she found out they were going to college and she said, 'I want that for my children.' And because she saw what Dr. King and the Civil Rights Act had done for those children and that any American family could take advantage, she sent us. We got somebody to help us fill out the forms. I went to college and then I helped my younger sister go to college, and because we went to college and we got jobs, we could move our mother out of sharecropping. And because we were white, we could move on ahead of those Negro children. They did OK. They became teachers, but I became the CEO. So we may not want to admit it, but we white folks have profited a whole lot from the Civil Rights movement." And then he said what I'm going to tell you all here. He said, "Give the Civil Rights movement a big round of applause." Let me ask you all to do that right now, would you please? [Sound of applause] We all have profited, if you weren't rich, because this is one of those drivers of transformation that changed the world for anybody who did not come from wealth, in the same way that the Social Security Act did in 1935.

Talk about transformation. Do you know that many in this country called FDR "a traitor to his class?" Because he came up with this notion of a way of helping poor old people to have some basic benefits in the '30s, because wealthy people said, if you're poor and you're old, you work, because there were no re-tirement benefits for people. They just said, "That's your problem." That's what wealthy people said. And FDR disagreed with his class. But now I have one question that makes my point about change and transformation that I want you to think about. What group of Americans, well-respected Americans, fought President Roosevelt when he proposed the GI Bill, which was designed to sup-port veterans going to college? What group of highly-respected Americans, in the early '40s, around 1943? I'll give you a hint. [Audience response: university presidents]. University presidents, starting with the president of the University of Chicago, to the president of Harvard University. They said, "If you let these veterans into our colleges, our institutions will become 'academic hobo jun-gles.'" And these were wonderfully educated, liberally educated men, good men. Don't miss my point. I am not being disparaging; I am simply telling you the truth. We must tell the truth in innovation; we cannot sweep the truth under the rug. We must tell the truth in order not to repeat history. What am I saying? I am saying that sometimes even our most educated people can want to keep things as they are because it's the way it's always been. And their view of the world was that college was for people of advantage. And they felt that's the way it should be. They said it would be really tragic for institutions to let the veterans in, and yet within several years, millions of veterans—hard-working people, mainly men, mainly white, but some women and some Blacks—went into those colleges, showed what hard work could do, and all of America realized regular people could get a college degree. That set the context for the '60s.

Why do I tell that story? Because it has a lesson for us today. We are living in a world in which more people than ever—we've seen an increase from 10% to 30%—have earned a bachelor's degree. Now we get to science, and here's what I want you to hear. I chaired the National Academies' study on underrepresenta-tion in STEM. It did not surprise us to find that only 20% of Blacks, Hispanics, Native Americans, and some [subgroups] of Asian Americans, those in under-educated groups, that underrepresented minorities who started in science and engineering were not completing STEM degrees. But people were shocked by the percentage of whites and Asian Americans who started with a major in

STEM, but did not graduate in those fields. Now here is my question: What percent of whites do you think, who started with a major in these areas, graduated in these areas? [Audience responses: 15%, 20%, 40%.] It's 31%. Now, the first response is usually, well, they didn't have a K–12 background. And here, if you look at the data, you'll be shocked at the percentage of those with perfect scores and 5s on AP exams. And this is the part that's interesting: Often, the higher the test scores, the more prestigious the university, the greater the probability that they start in science and they leave it within the first year or two. Some might say, "Oh, it's because of the money." No. If you go and look, what you'll see is this: They will tend to get a C. They'll go home and say, "I like something else." Because if you get a C in chemistry and A's in everything else, you're not going to go home saying, "I bombed out." You're just going to go home and say, "I love . . . [what you got the A's in]." I said this before, and my joke often is: People go to these wonderful places, but they just don't make it in science, particularly if they were reared in this country. Let's just be honest about it. Of any race. Because, of Asian Americans, it's only 41%.

And here is the point. I tend to say it, and I'm saying it jokingly, but I'm halfway serious. They start off in pre-med and they become great lawyers. All of you know people like that. They go to law school, because they read well, and they've got to do something, they want a profession, right? I said this at one of the most prestigious national agencies and the top lawyer there came to me and said afterwards, "You just told my story. I had perfect test scores. I had all 5s. I went to one of the best universities. I started off in pre-med, and here I am now, general counsel. You just told my story." I was speaking to the university presidents of one of the Midwestern states, and one of them said to me, "You just told my story. I started off in Double E (for the engineers in the room); I ended up in Single E." You get that one? From electrical [engineering] to English. She said, "I'm a great literary scholar, I love it. But I wanted to be and I was good!" We know what we call STEM first year courses. We call them "weed-out" courses. We know that.

And we are all working now on this issue, in different ways. On my campus, we are working to redesign courses. And we should be. At UMBC, we have the Chemistry Discovery Center, and we've got the Center for Active Learning. We all are doing these kinds of things. Congratulations to all of us. What we now have to do is look at the analytics. And we have to do it; we have to disaggregate the data. This is where we all should be at this point: looking at what difference we are making. The good news is that, in the most enlightened places, we all know students are bored in regular lectures, which I've said many times. It's an amazing lecturer who can go on for an hour and people are still fascinated. You see why I keep asking you questions. Did you notice that? I just recently asked the head of brain sciences at one of the major medical centers, "Is it really true that people can only concentrate for 20 minutes?" He said, "Freeman, we tell people that, but they really can't even concentrate that long!" [Laughs] But the point that I want to make, that I want you to think about is this. For UMBC, very quickly, we worked on starting with the Meyerhoff Program, which some of you have looked at. It involved the minority kids—the Black students first and then Hispanic students—and it worked so well that my white students said, "We want what they're getting." How often do white kids say they want what the Black kids are getting, right? I say this humorously, but am I telling the truth? When do your best white students ever say, "Oh my God! Look at how well they're doing." The success of the Meyerhoff program has to do with developing a sense of community among students, of inclusiveness, of teaching them a set of values that say, if you've got five students working together and four get A's and one gets a C, they all feel bad. Getting away from the cutthroat approach. We still teach students to be cutthroat in science, let's admit it. We still grade on the curve. If you grade on a curve, you're teaching students that you don't want them to work together. [Students think to themselves,] "I'm not going to tell you what I know, because I don't trust you, because if I tell you what I know and you're really worried about your grade, you're not going to tell me everything you know." So then, you're not teaching trust. And yet, when you go to companies, you see that they're working on projects together in groups. And so we've been able to work with employers from defense agencies to biotech companies to develop partnerships that help our students as well as the employers. We have over 100 companies on campus, biotech and IT companies; 25% of our CEOs are women, give us a big hand for that, please. [Applause] All right. And the notion is that our Meyerhoff Scholars Program, or our Center for Women in Technology, or the work that we're doing for the Sherman Scholars in getting more science and math majors to work in challenging schools where we have low-income kids, all include building community and encouraging group work. So whether we're talking about working with low-income kids of color or girls in technology, or having people thinking about the difficult, sticky questions of how to have more teachers, especially at the middle-school level, who know math, or of how we have more students succeeding in the first two years of science, the practice of group work is essential.

I'll give you one more practice that I think you should all look at if you haven't: I would not define earning a C in the first year of science as a success for any student of any race. Because if you do the analytics, what we have seen is that the probability of a student who earns a C in the first level of science actually going on to get a degree and finishing with a B average is very low. Go and look at the data on your own campuses. And yet all of our catalogs say, "If you get a D, you retake the course. If you get a C, you keep going." But the students who get Cs, they're struggling the whole time. And if they do graduate, they're going, "Whew! I made it." It's a real struggle, so we actually talk about success being at least a B in the first year work for students of any race. It's a high bar, but that is how we define success. It is a different game when you go to that level, but it means constantly working to supplement the academic and social experience in different ways. It means looking at the level of background students need in order to succeed. It doesn't mean that test scores have to be exactly the same. It means understanding what the gap can be and still expect that you can get it to a certain level. It means looking at what you do at the undergrad level in relation to what you do at the grad level.

Kelly Mack is here from the American Association of Colleges and Universities. Kelly had [an article] in the most recent issue of *Peer Review*, I believe. We're really pleased. Some of my women colleagues from my campus and others, including an assistant professor from the University of Puerto Rico who is a PhD from my campus in computer science, have written some articles involving Latina women in engineering and science, and one of the articles that has been very helpful at the grad and now undergrad levels involves a PhD candidate on my campus from Puerto Rico. She had gone to the University of Puerto Rico, an undergrad in chemical engineering, and she was doing so well in the PhD program, but her family—her husband—didn't understand why she had to work so late at night. Tragically, 10 years ago, he came to campus and killed her and then killed himself. This is one of those just unbelievable situations. Now part of our culture change has been to take that failure [and learn from it] because we all felt we didn't do enough for her, we didn't understand her cultural context well enough. I was out at the University of Colorado and I was talking about this, and some of the Latina engineering and science students came up to me to talk afterwards, and I was just so touched by their saying, "You told our story, in talking about how our families feel about our having to spend time in the evenings in the labs and what that means." We have taken that story and learned so much about what we could do to pull families of our students into the work—by holding celebrations at different times, for example—so they will better understand what it takes to get a PhD. It's the same thing we do with other students, but there are things we can do to help those families out. You know how, with a PhD, you never know when somebody's going to graduate? You're saying, "I just haven't gotten results yet," and the family is going, "Uh-huh, right. You just haven't gotten results." I often say about minorities that it's harder to get a person through a PhD program than it is to get a camel through a needle's eye. It goes on and on and on. But the response has been amazing to what we call the "Jessica Effect"—taking that tragedy and turning it into something that helps the whole cultural perspective change on our campus as we think about ways that we can make a difference in the lives, not just of Latinas on our campus, but also those who are engineering students, who have families, of women who have husbands and others who are worried about them being in labs at night, whatever it takes. It's about how we can bring the families into that work—whether they're Black, Hispanic, white women, whatever—that will let the families feel a part of the experience. Again, transforming the campus in a way that takes into account what those students need at the grad level, but learning things that can help at the undergrad level, too. That's the point I'm making.

And so, we are engaged in the work of culture change, whether it's about our Meyerhoff Scholars, supporting the families of graduate students, working with the Center for Women in Technology and what we're working to do to get more women into the technology area, working to prepare teachers in science and math, finding ways to help women faculty and men faculty understand what they can do or help, or white faculty to understand ways in which they are either supportive or not of students of color, helping people understanding what goes on with students who are first generation or have some disability, or even, I'm seeing now when talking about cultural changes, the way in which we can be terribly insensitive to veterans. We have had focus groups that have taught us how we have been amazingly insensitive to veterans in classrooms and in discussions, even in science classes.

And so I would urge you also to consider this thought: Any environment that is innovative is thinking about the analytics needed to specify groups to understand more deeply their experiences, to see how they're doing in the academic work, to see how they're doing through focus groups. Focus groups are really important for listening to their voices, to understanding if they feel somehow empowered. If you've not read the book chapter that my colleagues and I wrote, entitled, "Enhancing Representation, Retention, and Achievement of Minority Students in Higher Education: A Social Transformation Theory of Change"—the key point of the chapter is that the elements that worked not just for minority students, I would say for women students and for students in general, are those that develop empowering settings for student achievement, a larger institutional change process, multiple dimensions of organizational development change, transformation in organizational culture, assessment and evaluation, and then a change strategy that involves both implementation and sustainability. And it really does incorporate all that we've been doing for the past 20 years, so you can go and look at that.

But I want to leave you with just something that's from the heart, and it would be this: We are all here to help every student to dream about the possibilities.

When I was a student at Illinois, I would sit in classes as the only Black student, and I would think to myself, I wonder what the future holds? Will the day ever come when I won't have to be the only one sitting in a math class and looking around and seeing just white guys? I didn't see any women—there was one woman faculty member and she was not tenured. That was it. And so when I got a chance to be PI on our ADVANCE grant and I realized that we only had 12% women faculty—and here this is just 10 years ago—I said, "Wow!" I didn't even realize it. All the guys were saying we had so many more. But those women faculty were not tenured; they were not tenure-track. Institutional culture change. You get my point? And the women said to me, "Freeman, use your guiding principle. Don't use anecdotes. You see a lot of women over there, right? Are they in positions of power?" And when we did the analysis, it was 12%. And what we did, it was the same thing we had done with students. I would argue that you cannot separate these things. As we talk about student success at the undergrad level, we have to see how does that relate to what we do at the grad level, how are we doing with faculty, how are we doing with staff, how are we treating each of these groups? And how do we make it into some kind of ecosystem so we understand the kind of respect we give to those groups? How do we discuss them in a way that makes them related to each other? And most important, do we take the time to listen to the voices?

I remember thinking back 40 years ago to that grad student experience, to my experience in the Civil Rights Movement, and I just kept thinking, "Will this ever change?" And somehow, all of a sudden, I thought back to being in jail. And I remembered Dr. King coming with our parents to the jailhouse. And I was trying to keep the little kids from crying. It was awful. It was just so awful. And he said to the children, "What you do this day will have an impact on children who have not yet been born." And at the time, we did not understand the profundity of his words. And yet, I knew they were significant. And as I sat in those classrooms at Illinois, I kept thinking—somehow, something's going to happen. And even today, it's so clear to me, colleagues, that as difficult as things are, as much as I know that we have so far to go with poor children in our country, as much as I know for poor children—white, Black, Hispanic—I can't help but think that when we talk and focus on transformation, that the very activity here today should give us all hope. Because what we're saying is, "Not good enough! Not good enough." We cannot be satisfied that the brightest kids, the highest-achieving kids, the most advantaged kids are doing well. We know on my campus and on your campus, we can be much better. The language that we use says, "Success is never final." And so when you ask me what's the practice that makes the difference? It is the notion that we can be much better than we are. This is the idea. You are from some of the finest campuses. I would challenge you to produce leaders from every race, men and women, in STEM areas, because it is in these areas that we will solve the problems of tomorrow.

At the end of my mother's life, and I tell this over and over again, she said to me, "I know the end is near." And you never want to hear that. But it is the human experience that counts finally, not the PhD. And I said, "What's important to you?" She had developed dementia, but she knew I was close, and I'm an only child. And she said, "What's important? Relationships." And I was trying so hard not to cry, and she looked and me and said what she always said: "Just hold on to your faith, you'll be OK. Just hold on to your faith." She said, "Relationships. My relationship with my God." Then she said, "My relationship with my husband." She had forgotten my father had died twenty years before. But you never, you never, you just . . . OK. Then what she said shocked me [because] I'm an only child. She said, "You know, I have a son." And all of a sudden, all of my grief turned to anger. I'm thinking she's telling me she had a kid when she was a teenager. I am very angry. I'm thinking, TMI, as my students would say. Too much information. If I have not had a brother to this point, I do not want a brother now. Do not drop this bomb in my lap and die, no. And she says, "He's a college president." Thank God, she was talking about me. And then she gave me the gift that I give to you. This is the most powerful of all when I think about transformations. She said, "But you know, I understand that teachers touch eternity through their students." Teachers touch eternity through their students. Whatever I had to give, my sense of right and wrong, my lust for learning, my belief in [my students], my sense that they were important, I gave it to them, and I will live through them. And there's the point in all of this—that you are here to transform institutions because institutions are to teach people that they can do anything if they just dream it and work it and give it all they have. That's all we have to give: our sense of self and our belief in our students.

Watch your thoughts; they become your words. Watch your words; they become your actions. Watch your actions; they become your habits. Watch your habits; they become your character. Watch your character; it becomes your destiny. Dreams and values.

Thank you, all.